METAL IONS IN BIOLOGICAL SYSTEMS

VOLUME 25

Interrelations among Metal Ions, Enzymes, and Gene Expression

METAL IONS IN BIOLOGICAL SYSTEMS

Edited by

Helmut Sigel
Institute of Inorganic Chemistry
University of Basel
CH-4056 Basel, Switzerland

with Astrid Sigel

VOLUME 25

Interrelations among Metal Ions, Enzymes, and Gene Expression

MARCEL DEKKER, INC. New York and Basel

Library of Congress Cataloging-in-Publication Data

Interrelations among metal ions, enzymes, and gene expression / edited by Helmut Sigel with Astrid Sigel.
p. cm. -- (Metal ions in biological systems ; v.25)
Includes bibliographies.
Includes indexes.
ISBN 0-8247-8060-4 (alk. paper)
1. Metalloenzymes. 2. Metalloproteins. 3. Metal ions--Physiological effect. I. Sigel, Helmut. II. Sigel, Astrid. III. Series.
[DNLM: 1. Enzymes--metabolism. 2. Gene Expression Regulation--drug effects. 3. Ions--metabolism. 4. Metals--metabolism. W1 ME9611AT v. 25 / QU 130 I638]
QP532.M47 Vol. 25
[QP601.7]
574.19'214 s--dc19
[574.19'25]
DNLM/DLC
for Library of Congress 89-1077
CIP

This book is printed on acid-free paper.

MARCEL DEKKER, INC.
270 Madison Avenue, New York, New York 10016

Current printing (last digit):
10 9 8 7 6 5 4 3 2 1

PRINTED IN THE UNITED STATES OF AMERICA

Preface to the Series

Recently, the importance of metal ions to the vital functions of living organisms, hence their health and well-being, has become increasingly apparent. As a result, the long-neglected field of "bioinorganic chemistry" is now developing at a rapid pace. The research centers on the synthesis, stability, formation, structure, and reactivity of biological metal ion-containing compounds of low and high molecular weight. The metabolism and transport of metal ions and their complexes is being studied, and new models for complicated natural structures and processes are being devised and tested. The focal point of our attention is the connection between the chemistry of metal ions and their role for life.

No doubt, we are only at the brink of this process. Thus, it is with the intention of linking coordination chemistry and biochemistry in their widest sense that the series METAL IONS IN BIOLOGICAL SYSTEMS reflects the growing field of "bioinorganic chemistry." We hope, also, that this series will help to break down the barriers between the historically separate spheres of chemistry, biochemistry, biology, medicine, and physics, with the expectation that a good deal of the future outstanding discoveries will be made in the interdisciplinary areas of science.

Should this series prove a stimulus for new activities in this fascinating "field," it would well serve its purpose and would be a satisfactory result for the efforts spent by the authors.

Fall 1973

Helmut Sigel
Institute of Inorganic Chemistry
University of Basel
CH-4056 Basel, Switzerland

Preface to Volume 25

Nucleic acid polymerases and nucleases require metal ions for activity, indicating that the metabolism of nucleic acids and the action of the involved enzymes are closely interwoven with the properties of certain metal ions. In addition, critical changes in the intracellular and extracellular concentrations of heavy metals can be transduced by metal-responsive regulatory proteins into alterations in the expression of specific genes. For example, in many cells the synthesis of metallothioneins is induced at the transcriptional level by the same metals that subsequently bind to the expressed protein. Considering that metallothioneins, among other functions, can protect cells from the toxicity of certain metal ions, a very subtle interplay becomes evident.

In the first part of the book metal ion-induced mutagenesis, the role of metal ions in nucleases, metal-responsive switches governing gene expression, and related topics are covered. A further aspect treated in this volume is the opportunity to explore structure-function relationships in metalloproteins via site-directed mutagenesis and to gain insights into the mechanisms of metal ion-dependent enzymes. Such interrelations are considered, e.g., for Ca^{2+}-binding proteins, nucleases, heme proteins, alcohol dehydrogenase, carboxypeptidase A, alkaline phosphatase, and cytochrome c peroxidase. The chapters provide an authoritative view on this rapidly developing and fascinating field.

Helmut Sigel
Astrid Sigel

Contents

Chapter 6

"ZINC FINGERS": THE ROLE OF ZINC(II) IN TRANSCRIPTION FACTOR IIIA AND RELATED PROTEINS 235

Jeremy M. Berg

Chapter 7

SITE-DIRECTED MUTAGENESIS AND STRUCTURE-FUNCTION RELATIONS IN EF-HAND Ca^{2+}-BINDING PROTEINS 255

Sture Forsén, Torbjörn Drakenberg, Sara Linse, Peter Brodin, Peter Sellers, Charlotta Johansson, Eva Thulin, and Thomas Grundström

Chapter 8

GENETIC ALTERATION OF ACTIVE SITE RESIDUES OF STAPHYLOCOCCAL NUCLEASE: INSIGHTS INTO THE ENZYME MECHANISM 309

Albert S. Mildvan and Engin H. Serpersu

Chapter 9

ALCOHOL DEHYDROGENASE: STRUCTURE, CATALYSIS, AND SITE-DIRECTED MUTAGENESIS 335

Y. Pocker

Chapter 10

PROBING THE MECHANISM OF ACTION OF CARBOXYPEPTIDASE A BY INORGANIC, ORGANIC, AND MUTAGENIC MODIFICATIONS 359

David S. Auld, James F. Riordan, and Bert L. Vallee

Contributors

Numbers in parentheses indicate the pages on which the authors' contributions begin.

William M. Atkins Departments of Biochemistry and Chemistry, University of Illinois, 415 Roger Adams Laboratory, 1209 West California Street, Urbana, IL 61801, USA (417)

David S. Auld Center for Biochemical and Biophysical Sciences and Medicine, Harvard Medical School, Seeley G. Mudd Building, 250 Longwood Avenue, Boston, MA 02115, USA (359)

Jacqueline K. Barton Department of Chemistry, Columbia University, Havemeyer Hall, Box 953, 116th Street and Broadway, New York, NY 10027, USA (31)

Lena A. Basile Department of Chemistry, Columbia University, Havemeyer Hall, Box 953, 116th Street and Broadway, New York, NY 10027, USA (31)

Jeremy M. Berg Department of Chemistry, The Johns Hopkins University, Remsen Hall-Dunning Hall, 34th and Charles Streets, Baltimore, MD 21218, USA (235)

Peter Brodin Unit of Applied Cell and Molecular Biology, University of Umeå, S-901 87 Umeå, Sweden (255)

John E. Butler-Ransohoff* Laboratory of Bioorganic Chemistry and Biochemistry, The Rockefeller University, Box 113, 1230 York Avenue, New York, NY 10021-6399, USA (395)

Tauseef R. Butt Department of Molecular Pharmacology, Smith Kline and French Laboratories, P.O. Box 1539, King of Prussia, PA 19406, USA (147)

Joseph E. Coleman Department of Molecular Biophysics and Biochemistry, Yale University School of Medicine, Sterling Hall, Box 3333, 333 Cedar Street, New Haven, CT 06510-8024, USA (171)

---

*Present affiliation: Bayer AG, Pharmaforschungszentrum, 5600-Wuppertal, Federal Republic of Germany.

Stanley T. Crooke Department of Molecular Pharmacology, Smith Kline and French Laboratories, P.O. Box 1539, King of Prussia, PA 19406, USA (147)

Kathleen M. Downey Departments of Medicine and of Biochemistry and Molecular Biology, and the Center for Blood Diseases, University of Miami School of Medicine, P.O. Box 016 960, Miami, FL 33101, USA (1)

Torbjörn Drakenberg Physical Chemistry 2, Chemical Centre, Lund University, P.O. Box 124, S-221 00 Lund, Sweden (255)

David J. Ecker Department of Molecular Pharmacology, Smith Kline and French Laboratories, P.O. Box 1539, King of Prussia, PA 19406, USA (147)

Steven L. Edwards Department of Chemistry B-017, University of California, San Diego, La Jolla, CA 92093, USA (477)

Laurence A. Fishel Department of Chemistry B-017, University of California, San Diego, La Jolla, CA 92093, USA (477)

Sture Forsén Physical Chemistry 2, Chemical Centre, Lund University, P.O. Box 124, S-221 00 Lund, Sweden (255)

David P. Giedroc Department of Molecular Biophysics and Biochemistry, Yale University School of Medicine, Sterling Hall, Box 3333, 333 Cedar Street, New Haven, CT 06510-8024, USA (171)

Thomas Grundström Unit of Applied Cell and Molecular Biology, University of Umeå, S-901 87 Umeå, Sweden (255)

Charlotta Johansson Physical Chemistry 2, Chemical Centre, Lund University, P.O. Box 124, S-221 00 Lund, Sweden (255)

Emil Thomas Kaiser Laboratory of Bioorganic Chemistry and Biochemistry, The Rockefeller University, Box 113, 1230 York Avenue, New York, NY 10021-6399, USA (395)

Debra A. Kendall Laboratory of Bioorganic Chemistry and Biochemistry, The Rockefeller University, Box 113, 1230 York Avenue, New York, NY 10021-6399, USA (395)

Joseph Kraut Department of Chemistry B-017, University of California, San Diego, La Jolla, CA 92093, USA (477)

Sara Linse Physical Chemistry 2, Chemical Centre, Lund University, P.O. Box 124, S-221 00 Lund, Sweden (255)

J. Matthew Mauro Department of Chemistry B-017, University of California, San Diego, La Jolla, CA 92093, USA (477)

Albert S. Mildvan Department of Biological Chemistry, The Johns Hopkins University School of Medicine, 725 N. Wolfe Street, Baltimore, MD 21205, USA (309)

Mark A. Miller Department of Chemistry B-017, University of California, San Diego, La Jolla, CA 92093, USA (477)

Thomas V. O'Halloran Department of Chemistry and Department of Biochemistry, Molecular Biology, and Cell Biology, Northwestern University, 2145 Sheridan Road, Evanston, IL 60201, USA (105)

Y. Pocker Department of Chemistry, BG-10, University of Washington, Seattle, WA 98195, USA (335)

James F. Riordan Center for Biochemical and Biophysical Sciences and Medicine, Harvard Medical School, Seeley G. Mudd Building, 250 Longwood Avenue, Boston, MA 02115, USA (359)

Peter Sellers Physical Chemistry 2, Chemical Centre, Lund University, P.O. Box 124, S-221 00 Lund, Sweden (255)

Engin H. Serpersu Department of Biological Chemistry, The Johns Hopkins University School of Medicine, 725 N. Wolfe Street, Baltimore, MD 21205, USA (309)

Stephen G. Sligar Departments of Biochemistry and Chemistry, University of Illinois, 415 Roger Adams Laboratory, 1209 West California Street, Urbana, IL 61801, USA (417)

Antero G. So Departments of Medicine and of Biochemistry and Molecular Biology, and the Center for Blood Diseases, University of Miami School of Medicine, P.O. Box 016 960, Miami, FL 33101, USA (1)

Barry A. Springer Departments of Biochemistry and Chemistry, University of Illinois, 415 Roger Adams Laboratory, 1209 West California Street, Urbana, IL 61801, USA (417)

Patrick S. Stayton Departments of Biochemistry and Chemistry, University of Illinois, 415 Roger Adams Laboratory, 1209 West California Street, Urbana, IL 61801, USA (417)

Eva Thulin Physical Chemistry 2, Chemical Centre, Lund University, P.O. Box 124, S-221 00 Lund, Sweden (255)

Bert L. Vallee Center for Biochemical and Biophysical Sciences and Medicine, Harvard Medical School, Seeley G. Mudd Building, 250 Longwood Avenue, Boston, MA 02115, USA (359)

Jimin Wang Department of Chemistry B-017, University of California, San Diego, La Jolla, CA 92093, USA (477)

Contents of Other Volumes

*Out of print

*Out of print

Volume 11. Metal Complexes as Anticancer Agents

Volume 12. Properties of Copper

Volume 13. Copper Proteins

Volume 14. Inorganic Drugs in Deficiency and Disease

Volume 15. Zinc and Its Role in Biology and Nutrition

Volume 16. Methods Involving Metal Ions and Complexes in Clinical Chemistry

Volume 17. Calcium and Its Role in Biology

Volume 18. Circulation of Metals in the Environment

Volume 19. Antibiotics and Their Complexes

Volume 20. Concepts on Metal Ion Toxicity

Volume 21. Applications of Nuclear Magnetic Resonance to Paramagnetic Species

Volume 22. ENDOR, EPR, and Electron Spin Echo for Probing Coordination Spheres

Volume 23. Nickel and Its Role in Biology

Volume 24. Aluminum and Its Role in Biology

Other volumes are in preparation.

Comments and suggestions with regard to contents, topics, and the like for future volumes of the series would be greatly welcome.

The following Marcel Dekker, Inc. book is also of interest for any reader dealing with metals or other inorganic compounds:

HANDBOOK ON TOXICITY OF INORGANIC COMPOUNDS edited by Hans G. Seiler and Helmut Sigel, with Astrid Sigel

METAL IONS IN BIOLOGICAL SYSTEMS

VOLUME 25

Interrelations among Metal Ions, Enzymes, and Gene Expression

1

Metal Ion-Induced Mutagenesis In Vitro: Molecular Mechanisms

Kathleen M. Downey and Antero G. So
Departments of Medicine and of Biochemistry and Molecular Biology, and the Center for Blood Diseases
University of Miami School of Medicine
P.O. Box 016 960
Miami, FL 33101

1. INTRODUCTION

All DNA polymerases require a divalent cation for catalytic activity. Mg^{2+} is considered to be the physiological activator; however, other divalent cations such as Mn^{2+}, Co^{2+}, Be^{2+}, and Ni^{2+} can substitute for Mg^{2+} as metal activators in vitro [1]. Of these metal ions, Mn^{2+} has been the most extensively studied. It is known to be mutagenic and carcinogenic in vivo [2,3], and substitution of Mn^{2+} for Mg^{2+} in vitro results in diminished fidelity of DNA synthesis. Increased error frequency in the presence of Mn^{2+} has been observed with *E. coli* DNA polymerase I [4-6], T_4 DNA polymerase [7-10], mammalian DNA polymerases α and β [11,12], and avian myoblastosis virus (AMV) reverse transcriptase [13].

DNA polymerase I of *E. coli* is the most extensively studied and best characterized DNA polymerase; the complete primary structure has been determined [14,15] and the crystal structure of the large fragment (Klenow fragment) elucidated [16,17]. In many respects DNA polymerase I is a prototype DNA polymerase, and it is likely that findings obtained with this enzyme will be applicable to other DNA polymerases.

In order to understand the molecular mechanism of metal ion-induced mutagenesis, it is necessary to understand the mechanisms by which the fidelity of DNA synthesis is maintained. In this review, we will focus on the molecular mechanisms of base selection by the polymerase activity and proofreading by the 3'- to 5'-exonuclease activity of the Klenow fragment of DNA polymerase I of *E. coli,* and the effects of Mn^{2+} on these error discrimination steps.

2. FIDELITY OF DNA SYNTHESIS

In order to maintain viability and species identity, DNA must be replicated and repaired with great fidelity, whereas evolution is only possible if genetic variability is generated. These opposing demands have resulted in DNA replication and repair processes with low but finite error frequencies. Measurements of spontaneous mutation frequencies in prokaryotes suggest that the average frequency of substitution mutation is on the order of 10^{-7} to 10^{-10} misincorporations per base pair replicated [18,19]. It is now believed that this level of accuracy is the result of a multistep process: (1) base selection in the initial incorporation of a complementary nucleotide at the 3' end, i.e., primer terminus, of a growing DNA chain, (2) exonucleolytic excision of a newly added mismatched nucleotide at the primer terminus, i.e., proofreading, and (3) postreplicative excision of mismatched nucleotides by a methyl-directed endonuclease [20]. The first two steps in error discrimination, selection of a deoxynucleoside triphosphate (dNTP) that is complementary to the corresponding nucleotide in the template strand for incorporation into the growing primer chain and hydrolysis of a mismatched nucleotide at the primer terminus prior to chain extension, are both functions of DNA polymerases.

The ability of DNA polymerase to discriminate between nucleotide substrates that are complementary and noncomplementary to the corresponding nucleotide in the template is thought to be based on the difference in free energy between correct and incorrect base pairings. This free-energy difference, which has been estimated to be 1 to 3 kcal from measurements of nucleotide interactions and the stability of polynucleotide helices in solution, would predict an error frequency for insertion of a nucleotide opposite a template base of 1/100 [21]. However, in vitro measurements of error frequency by DNA polymerases in the absence of exonucleolytic proofreading are one to two orders of magnitude lower than predicted [22], suggesting that the DNA polymerase enhances the accuracy of base pairing at the base selection step. Proofreading by the 3'- to 5'-exonuclease activity of the DNA polymerase further enhances fidelity [20].

3. MECHANISM OF BASE SELECTION

The mechanisms by which the high specificity of base selection is achieved by DNA polymerases are not well understood and still remain the subject of much controversy. The ability of a DNA polymerase to discriminate between nucleotide substrates that are complementary or noncomplementary to the corresponding nucleotide in the template is based, at least in part, on the difference in free energy between correct and incorrect base pairings. It has been demonstrated in several laboratories that the apparent K_m values for mismatched nucleotide substrates are at least one to two orders of magnitude higher than the K_m values for matched nucleotide substrates [6,23]. However, whether the polymerase further discriminates against a mismatched nucleotide by lowering its rate of insertion has not been fully resolved.

3.1. Passive Polymerase Model

The passive polymerase model for nucleotide insertion, which is also known as the K_m discrimination model, suggests that DNA polymerases play no active role in base selection but rather they insert any dNTP resident at their active sites [24,25]. It has been proposed that kinetic "on rates" for dNTPs are diffusion-controlled and equivalent for matched and mismatched nucleotides, whereas "off rates" are dependent on base-pairing interactions with the template and are therefore proportional to base pair stability, possibly amplified by the imposition of stereochemical constraints and base stacking [10,25]. This hypothesis predicts that error frequency is determined primarily by K_m differences between complementary and noncomplementary nucleotide substrates, and is supported by the findings that the sixfold preference of T_4 DNA polymerase for incorporation of adenine over 2-aminopurine (AP) can be entirely accounted for by the free-energy difference between A:T and AP:T base pairs [10]. Similarly, the 22-fold preference of DNA polymerase α for insertion of dTMP rather than dCMP across from a template AP can be accounted for by the 25-fold difference in the K_m values for dCTP and dTTP determined with AP-containing templates [26]. As predicted by the passive polymerase model, the V_{max} values for insertion of dTTP or dCTP across from a template AP were found to be identical, as were the V_{max} values for insertion of either 2-aminopurine deoxynucleoside triphosphate (dAPTP) or dATP across from a template T [10,26]. More recently, it was suggested that the free-energy differences between complementary and noncomplementary base pairs can be amplified by the exclusion of water at the polymerase active site, thus enabling the polymerase to achieve high fidelity of DNA synthesis without undergoing any conformational changes [27].

3.2. Active Polymerase Model

The active polymerase model, also known as the V_{max} discrimination model, suggests that the DNA polymerase actively discriminates against a mismatched nucleotide, i.e., that in addition to K_m discrimination, the DNA polymerase reads each template base and undergoes conformational changes which result in insertion of noncomplementary dNTPs at a slower rate than complementary dNTPs [8,20,21]. In studies with the homopolymer duplex poly(dA)/poly(dT) as template/primer, Gillin and Nossal [23] reported a 10-fold higher apparent K_m for insertion of mismatched dGTP vs. matched dATP, a 300-fold higher apparent K_m for mismatched dCTP vs. matched dTTP, and apparent V_{max} values approximately 10-fold higher for the complementary vs. the noncomplementary nucleotide with T_4 DNA polymerase. Fersht et al. [6] found a 400-fold apparent K_m difference between dATP and dGTP for insertion into the poly(dA-dT) alternating copolymer by *E. coli* DNA polymerase I. They further reported V_{max} differences of 20- to 30-fold between complementary dATP (A:T pair) and mismatched dGTP (G:T pair). These studies suggest that both DNA polymerase I and T_4 DNA polymerase may utilize V_{max} differences in addition to K_m differences in error discrimination.

It is interesting to note that the data in support of V_{max} discrimination have come from studies with naturally occurring nucleotides, whereas data in support of K_m discrimination have come from studies with base analogs. Thus it is possible that the discrepancies are the result of differences in the ability of DNA polymerases to discriminate against base analogs rather than noncomplementary bases. However, another possibility is that the assays may not have measured the rate of phosphodiester bond formation solely as function of dNTP concentration.

In the absence of an assay sensitive enough to measure the rate of formation of a single phosphodiester bond, the rate of phosphodiester bond formation has, of necessity, been determined by measuring the rate of formation of many bonds and assuming (1) that the rate of formation of each phosphodiester bond is the same and (2) that phosphodiester bond formation is always rate determining. The formation of a phosphodiester bond between a primer terminus and a dNTP is dependent on the binding of both substrates, i.e., primer terminus and dNTP, at the active site of the polymerase. Thus, measurements of the rate of phosphodiester bond formation as a function of dNTP concentration are only valid if the primer terminus is held constant for each phosphodiester bond formed. Since the dNTP substrate for the formation of one phosphodiester bond becomes the dNMP primer terminus for the formation of the next phosphodiester bond, measurements of the rates of phosphodiester bond formation as a function of dNTP concentration are influenced both by the affinity of dNTP for the triphosphate site of the polymerase and by the affinity of the incorporated dNMP for the primer terminus site of the polymerase. Consequently, even with the same template/primer, measurements of the rates of incorporation of different dNTPs cannot be interpreted as measurements of the rates of phosphodiester bond formation, since two different dNTPs will necessarily generate two different primer termini. Furthermore, with a synthetic homopolymer duplex template/primer there are two primer termini available, and thus it is difficult to delineate whether the apparent Michaelis constants obtained represent insertion events across from purine or pyrimidine template bases or both, particularly under mutagenic conditions. Another problem arises when one is attempting to measure rates of phosphodiester bond formation with noncomplementary nucleotides, particularly with a DNA polymerase that does not have a proofreading 3'- to 5'-exonuclease activity, such as mammalian DNA polymerases α, β, and γ. Insertion of a noncomplementary nucleotide results in the formation of a primer terminus with an extremely low probability of hydrogen bonding with the template base, and therefore insertion of additional noncomplementary nucleotides is extremely unlikely. It has been estimated that DNA polymerase α incorporates a complementary dNTP into a mismatched primer terminus at a rate which is approximately 5% of that with a matched primer terminus [28], and presumably the extension of a mismatched primer terminus with a noncomplementary dNTP would be significantly slower.

To overcome this apparent dilemma, it is possible to measure the rates of phosphodiester bond formation with a noncomplementary nucleotide by using a polymerase that contains a 3'- to 5'-exonuclease activity that is capable of proofreading. This is based on the observation that the affinity of a mismatched nucleotide at the primer terminus for binding at the active site of polymerase activity is so low that it preferentially binds to the active site of the exonuclease activity and is excised, and that the rate of hydrolysis of a mismatched nucleotide is much faster than its rate of extension [29]. Thus, the rate of turnover of a noncomplementary substrate, i.e., conversion of dNTP to dNMP, would be limited by the rate of phosphodiester bond formation and would correspond to the rate of insertion of a noncomplementary nucleotide.

Based on these considerations, we recently developed a method to determine the rate of phosphodiester bond formation as a function of dNTP concentrations for both complementary and noncomplementary dNTPs by (1) measuring the rate of incorporation of a complementary dNTP with a homopolymer template under conditions where the rate of

```
 ,C-C-C-C-C-C-C-C-C-C-C-C-C-C-C-C-C-C  5'
C : : : : : :
 `G-G-G-G-G-G  3'


 ,T-T-T-T-T-T-T-T-T-T-T-T-T-T-T-T-T-T  5'
T : : : : : :
 `A-A-A-A-A-A  3'
```

FIG. 1. Schematic representation of hook polymer template/primers used to determine rates of phosphodiester bond formation with complementary and noncomplementary nucleotide substrates.

turnover is minimal, and (2) measuring the rate of turnover of a noncomplementary dNTP where the rate of hydrolysis is rapid.

The template/primers used in these studies were primed homopolymers in which the primer is covalently linked to the template, i.e., hook polymers (Fig. 1), in order to maintain a constant primer terminus and avoid any ambiguity that could arise when more than one primer terminus is available for phosphodiester bond formation. By using hook homopolymers as templates, one can measure the rate of phosphodiester bond formation between a constant primer terminus as nonvariable substrate and either a complementary or noncomplementary dNTP as variable substrate. With complementary dNTP as variable substrate, incorporation of dNMP generates a primer terminus that is identical to the original, so that the nature of the primer terminus is held constant for every phosphodiester bond formed. Furthermore, the anchoring of the primer by covalent attachment to the 3' end of the template strand provides a long stretch of single-stranded template for replication, so that available template space does not limit the reaction and affect the kinetic constants for nucleotide insertion. With a noncomplementary dNTP as variable substrate, incorporation of dNMP into the primer terminus results in a mismatched primer terminus which is rapidly hydrolyzed to free dNMP by the proofreading 3'- to 5'-exonuclease of DNA polymerase I, regenerating the original primer terminus. The rate of insertion of either complementary or noncomplementary dNTPs is determined by measuring the rate of total utilization of dNTP, i.e., the sum of dNMP incorporated into polynucleotide and dNMP incorporated and released as free dNMP (turnover).

Our studies on the mechanism of base selection by the Klenow fragment of DNA polymerase I are based on the premise that the template-dependent conversion of noncomplementary dNTP to dNMP is the result of alternate incorporation of dNMP at the primer terminus by the polymerase activity followed by hydrolysis to free dNMP by the 3'- to 5'-exonuclease activity, and that the rate of turnover of a noncomplementary nucleotide is limited by the rate of phosphodiester bond formation. Consequently, it is important to determine whether any alternate pathway exists for turnover and to demonstrate that the rate of turnover of noncomplementary nucleotide is not limited by the rate of hydrolysis.

3.2.1. Preinsertion Proofreading

A general mechanism for error discrimination by DNA polymerase has been proposed by Hopfield [30], in which proofreading occurs prior to phosphodiester bond formation by the hydrolysis of an unstable enzyme-dNTP intermediate to pyrophosphate and dNMP. In this error discrimination mechanism, the hydroxyl group of H_2O competes with the hydroxyl group of the primer terminus in a nucleophilic attack on the incoming dNTP [31]. Consequently, it is expected that, in the absence of a hydroxyl group at the primer terminus, the rate of preinsertion proofreading would be significantly enhanced. In contrast, we have found that the conversion of dNTP to dNMP is dependent on the presence of a 3'-hydroxyl group at the primer terminus. Further evidence for the lack of contribution of preinsertion proofreading to the template-dependent generation of dNMP is the demonstration that the turnover reaction can be essentially abolished by inhibiting the 3'- to 5'-exonuclease activity by dNMP. The active site for the 3'- to 5'-exonuclease activity is functionally and spatially distinct from the polymerase active site and the 3'- to

5'-exonuclease activity can be selectively inhibited by dNMP [16,32]. Since there is only a single binding site for dNMP in the polymerase [33], inhibition of turnover of dNTP to dNMP by 5'-dNMP would be attributable to inhibition of proofreading by the 3'- to 5'-exonuclease activity of DNA polymerase and not to inhibition of the hydrolysis of enzyme-dNTP intermediates. These findings suggest that template-dependent conversion of noncomplementary dNTP to dNMP requires an active 3'- to 5'-exonuclease activity and argues against kinetic proofreading as an important contributor to the high fidelity of DNA polymerase I.

3.3. Factors that Affect the Kinetics of Nucleotide Insertion

3.3.1. Rate of Hydrolysis of Mismatched Nucleotides at the Primer Terminus

Since the rate of formation of a phosphodiester bond between a primer terminus and a noncomplementary dNTP is determined by measuring the rate of template-dependent turnover of dNTP to dNMP, it is important to establish that the rate-limiting step in the turnover reaction is the rate of phosphodiester bond formation and not the rate of hydrolysis of a mismatched primer terminus. Comparisons of the rates of turnover of mismatched nucleotides with the rates of hydrolysis of mismatched primer termini revealed that the rate of hydrolysis of the mismatched nucleotide dAMP from poly(dC)-oligo(dG)-(dA) is 30-fold higher than the maximal rate of insertion of dATP as measured by the rate of template-dependent turnover of dATP to dAMP and the rate of hydrolysis of mismatched nucleotide dGMP from poly(dT)-oligo(dA)-(dG) is eightfold higher than the maximal catalytic rate for insertion of dGTP into poly(dT)-oligo(dA)-(dG). Thus the rates of turnover of noncomplementary nucleotides are not limited by their rates of hydrolysis and are most likely limited by rates of phosphodiester bond formation [34,35].

3.3.2. *Influence of Available Template Space on V_{max} Values for Insertion of Complementary Nucleotides*

The rate of template-dependent turnover of a complementary nucleotide is limited by the rate of hydrolysis of a matched primer terminus and not by the rate of phosphodiester bond formation. As a consequence, with template/primers which contain little template space, e.g., homopolymer duplexes or alternating copolymers, the V_{max} for insertion of a complementary dNTP is significantly lower than with a template/primer which contains ample template space, e.g., hook polymers [36]. For example, the apparent K_m values for insertion of dGTP by DNA polymerase I are identical with either activated calf thymus DNA, poly(dG-dC), or poly(dC)-oligo(dG) as template/primer; however, the V_{max} values for dGTP insertion are approximately five- to sevenfold lower with either activated calf thymus DNA or poly(dG-dC) as template/primer as compared to the poly(dC)-oligo(dG) hook polymer (Table 1). As might be expected, the rate of template-dependent

TABLE 1

Effect of Template Space on the V_{max} for Insertion of dGTP

	dGTP Insertion	
Template/primer	K_m (μM)	V_{max} (μmol/30'/mg)
Activated calf thymus DNA	4.1 ± 1.2	216 ± 13
Poly(dG-dC)	4.6 ± 1.9	160 ± 16
Poly(dC)-oligo(dG)	3.7 ± 0.4	1120 ± 53

Note: Concentration of dGTP was varied over a 150-fold concentration range (0.72-107 μM). When poly(dG-dC) was used as template/primer, dCTP was present at 50 μM. When activated calf thymus DNA was used as template/primer, dCTP, dATP, and dTTP were present at 50 μM. Apparent K_m and V_{max} values were determined and assay conditions were as described in Ref. 34.

turnover of complementary dNTP to dNMP is significant with alternating copolymer or activated DNA template/primers but negligible with hook polymer template/primers. The rate of dGTP turnover represents approximately 5-10% of the total insertion rate with DNA template/primers that have little replicating space, e.g., activated calf thymus DNA and poly(dG-dC). In contrast with the hook polymer poly(dC)-oligo(dG) as template/primer, insertion of the complementary substrate dGTP resulted in greater than 99% of dGMP incorporated into polynucleotide and less than 1% released as free dGMP. Thus, the V_{max} value for insertion of a complementary nucleotide substrate can be greatly affected by the availability of template space.

3.4. Kinetic Constants for dNTP Insertion

With poly(dC)-oligo(dG) as template/primer, the K_m for insertion of the complementary substrate dGTP (G:C pair) was found to be approximately 20-fold lower than that for the noncomplementary substrate dATP (A:C mispair), whereas the V_{max} for dGTP insertion was found to be approximately 5,000-fold higher than that for dATP insertion (Table 2). With poly(dT)-oligo(dA) as template/primer, the K_m for the complementary substrate dATP (T:A pair) was found to be approximately tenfold lower than that for the noncomplementary substrate dGTP (T:G mispair), whereas the V_{max} for dATP insertion was approximately 1,600-fold higher than that for dGTP insertion. With either hook polymer, insertion of noncomplementary pyrimidine nucleotides was below the level of detection.

The specificity of DNA polymerase I for inserting a G opposite a template C ($k_{cat}/K_m = 1.2 \cdot 10^7$) rather than an A opposite a template C ($k_{cat}/K_m = 10^2$) is of the order of $1 \cdot 10^5$ (Table 2), and the specificity for inserting an A opposite a template T ($k_{cat}/K_m = 6 \cdot 10^5$) vs. a G opposite a template T ($k_{cat}/K_m = 6 \cdot 10^2$) is of the order of $1 \cdot 10^4$. These values are comparable to error frequencies observed in vitro and consistent with the predicted specificity values for a single base-pairing selection step.

These results indicate that the high specificity of base selection by DNA polymerase I is achieved by utilization of both

TABLE 2

Apparent Michaelis-Menten Constants for the Insertion of Complementary and Noncomplementary Nucleotides into the Hook Polymers Poly(dC)-oligo(dG) and Poly(dT)-oligo(dA)

Substrate	K_m (μM)	V_{max} (μmol/30'/mg)	k_{cat} (sec^{-1})	k_{cat}/K_m ($sec^{-1}\ M^{-1}$)
	Template/Primer: $(dC)_{194}$-$(dG)_{12}$			
dGTP	3.7 ± 0.5	1120 ± 53	42.4	$1.2 \cdot 10^7$
dATP	63.7 ± 11	0.17 ± 0.03	0.0064	$1.0 \cdot 10^2$
	Template/Primer: $(dT)_{350}$-$(dA)_{14}$			
dATP	8.7 ± 1.2	1413 ± 42	53.4	$6.1 \cdot 10^6$
dGTP	51.5 ± 10	0.860 ± 0.05	0.033	$6.4 \cdot 10^2$

Note: When $(dC)_{194}$-$(dG)_{12}$ was template/primer, the concentration of Mg^{2+} was 5 mM and the concentrations of dNTPs were varied as follows: dGTP (0.72-107 μM) and dATP (10 μM to 2 mM). Apparent K_m and V_{max} values are reported as mean values ± SEM. Assay conditions were as described in Ref. 33. When $(dT)_{350}$-$(dA)_{14}$ was template/primer, the concentration of Mg^{2+} was 1 mM and the concentrations of dNTPs were varied as follows: dATP (1.7-202 μM) and dGTP (5.5 μM to 1 mM). The assay conditions were as described in Ref. 35.

K_m and V_{max} differences between complementary and noncomplementary nucleotide substrates, and suggest that the DNA polymerase plays an active role in error discrimination during base selection, i.e., DNA polymerase I actively discriminates between complementary and noncomplementary nucleotides resident at the active site, possibly by assuming a different conformation upon binding of a nucleotide substrate that is complementary to the corresponding nucleotide in the template. These results are in contrast to the results of studies with T_4 DNA polymerase [10] in which the use of base analogs as model noncomplementary substrates has suggested that DNA polymerases do not actively discriminate against noncomplementary substrates, but rather that they insert any nucleotide resident at the active site into a growing chain, residence time being reflected in apparent K_m. Whether these two DNA polymerases differ in the mechanisms by which they discriminate between complementary and noncomplementary nucleotide substrates, or whether the stereochemistries of base pairing between 2-aminopurine and adenine with thymidine in the template are very similar and therefore the differences between these bases are not discerned by the DNA polymerase is not clear at present.

4. MECHANISM OF PROOFREADING

The recent elucidation of the crystal structure of the Klenow fragment of DNA polymerase I [16,17] in addition to inhibitor studies with this enzyme [32,37,38] have suggested that the active sites of the polymerase activity and the 3'- to 5'-exonuclease activity are structurally and functionally separate and distinct.

The three-dimensional structure of the Klenow fragment revealed that the 60-kD polypeptide is folded into two structural domains [17]. The polymerase active site is located at the larger C-terminal domain and the 3'- to 5'-exonuclease active site is confined to the N-terminal domain, the two active sites being separated by about 30 Å. That the active sites are separate is further supported by the recent cloning and expression of the DNA coding for the large domain, which demonstrated that the C-terminal domain possesses polymerase activity but is devoid of measurable 3'- to 5'-exonuclease activity [39].

Previous kinetic studies demonstrated that the 3'- to 5'-exonuclease activity of *E. coli* DNA polymerase I can be selectively inhibited by nucleoside 5'-monophosphates (5'-NMP) whereas the polymerase activity is not affected, and that 5'-NMPs are competitive inhibitors of the 3'- to 5'-exonuclease [32]. The inhibitor constants for several 5'-NMPs were found to correlate with the binding constants for these nucleotides determined by Huberman and Kornberg [33]. Based on these findings, it was proposed that the active sites for the polymerase and exonuclease activities are distinct and separate, and a model was proposed for the coordination of the

3'- to 5'-exonuclease and polymerase activities in maintaining the fidelity of DNA synthesis [29,32]. In this model it was suggested that a primer terminus can bind to one of the two active sites of the DNA polymerase and that the fate of the primer terminus, either extension or hydrolysis, depends on its ability to base-pair with the base at the template, the stability of the base pair formed, and the rate of subsequent chain extension. A mismatched nucleotide at the primer terminus would therefore have two effects: (1) since it could not base-pair with the template base, its binding to the exonuclease active site is favored and it would become a substrate for hydrolysis, and (2) because of its inability to form a stable base pair, the rate of formation of the next phosphodiester bond would be slow and this in turn would favor its binding at the exonuclease active site, resulting in its hydrolysis.

Evidence that the 3'- to 5'-exonuclease activity serves a proofreading function was first demonstrated by Brutlag and Kornberg [40], who suggested that the specificity of the exonuclease activity for a mismatched or frayed primer terminus would allow error correction prior to further chain extension. Further evidence for the role of the exonuclease in maintaining replication fidelity has come from studies with mutant T_4 DNA polymerases which exhibit either mutator or antimutator phenotypes. Examination of the ratios of exonuclease to polymerase of the partially purified DNA polymerases from these mutants has shown that mutator strains have a lower ratio of exonuclease to polymerase than the wild type, whereas anitmutator strains have a higher ratio [41].

Additional evidence that proofreading by the 3'- to 5'-exonuclease activity of DNA polymerase has a role in maintaining the fidelity of DNA synthesis has come from the studies in which selective inhibition of 3'- to 5'-exonuclease activity by 5'-NMP was shown to diminish the fidelity of DNA synthesis by DNA polymerase I on synthetic polynucleotide template/primers [42] and from similar studies by Kunkel et al. [38], in which the inhibition of exonuclease activity by 5'-NMP was found to increase error frequency with a natural DNA template/primer. Similarly, selective inhibition of 3'- to 5'-exonuclease activity of DNA polymerase III holoenzyme by 5'-NMP was also found to result in a decrease in the fidelity of DNA synthesis with either synthetic polynucleotide or natural DNA template/primers [43]. Increases in error frequency were also found to result when deoxynucleoside thiotriphosphates were substituted for dNTPs as substrates for DNA polymerase I [44]. These nucleotide analogs, which contain a sulfur atom in place of an oxygen on the α-phosphate, are incorporated into DNA by the polymerase activity but are not effectively excised by the 3'- to 5'-exonuclease activity of DNA polymerase I [44-46]. The effect of high concentrations of the next correct nucleotide in decreasing the fidelity of DNA synthesis with DNA polymerases containing an associated 3'- to 5'-exonuclease activity, but not with DNA polymerases devoid of proofreading activity, is also consistent with a proofreading function for the 3'- to 5'-exonuclease activity, as the rapid addition of the next correct nucleotide after incorporation of a mismatched nucleotide would allow less opportunity for the enzyme to excise the mismatched nucleotide [38,43].

5. MANGANESE MUTAGENESIS: EFFECTS ON ERROR DISCRIMINATION

The mechanism by which Mn^{2+} decreases the fidelity of DNA synthesis is not fully understood. For a DNA polymerase with proofreading 3'- to 5'-exonuclease activity, Mn^{2+} could exert its effects on either or both of the two base-pairing selection steps involved in discrimination between complementary and mismatched nucleotides: (1) insertion of a properly base-paired nucleotide into the primer terminus by the polymerase activity (base selection) and/or (2) removal of a non-base-paired nucleotide at the primer terminus by the 3'- to 5'-exonuclease activity prior to chain extension (proofreading).

5.1. Effects of Mn^{2+} on Base Selection

The effects of Mn^{2+} on base selection have been investigated by determining the apparent K_m and V_{max} values for insertion of either complementary or noncomplementary nucleotide substrates into the hook polymers poly(dC)-oligo(dG) [34] and poly(dT)-oligo(dA) [34] by the large fragment of DNA polymerase I in the presence of either Mg^{2+} or Mn^{2+} or both. With both hook polymers the substitution of Mn^{2+} for Mg^{2+} as divalent cation or addition of Mn^{2+} to optimal levels of Mg^{2+} resulted in an increased V_{max} for insertion of the noncomplementary nucleotide. In contrast, Mn^{2+} had no significant effect on the K_m values for either complementary or noncomplementary nucleotide substrates. With poly(dC)-oligo(dG) as template/primer (Table 3), the V_{max} for insertion of the noncomplementary substrate dATP was increased approximately 50-fold in the presence of Mn^{2+} vs. Mg^{2+} while the V_{max} for insertion of the complementary substrate dGTP was decreased approximately twofold. With poly(dT)-oligo(dA) as template/primer (Table 4), the substitution of Mn^{2+} for Mg^{2+} increased the V_{max} for insertion of the noncomplementary nucleotide dGTP approximately 40-fold and had no effect on the V_{max} for insertion of the complementary substrate dATP. With either poly(dC)-oligo(dG) or poly(dT)-oligo(dA) as template/primer, substitution of Mn^{2+} for Mg^{2+} resulted in a marked increase in the insertion of noncomplementary pyrimidine nucleotides, from nondetectable levels in the presence of Mg^{2+} to easily detectable levels in the presence of Mn^{2+}. Furthermore, the effects of Mn^{2+} predominated when both cations were present, i.e., the increased V_{max} values for insertion of noncomplementary nucleotide substrates observed in the presence of Mn^{2+} were not significantly different in the presence of both Mn^{2+} and Mg^{2+}.

With both hook polymers studied, substitution of Mn^{2+} for Mg^{2+} had no significant effect on the apparent K_m values for either complementary or noncomplementary substrates. These results are in contrast to those of Fersht et al. [6], who found that substitution of Mn^{2+} for

TABLE 3

Effects of Mn^{2+} on the Specificity of Insertion of Either Complementary or Mismatched Nucleotides into Poly(dC)-oligo(dG)

Divalent cation	K_m (μM)	V_{max} (μmol/30'/mg)	k_{cat} (sec^{-1})	k_{cat}/K_m (1/mol/sec)
	Insertion of dGTP (G:C pair)			
5.0 mM Mg^{2+}	3.7 ± 0.5	1120 ± 53	42.4	$1.2 \cdot 10^7$
0.1 mM Mn^{2+}	3.5 ± 1.2	515 ± 66	19.5	$5.6 \cdot 10^6$
0.3 mM Mn^{2+}	2.5 ± 0.7	315 ± 32	11.9	$4.8 \cdot 10^6$
5.0 mM Mg^{2+} + 1.0 mM Mn^{2+}	1.9 ± 0.5	222 ± 14	8.39	$4.5 \cdot 10^6$
	Insertion of dATP (A:C mispair)			
5.0 mM Mg^{2+}	63.7 ± 11	0.17 ± 0.03	0.0064	$1.0 \cdot 10^2$
0.3 mM Mn^{2+}	78.9 ± 32	8.02 ± 1.8	0.303	$3.8 \cdot 10^3$
5.0 mM Mg^{2+} + 0.5 mM Mn^{2+}	58.0 ± 10	6.62 ± 0.36	0.250	$4.3 \cdot 10^3$
	Insertion of dTTP (T:C mispair)			
5.0 mM Mg^{2+}		(a)		
0.5 mM Mn^{2+}	119 ± 30	7.34 ± 0.8	0.278	$2.3 \cdot 10^3$

(a) Below detection limit.

Note: The concentration of dGTP was varied over a 150-fold concentration range (0.72-107 μM). dATP and dTTP concentrations were varied over a 200-fold concentration range (10 μM to 2 mM). Apparent K_m and V_{max} values are reported ±1 SEM. Assay conditions were as described in Ref. 34.

TABLE 4

Effects of Mn^{2+} on the Specificity of Insertion of Either Complementary or Mismatched Nucleotides into Poly(dT)-oligo(dA)

Divalent cation	K_m (μM)	V_{max} (μmol/30'/mg)	k_{cat} (sec^{-1})	k_{cat}/K_m (1/mol/sec)
		Insertion of dATP (A:T pair)		
1.0 mM Mg^{2+}	8.74 ± 1.2	1413 ± 43	53.4	$6.1 \cdot 10^6$
5.0 mM Mg^{2+}	3.27 ± 0.42	139 ± 3.2	5.26	$1.6 \cdot 10^6$
0.2 mM Mn^{2+}	3.46 ± 0.48	1075 ± 29	40.6	$1.2 \cdot 10^7$
1.0 mM Mn^{2+}	13.6 ± 2.0	151 ± 5.3	5.71	$4.2 \cdot 10^5$
		Insertion of dGTP (G:T mispair)		
1.0 mM Mg^{2+}	51.5 ± 10	0.86 ± 0.05	0.033	$6.4 \cdot 10^2$
5.0 mM Mg^{2+}	71.0 ± 21	0.81 ± 0.08	0.031	$4.4 \cdot 10^2$
0.3 mM Mn^{2+}	62.9 ± 3	34.7 ± 0.5	1.31	$2.1 \cdot 10^4$
1.0 mM Mn^{2+}	44.6 ± 21	3.86 ± 0.42	0.146	$3.3 \cdot 10^3$
		Insertion of dCTP (C:T mispair)		
1.0 mM Mg^{2+}		(a)		
0.3 mM Mn^{2+}	213 ± 16	18.2 ± 0.4	0.69	$3.2 \cdot 10^4$

(a) Below detection limit.

Note: Substrate concentrations were varied as follows: dATP (1.7-202 μM), dGTP (5.5 μM to 1 mM), and dCTP (2.1 μM to 2 mM). Apparent K_m and V_{max} values are reported as mean values ± SEM. In Mn^{2+}-activated reactions, dNTP was added as a 1:1 complex with cation. Assay conditions were as described in Ref. 35.

Source: Reprinted from [35] by permission of the American Chemical Society.

Mg^{2+} resulted in markedly lower apparent K_m values for noncomplementary dNTPs with DNA polymerase I. This discrepancy may be explained by the effects of the different buffers that were used in these two studies.

In the studies of Fersht et al. [6], apparent K_m and V_{max} values for complementary and noncomplementary nucleotide substrates were determined in phosphate buffer when Mg^{2+} was the divalent cation and in Tris-HCl buffer when Mn^{2+} was the divalent cation. When the apparent K_m values for complementary and noncomplementary nucleotides were determined in both Hepes buffer and phosphate buffer with Mg^{2+} as divalent cation, it was found that the K_m values for complementary nucleotides were essentially the same in both buffers. However, the K_m values for noncomplementary nucleotides were markedly increased in phosphate buffer as compared to Hepes buffer, i.e., 10- to 20-fold [35]. The marked effect of phosphate ion on the kinetic constants for insertion of noncomplementary nucleotides but not complementary nucleotides suggested that phosphate ion might be affecting the 3'- to 5'-exonuclease activity of DNA polymerase I, since the rate of insertion of a noncomplementary nucleotide is determined by measuring the rate of template-dependent turnover of dNTP to dNMP, which involves alternate incorporation by the polymerase activity and hydrolysis by the 3'- to 5'-exonuclease activity, while the rate of insertion of complementary nucleotides is essentially equivalent to the rate of incorporation of dNMP into polynucleotide.

The effects of phosphate ion on the rate of turnover of mismatched dGTP to dGMP with poly(dA-dT) as template/primer and the rate of hydrolysis of poly(dT) were determined. It was found that both the rate of hydrolysis of poly(dT) and the rate of turnover of a noncomplementary nucleotide were extremely sensitive to inhibition by phosphate ion, 50% inhibition occurring at 25 mM and greater than 90% inhibition at 80 mM phosphate ion. In contrast, the polymerase activity of DNA polymerase I was not inhibited by phosphate ion; in fact, the rate of incorporation of complementary dATP and dTTP into poly(dA-dT) was somewhat stimulated by the addition of phosphate ion [35]. This suggests that phosphate ion, by selectively inhibiting the proofreading activity of DNA polymerase I, can affect the kinetic constants for insertion of noncomplementary nucleotides but not complementary nucleotides. It is likely, therefore, that in the previous studies by Fersht et al. [6], the marked decrease in apparent K_m for noncomplementary nucleotides that was observed when Mn^{2+} in Tris-HCl buffer was substituted for Mg^{2+} in phosphate buffer was the result of the change in buffer rather than the change in divalent cation. When the buffer is held constant, the substitution of Mn^{2+} for Mg^{2+} does not significantly affect the apparent K_m values for either complementary or noncomplementary nucleotides. V_{max} values for insertion of noncomplementary nucleotides are markedly increased whereas the V_{max} values for insertion of complementary nucleotides are only slightly affected. Thus, Mn^{2+} alters the specificity of base selection primarily by increasing the catalytic rate for insertion of a noncomplementary nucleotide substrate.

Studies on the effects of substitution of Mn^{2+} for Mg^{2+} on the kinetic parameters for insertion of 2-aminopurine deoxynucleoside triphosphate (dAPTP) as a mismatched nucleotide with T_4 DNA polymerase [10] suggested that the major effect was a differential decrease in the K_m for the insertion of dAPTP relative to dATP. The V_{max} for the insertion of dAPTP was identical to that for dATP in either Mg^{2+} or Mn^{2+}, and when Mn^{2+} was substituted for Mg^{2+}, both reactions were inhibited to an equivalent extent, i.e., 40-fold. In contrast, substitution of Mn^{2+} for Mg^{2+} resulted in a 10-fold reduction in the K_m for the "mismatched" nucleotide dAPTP and a relatively smaller (two- to threefold) reduction in the K_m for the complementary nucleotide dATP. At present it is not clear whether the different effects of Mn^{2+} observed in the two studies represent true differences between the two DNA polymerases or differences in the assay systems.

5.2. Effects of Mn^{2+} on Proofreading

To determine whether Mn^{2+} affects the specificity of proofreading by the 3'- to 5'-exonuclease activity of DNA polymerase I, the rates

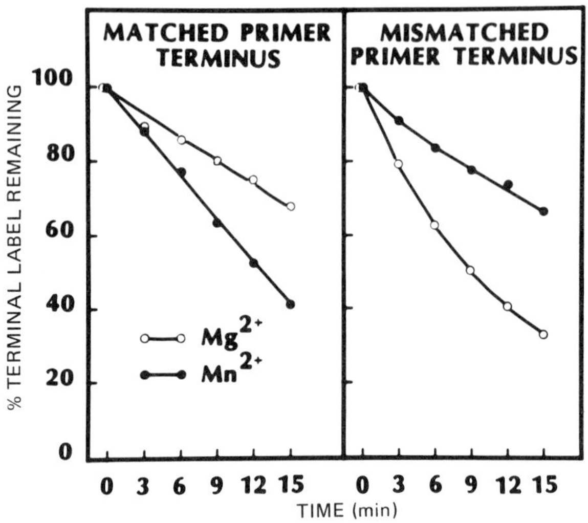

FIG. 2. Effect of 5 mM Mg^{2+} and 0.5 mM Mn^{2+} on the rates of hydrolysis of either complementary dGMP or noncomplementary dAMP from the primer terminus of poly(dC)-oligo(dG). For hydrolysis of the complementary primer terminus (panel A), the concentration of enzyme was 10 times that for hydrolysis of the noncomplementary primer terminus (panel B). (Reprinted from Ref. 34 with permission.)

of hydrolysis of matched and mismatched primer termini were measured in the presence of either Mg^{2+} or Mn^{2+}. With poly(dC)-oligo(dG) containing either a complementary (dGMP) or noncomplementary (dAMP) nucleotide at the 3' terminus, the rate of hydrolysis of the mismatched dAMP terminus was found to be two- to threefold greater in Mg^{2+} than in Mn^{2+}, while the rate of hydrolysis of the matched dGMP terminus was approximately twofold higher in Mn^{2+} than in Mg^{2+} (Fig. 2). Similarly, with poly(dT)-oligo(dA) containing either a complementary (dAMP) or noncomplementary (dGMP) nucleotide at the 3' terminus, the rate of hydrolysis of the mismatched dGMP terminus was approximately twofold higher in Mg^{2+} than in Mn^{2+}, whereas the rate of hydrolysis of the base-paired dAMP terminus was two- to threefold higher in Mn^{2+} than in Mg^{2+} [35]. Thus, the specificity of the proofreading 3'- to 5'-exonuclease for a mismatched primer terminus is altered in the presence of Mn^{2+}.

Previous studies on the effects of Mn^{2+} on the 3'- to 5'-exonuclease activity of DNA polymerase I suggested that inhibition of proofreading does not play a role in Mn^{2+} mutagenesis, particularly since mutagenesis induced by Mn^{2+} can occur in DNA polymerases lacking proofreading exonuclease activity, i.e., AMV reverse transcriptase and DNA polymerases α and β [1]. Sirover et al. [47] demonstrated no change in the extent of hydrolysis of a mismatched primer terminus when Mg^{2+} was replaced by Mn^{2+}, and Miyaki et al. [48] showed that the rate of hydrolysis of a properly base-paired primer terminus was stimulated by Mn^{2+}. The results of our studies with hook polymers either poly(dC)-oligo(dG) or poly(dT)-oligo(dA) show that, whereas the rate of hydrolysis of a properly base-paired primer terminus is stimulated in the presence of Mn^{2+}, the rate of hydrolysis of a mismatched primer terminus is inhibited. Thus, substitution of Mn^{2+} for Mg^{2+} alters the specificity of hydrolysis by the 3'- to 5'-exonuclease activity. Furthermore, the decreased ability of the enzyme to hydrolyze a mismatched nucleotide at the primer terminus when Mg^{2+} is replaced by Mn^{2+} could allow extension of mismatched primer termini, resulting in increased error frequency.

The substitution of Mn^{2+} for Mg^{2+} has also been found to affect the sensitivity of the 3'- to 5'-exonuclease to inhibition by nucleoside monophosphates (5'-NMP). Previous studies in our laboratory showed that 5'-NMPs selectively inhibit the 3'- to 5'-exonuclease activity but not the polymerase activity of DNA polymerase I by competing with the primer terminus for binding at the active site of the exonuclease [32,42]. We have found that, in the presence of Mn^{2+} vs. Mg^{2+}, the concentration of 5'-NMP required to inhibit the exonuclease activity 50% was decreased five- to 10-fold, depending on the nature of the 5'-NMP [35].

The fact that Mn^{2+} alters not only the specificity of base selection by the polymerase activity and the specificity of proofreading by the 3'- to 5'-exonuclease activity, but also the sensitivity of the exonuclease activity to inhibition by 5'-NMP suggests that the mutagenicity of Mn^{2+} may be due to its interaction with the enzyme-template complex, resulting in an altered conformation at the active site of both the polymerase and exonuclease activities.

The effects of Mn^{2+} on the specificity of base selection and proofreading by DNA polymerase I with both G:C-containing templates and A:T-containing templates are at least qualitatively the same, suggesting that base-specific effects of Mn^{2+} are unlikely to account for its mutagenic effect [34,35]. On the other hand, Hillebrand and Beattie [49] found that with natural DNA templates the fidelity of DNA polymerase I in the presence of Mn^{2+} vs. Mg^{2+} is sequence-dependent. In these studies, Mn^{2+}-induced misincorporation was found to correlate with the occurrence of G:C base pairs 5' to the misincorporation. Since a G:C base pair at the primer terminus stabilizes the primer terminus to excision by the exonuclease activity [50,51], it is possible that the effects of Mn^{2+} on the specificity of proofreading may account for the sequence-dependent effects seen with natural DNA templates.

6. MANGANESE MUTAGENESIS: POTENTIAL SITES OF ACTION

Since Mn^{2+} can bind to either the DNA polymerase, the dNTP substrates, or the template/primer, it can exert its mutagenic effects through its interaction with one or more of these molecules. The binding of both dNTPs and dNMPs to DNA polymerase I in the absence of template/primer has been shown to require divalent cation, and each has been shown to bind to a distinct site on the polymerase [52].

NMR studies with DNA polymerase I have suggested that the enzyme, in the absence of template/primer, immobilized dNTP substrates in a β-like conformation, and it has been suggested that enhanced misincorporation in the presence of Mn^{2+} could be the result of its interaction with dNTP substrates so as to change their base-pairing specificities [53]. However, studies with both *E. coli* and T_4 DNA polymerases have shown that addition of increasing concentrations of Mg^{2+} to Mn^{2+}-containing reaction mixtures, up to a ratio of 10:1, does not reduce the error frequency observed in the presence of Mn^{2+} alone. Since the binding affinities of Mn^{2+} and Mg^{2+} for dNTPs are very similar, the lack of strong competition by Mg^{2+} in fidelity experiments argues against the idea that Mn^{2+} effects are exerted solely through its interaction with dNTPs [9,34,35].

Studies by Slater et al. [54] on the binding of Mn^{2+} to DNA polymerase I demonstrated that there is one tight binding site with a dissociation constant of 1 μM, four intermediate binding sites with dissociation constants of 29 μM, and approximately 20 weak binding sites with dissociation constants of about 830 μM. It was proposed that the tight binding site corresponds to the polymerase active site, and at concentrations of Mn^{2+} where the active site would be occupied (less than 10 μM), a threefold increase in error rate was observed.

Beckman et al. [55] determined the fidelity of DNA synthesis with DNA polymerase I as a function of free Mn^{2+} concentration and correlated this with measured dissociation constants of Mn^{2+} from enzyme, template, and dNTP substrates. At very low Mn^{2+} concentrations (less than 1 μM), where binding of Mn^{2+} to the DNA polymerase occurs, the overall fidelity of DNA synthesis by DNA polymerase I was found to be similar to that of the Mg^{2+}-activated enzyme, whereas at Mn^{2+} concentrations between 1 and 100 μM, where binding of Mn^{2+} to the DNA template occurs, fidelity of DNA synthesis was markedly affected. The results of this study differ somewhat from those of previous studies by Slater et al. [54], who reported that a large increase in misincorporation with DNA polymerase I occurs at free Mn^{2+} concentrations between 100 and 140 μM, where the binding of Mn^{2+} to the template is increasing most rapidly. The results suggested to these authors that Mn^{2+}-induced modification of the template may be the primary mechanism of Mn^{2+}-induced mutagenesis, i.e., that the binding of Mn^{2+} to the template causes changes in base-pairing specificity. However, the interpretation that Mn^{2+}-induced mutagenesis is due to changes in base-pairing specificity is not consistent with our findings of marked effects of low Mn^{2+} concentrations on the individual error discrimination steps, i.e., base selection and proofreading [34, 35]. Furthermore, substitution of Mn^{2+} for Mg^{2+} was found to result in a large increase in the V_{max} for insertion of noncomplementary nucleotides by DNA polymerase I, with little or no effect on the apparent K_m values for insertion of either complementary or noncomplementary nucleotides [34,35]. One would expect apparent K_m changes rather than V_{max} changes if the effect of Mn^{2+} is on altering the specificity of base pairing, whether mediated through enzyme or template effects. However, the fact that Mn^{2+} alters not only the specificity of base selection by the polymerase activity and the specificity of proofreading by the 3'- to 5'-exonuclease activity, but also the sensitivity of the exonuclease activity to inhibition by 5'-NMP suggests that the mutagenicity of Mn^{2+} may be due to its interaction with the enzyme-template complex, resulting in an altered conformation at the active sites of both the polymerase and exonuclease activities. Whether the effects of Mn^{2+} are primarily the result of an altered enzyme conformation or whether they are secondary to changes in DNA structure must await further studies.

Recently, more definitive studies on the binding of metals to specific sites on the Klenow fragment of DNA polymerase I were performed [39]. Crystal structure analysis of a Mn^{2+}-dNMP-enzyme complex revealed that the smaller N-terminal domain, which contains the 3'- to 5'-exonuclease active site, contains two binding sites (A and B) for divalent metal ion. By means of site-directed mutagenesis, two mutant proteins were produced, both of which had high levels of polymerase activity but essentially no exonuclease activity. Crystallographic analysis of the mutant proteins demonstrated that mutation at site B resulted in a loss of catalytic activity but not the ability of the enzyme to bind dNMP (a substrate analog) and metal ion at site A, suggesting that the metal ion that binds at site B plays a direct role in catalysis. In contrast, mutation at site A led to a loss of the ability of the enzyme to bind metal ion either at site A or B or to bind dNMP, suggesting that metal ion A is involved in substrate binding.

7. GENERAL CONCLUSIONS

The mechanisms by which the mutagenic metal ion Mn^{2+} diminishes the fidelity of DNA synthesis have been studied by multiple laboratories over a number of years. Although many approaches have been used to resolve this question, a clear understanding of the effects of Mn^{2+} on the individual error discrimination steps carried out by DNA polymerases is not yet at hand. Some of the discrepancies and controversies that have arisen can be ascribed to differences in assay conditions used by different laboratories, e.g., buffer effects, and some may be the result of genuine differences among different DNA polymerases. The recent availability of high-resolution structural information on the Klenow fragment of DNA polymerase I, as well as the wealth of biochemical information on this enzyme obtained over a number of years, makes it an excellent model system for resolving these questions. The ultimate resolution of the pleiotropic effects of Mn^{2+} on error discrimination will clearly require a multifaceted approach, ideally involving biochemical, structural, and genetic studies.

ACKNOWLEDGMENT

The work from our laboratories reviewed here was supported by grant GM33507 from the National Institutes of Health. We thank our coworkers, listed on the publications cited from our laboratories, for their many contributions.

REFERENCES

1. L. A. Loeb and A. S. Mildvan, in "Metal Ions in Genetic Information Transfer" (G. L. Eichhorn and L. G. Marzilli, eds.), Elsevier North-Holland, New York, 1981, p. 125 ff.
2. A. Orgel and L. E. Orgel, *J. Mol. Biol., 14,* 453 (1965).

3. G. D. Stoner, M. G. Shinkin, M. C. Troyell, T. L. Thompson, and L. S. Terry, *Cancer Res.*, *36*, 1744 (1976).

4. T. A. Trautner, M. N. Swartz, and A. Kornberg, *Proc. Natl. Acad. Sci. USA*, *48*, 499 (1962).

5. M. A. Sirover, D. K. Dube, and L. A. Loeb, *J. Biol. Chem.*, *254*, 107 (1979).

6. A. R. Fersht, J.-P. Shi, and W.-C. Tsui, *J. Mol. Biol.*, *165*, 655 (1983).

7. Z. W. Hall and I. R. Lehman, *J. Mol. Biol.*, *36*, 321 (1968).

8. F. D. Gillin and N. G. Nossal, *J. Biol. Chem.*, *251*, 5225 (1976).

9. U. Hibner and B. M. Alberts, *Nature*, *205*, 300 (1980).

10. M. F. Goodman, S. Keener, S. Guidotti, and E. W. Branscomb, *J. Biol. Chem.*, *258*, 3469 (1983).

11. S. Linn, M. Kairis, and N. Holliday, *Proc. Natl. Acad. Sci. USA*, *73*, 2818 (1976).

12. T. S.-F. Wang, D. L. Eichler, and D. Korn, *Biochemistry*, *16*, 4927 (1977).

13. M. A. Sirover and L. A. Loeb, *J. Biol. Chem.*, *252*, 3605 (1977).

14. W. E. Brown, K. H. Stamp, and W. S. Kelley, *J. Biol. Chem.*, *257*, 1965 (1982).

15. C. M. Joyce, W. S. Kelley, and N. D. F. Grindley, *J. Biol. Chem.*, *257*, 1958 (1982).

16. D. L. Ollis, P. Brick, R. Hamlin, N. G. Xuong, and T. A. Steitz, *Nature*, *313*, 762 (1985).

17. C. M. Joyce and T. A. Steitz, *Trends Biochem. Sci.*, *12*, 288 (1987).

18. J. W. Drake, *Nature*, *221*, 1132 (1969).

19. E. C. Cox, *Ann. Rev. Genet.*, *10*, 135 (1976).

20. L. A. Loeb and T. A. Kunkel, *Ann. Rev. Biochem.*, *51*, 429 (1982).

21. A. S. Mildvan, *Ann. Rev. Biochem.*, *43*, 357 (1974).

22. T. A. Kunkel and L. A. Loeb, *Science*, *213*, 765 (1981).

23. F. D. Gillin and N. G. Nossal, *Biochem. Biophys. Res. Commun.*, *64*, 457 (1975).

24. D. J. Galas and E. W. Branscomb, *J. Mol. Biol.*, *124*, 653 (1978).

25. L. K. Clayton, M. G. Goodman, E. W. Branscomb, and D. J. Galas, *J. Biol. Chem.*, *254*, 1902 (1979).

26. S. M. Watanabe and M. G. Goodman, *Proc. Natl. Acad. Sci. USA*, *79*, 6429 (1982).

27. J. Petruska, L. C. Sowers, and M. F. Goodman, *Proc. Natl. Acad. Sci. USA*, *83*, 1559 (1986).

28. B. Reckman, F. Grosse, and G. Krauss, *Nucleic Acids Res., 11,* 7251 (1983).

29. B. G. Que, K. M. Downey, and A. G. So, *Biochemistry, 18,* 2064 (1979).

30. J. J. Hopfield, *Proc. Natl. Acad. Sci. USA, 71,* 4135 (1974).

31. A. Fersht, "Enzyme Structure and Mechanism", W. H. Freeman, New York, 1985, p. 366.

32. B. G. Que, K. M. Downey, and A. G. So, *Biochemistry, 17,* 1603 (1978).

33. J. A. Huberman and A. Kornberg, *J. Biol. Chem., 236,* 864 (1970).

34. W. S. El-Deiry, K. M. Downey, and A. G. So, *Proc. Natl. Acad. Sci. USA, 81,* 7378 (1984).

35. W. S. El-Deiry, A. G. So, and K. M. Downey, *Biochemistry, 27,* 546 (1988).

36. W. S. El-Deiry, A. G. So, and K. M. Downey, unpublished observations.

37. H. Klenow and K. Overgaard-Hansen, *Biochim. Biophys. Acta, 665,* 187 (1981).

38. T. A. Kunkel, R. M. Schaaper, R. A. Beckman, and L. A. Loeb, *J. Biol. Chem., 256,* 9883 (1981).

39. V. Derbyshire, P. S. Freemont, M. R. Sanderson, L. Beese, J. Freidman, C. M. Joyce, and T. A. Steitz, *Science, 240,* 199 (1988).

40. D. Brutlag and A. Kornberg, *J. Biol. Chem., 247,* 241 (1972).

41. N. Muzyczka, R. L. Poland, and M. J. Bessman, *J. Biol. Chem., 247,* 7116 (1972).

42. J. J. Byrnes, K. M. Downey, B. G. Que, M. Y. W. Lee, V. Black, and A. G. So, *Biochemistry, 16,* 3740 (1977).

43. A. R. Fersht and J. W. Knill-Jones, *J. Mol. Biol., 165,* 655 (1983).

44. T. A. Kunkel, F. Eckstein, A. S. Mildvan, R. M. Koplitz, and L. A. Loeb, *Proc. Natl. Acad. Sci. USA, 78,* 6734 (1981).

45. H. P. Vosberg and F. Eckstein, *Biochemistry, 16,* 3633 (1977).

46. P. M. J. Burgers and F. Eckstein, *J. Biol. Chem., 254,* 6889 (1979).

47. M. A. Sirover, D. K. Dube, and L. A. Loeb, *J. Biol. Chem., 254,* 107 (1979).

48. M. Miyaki, I. Murata, M. Osabe, and T. Ono, *Biochem. Biophys. Res. Commun., 77,* 854 (1977).

49. G. G. Hillebrand and K. L. Beattie, *Nucleic Acids Res., 12,* 3173 (1984).

50. R. C. Pless, L. M. Levitt, and M. J. Bessman, *Biochemistry, 20,* 6235 (1981).

51. J. E. Patten, A. G. So, and K. M. Downey, *Biochemistry, 23,* 1613 (1984).

52. A. Kornberg, "DNA Replication", W. H. Freeman and Company, New York, 1980.

53. D. L. Sloan, L. A. Loeb, A. S. Mildvan, and R. J. Feldman, *J. Biol. Chem., 250,* 8913 (1975).

54. J. P. Slater, I. Tamir, L. A. Loeb, and A. S. Mildvan, *J. Biol. Chem., 247,* 2784 (1972).

55. R. A. Beckman, A. S. Mildvan, and L. A. Loeb, *Biochemistry, 24,* 5810 (1985).

2

Metallonucleases: Real and Artificial

Lena A. Basile and Jacqueline K. Barton
Department of Chemistry
Columbia University
Havemeyer Hall, Box 953
116th Street and Broadway
New York, NY 10027

1. INTRODUCTION

Nucleases, molecules which cleave DNA either at specific sites along the strand or in an indiscriminant, nonspecific fashion, have become the molecular scalpels of the biochemist. They are the indispensable tools for analyzing DNA structure, sequencing DNA molecules, and isolating and cloning genes. Although the use of nucleases in modern-day molecular biology has become quite commonplace, an understanding of the chemistry behind these tools has been less clearly developed.

The reactions performed by both site-specific and nonspecific cleaving molecules are quite remarkable and intriguingly complex. These reactions can be broken down in terms of two distinct chemical events: (1) the binding step, involving recognition of a particular aspect of DNA base sequence or structure; and (2) the cleaving step, involving a series of transformations which lead either directly or indirectly to hydrolysis of the phosphodiester linkage and scission of the DNA backbone. It is curious to consider how a restriction endonuclease, for example, first is able to search out and find a particular sequence of four to eight bases along the endless helical strand, and second is able then to effect double-stranded endonucleolytic cleavage, generally a difficult nucleophilic reaction which must here be performed along a polyanion. Recently, small synthetic DNA cleaving molecules have also been developed, in part, as a means of dissecting the elements necessary to obtain site-specific reactions along the strand, and studies with these cleaving molecules have been helpful in pinpointing key features needed for specificity and reactivity along the helix. These small inorganic complexes furthermore are themselves now providing valuable tools for molecular biologists.

In this chapter we describe a variety of nucleolytic reactions, using as examples both true enzymes and synthetic molecules, and all of which require metal ions for activity. For the purpose of this chapter, we define a "true metallonuclease" as an enzyme which degrades nucleic acids and, in doing so, has an absolute requirement for metal ions; we define an "artificial metallonuclease" as a synthetic molecule which also degrades nucleic acids and depends on metal ions for activity, but does not react catalytically in the true sense. For naturally occurring nucleases, numerous examples exist in which metal ions have been shown to be essential to the reactivity of the enzyme. Only in a few cases, however, has the role for the metal ion involved been elucidated. For synthetic nucleases, the primary role of the metal has been generally to serve as a center for the generation of metal-mediated redox reactions in close proximity to the DNA helix leading to scission of the DNA strand. The reactions mediated by these redox-based nucleases, therefore, are clearly quite different from the types of reactions performed by true DNA-cleaving enzymes.

For what roles might a metal cation be particularly advantageous? With respect to aspects of recognition, the metal center may serve as a structural center for the DNA-binding moiety, and it is becoming increasingly clear that the recognition of structural elements determined by base sequence may be critical to site specificity. Some examples of this notion in the case of recognition by small synthetic molecules are described below. With respect to aspects of endonucleolytic cleavage, the need to carry out a nucleophilic reaction at the negatively charged DNA phosphodiester introduces an important function for metal cations. Indeed, most of the phosphate chemistry carried out in nature appears to be catalyzed by metalloenzymes, and we will see many such examples in this chapter. In addition, the metal ion may serve a structural role in the catalytic step, since metal ion coordination can serve to orient catalytically important groups for reactivity.

Shown in Fig. 1 is the substrate for these nucleolytic cleavage reactions and, in some respect, the ligands for the metal ions, double-helical DNA in its various double-stranded conformations (for a review, see [1]). The structures of DNA vary substantially, and in particular with regard to helical pitch, groove dimensions, and even helix chirality. The B form of DNA, most common, is characterized by a regular right-handed helical structure, with base pairs stacked perpendicular to the helical axis. (Recent crystallographic results, however, have indicated a propeller twisting of the bases paired one to another.) Distinctive helical grooves are apparent in a B-form helix with average widths of 17 and 12 Å for major and minor grooves, respectively. The A-form helix is also of right-handed helical chirality, but here the bases are tilted by as much as 15 Å with respect to the helix axis and skewed outward toward the direction of the minor groove. What results for the A conformation is therefore more of a helical barrel, with the minor groove acting as the barrel's surface and what was the major groove for the B-form helix now pulled deeply into the helix interior. The Z-form helix is of course most dramatically different. The long and slender helix spirals in a left-handed sense and in an alternating zig-zag fashion, owing to the alternation in both sugar pucker and base orientation about the glycosidic bond (syn vs. anti) up the strand. Z-DNA is characterized by a wide and shallow major groove, really more of a convex surface, and a minor groove narrowed down into a crevice between major groove surfaces. These different helical structures are representative of the diversity of helical conformations that double-stranded DNA may adopt. Other conformations exist as well, including bent sites, hairpin loops, and cruciforms. In considering how nucleases, natural and synthetic, both recognize and react at sites along the DNA strand,

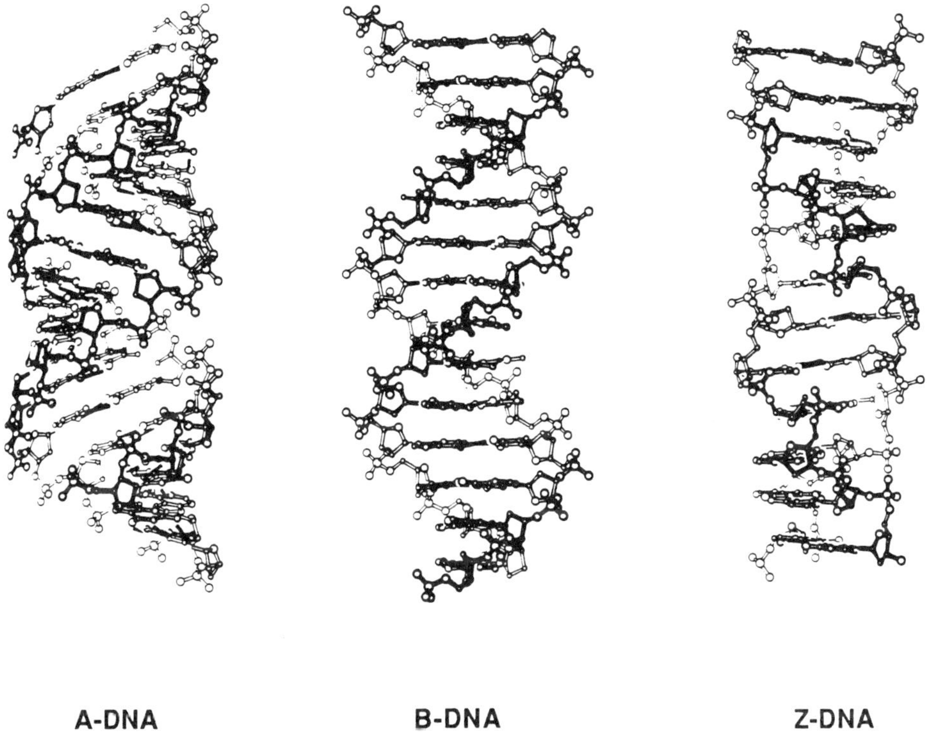

FIG. 1. Structural conformers of double-helical DNA. (Adapted from Ref. 1.)

it is important to keep in mind this polymorphic character of DNA, and how the variations in DNA structure may affect both binding selectivity and reactivity.

An examination of metal-mediated chemistry of nucleases, both enzymatic and synthetic systems, becomes quite helpful in understanding the DNA binding and cleavage reactions carried out in nature and in the future design and applications of new artificial nucleases. Those features which are common to or even distinctive among DNA binding metalloenzymes and small synthetic coordination complexes, furthermore, provide a basis for the development of still more efficient and specific schemes. This chapter, which describes a variety of such nucleolytic reactions, is intended not as an exhaustive review of the literature but instead, hopefully, as an illustrative account, one which may highlight some of the intriguing chemistry carried out by these different molecules and one which may underscore points where the metal may hold an essential role in activity.

2. NONSPECIFIC DNA CLEAVAGE

We examine first molecules which cleave DNA in a random, nonspecific fashion. These molecules are often quite useful as tools to examine other molecules specifically bound to DNA, and therefore the lack of inherent specificity in these reagents themselves is essential. To cleave DNA in a nondiscriminant manner requires both a binding interaction which involves no site or sequence selectivity and a cleaving reaction which shows no dependence on base composition or conformation. Nonspecific nucleases in fact usually react with little, if any, base specificity but instead appear to recognize changes in the local conformation of DNA, such as groove widths and orientations about the phosphate backbone; therefore some sequence or structural selectivity is encountered. Metal ions facilitate the nonspecific cleavage of DNA. With respect to sequence neutrality in the cleavage reaction, the ideal reaction might be that involving hydrolysis of the phosphodiester. One obvious role for metal ions in hydrolysis of DNA is to function as electrophilic catalysts, i.e., in polarizing bonds and reducing the negative charge on the phosphate backbone, rendering the phosphodiester linkage more vulnerable to nucleophilic attack. We encounter several such examples below. However, model systems have elucidated also the importance of metal ions as a source of metal-bound hydroxide ions, a potent nucleophile at neutral pH. Curiously, the highest degree of nonspecific or sequence-neutral cleavage of DNA with synthetic molecules, as described below, is found through an innocent delivery of hydroxyl radicals to the DNA helix. All told, what is required for the design of a nonspecific nuclease is sequence-independent binding, if binding at all, and sequence-independent reactivity, requirements perhaps as complex to obtain as those necessary for sequence specificity.

2.1. Enzymatic Cleavage

2.1.1. S1 Nuclease and Other Single-Strand Specific Metallonucleases

Metallonucleases that preferentially hydrolyze single-stranded nucleic acids, such as endonuclease S1 [2], mung bean nuclease [3], *Neurospora crassa* endonuclease [4], *Physarum polycephalum* nuclease [5,6], and SC nuclease [7], are among enzymes of a class (EC 3.1.30.1) that are widely used as probes of nucleic acid secondary structure. From inspection of Table 1 (adapted from Ref. 8), it can be seen that nucleases of this class are active at either neutral or acid pH. These DNA-binding enzymes generally require either Zn^{2+} or Co^{2+} for enzymatic activity. The biological function of these enzymes is unknown. However, some nucleases have been implicated in such processes as recombination and DNA repair in vivo [9,11,12,15]. Characteristic features of nucleases of this class include the preference for single-stranded nucleic acid substrates over double-stranded polynucleotides, a lack of discrimination in binding between ribonucleic acids and deoxyribonucleic acids, the endonucleolytic hydrolysis of single-strand substrates, and generation of products with 5'-phosphoryl and 3'-hydroxyl termini. In a comparative study of S1, mung bean, *Neurospora*, and SC nucleases with double-strand vs. single-strand substrates, it was determined that mung bean nuclease has the highest preference for single-stranded DNA, followed in order by S1 nuclease, SC nuclease, and *Neurospora* nuclease [8]. This distinction may also reflect the effect of pH on substrate conformation rather than an absolute ability of these enzymes to discriminate higher order structure.

S1 nuclease has been widely used as a structural probe of DNA conformation. This nuclease, which is most active at pH 4-5 and has a strict requirement for zinc ion (or cobaltous ion), has been shown

TABLE 1

Properties of Metallonucleases Belonging to Enzyme Class EC 3.1.30.1

Nuclease	Activity ratio ssDNase/dsDNase	M_r (x 10^{-3})	Optimal pH range	Metal required[a]	Ref.
Neurospora (endoexonuclease)	20	80	Neutral	Co^{2+}	9
Wheat seedling	20	43	Acid	Zn^{2+}	10
Ustilago	200	42	Neutral	Zn^{2+} Co^{2+}	11,12
Penicillium (P1)	250	44	Acid	Zn^{2+}	13,14
Neurospora (55,000)	250-4,000	55	Neutral	Co^{2+}	4
Schizophyllum (SC)	20-5,000	34	Neutral	Zn^{2+} Co^{2+}	7,8
Mung bean	10,000	39	Acid	Zn^{2+}	3
Aspergillus (S1)	30,000	32	Acid	Zn^{2+} Co^{2+}	2
Physarum	b	32	Neutral	Zn^{2+}	5,6

[a]When two metals are listed, the requirement is either for one or the other, but not both.
[b]Activity on double-stranded DNA was not detectable.
Source: Adapted from Ref. 8.

to detect locally unpaired bases which may exist in supercoiled plasmids at the loops of cruciforms [18] and between right- and left-handed helical segments [19]. It is interesting to speculate as to whether the metal helps to promote or recognize such conformational variations because the binding of metal ions alone to DNA can alter its conformation by locally unwinding the helix. Moreover, zinc ion, in particular, has been found to allow for a facile, reversible unwinding and rewinding of DNA (by heating and cooling, respectively) [16].

Recently, studies were conducted to determine the structural features of DNA recognized by this metallonuclease [17]. These studies involved the synthesis of DNA oligomers which could form either a hairpin loop or a double-stranded segment of DNA. It appears from the reaction of the S1 nuclease with different DNA structures of the same sequence that a string of exposed phosphates may be the structural feature recognized by the enzyme, since three to four phosphates in succession, found in the loops and helix termini of an oligomer (20-mer), were found to be required to allow single-strand access to the nuclease. In a double helix, it is suggested that the two sugar-phosphate strands may protect one another from attack by the nuclease because of their proximity, and that only in a loop or at a helical terminus does single-strand access become possible. At present, little is known concerning the DNA structural features recognized by other enzymes of the same class to which S1 nuclease belongs, but an interesting correlation has been apparent between sites recognized by mung bean nuclease and by a simple coordination complex described below, tris(4,7-diphenyl-1,10-phenanthroline)cobalt(III) (see Sect. 3.2.3), both recognizing conformationally distinct sites which exist in noncoding regions of the plasmid pBR322 [20,21].

The cleavage reaction for nucleases of this class generally appears to depend on the presence of metal ions. The activity of many of these enzymes can be inhibited by metal chelators, and reversal of this inhibition can be observed by the readdition of metal ions (in excess to chelator concentrations). The requirement of many of these enzymes for zinc or cobalt for activity is understandable in terms of the DNA strand scission reaction which they perform, yet specific information concerning the role of the metal has not been forthcoming. Clearly, the mechanism of action of these metallonucleases needs to be examined in terms of the metal, and perhaps a unified picture of metal-mediated phosphate hydrolysis can then emerge for the enzymes of this class.

2.1.2. Deoxyribonuclease I (DNase I)

DNase I (M_r = 31,000) from bovine pancreas, an endonuclease which degrades double-stranded DNA in a nonspecific but sequence-dependent manner, has been used as a biochemical scissors in numerous reactions that probe DNA structure and protein-DNA interactions (for reviews, see Refs. 22, 23). In the presence of metal ions, this enzyme cleaves duplex DNA through nucleophilic attack on the P-O3' bond, which is severed, yielding 5'-phosphooligonucleotides. Single-stranded DNA is also degraded by DNase I, but at a rate four orders of magnitude slower than its double-stranded counterpart [22]. Additionally, the divalent metal ion requirement of this endonuclease for hydrolysis of DNA shows maximal activity in the presence of Ca^{2+} and Mg^{2+} or Ca^{2+} and Mn^{2+} [24]. The mode of action and specificity of the enzyme, as well as the rate of hydrolysis, also appear to depend on the type of metal cation used in the reaction [17,25]. It has been found that in the presence of Ca^{2+} and Mg^{2+}, DNase I causes single-strand breaks in duplex DNA, whereas double-strand breaks in a supercoiled plasmid have been reported in the presence of Mn^{2+} [26].

The structure of DNase I has been solved to high resolution, and the binding of calcium deoxythymidine-3',5'-diphosphate (Ca-pTp) at the active site has been recently investigated [27-29]. From the crystallographic data obtained, a mechanism of action is proposed for this glycoprotein and also a model for its interaction with the minor groove of DNA. Moreover, the metals in this system appear to have several functions: both structural and catalytic roles are proposed for calcium ions observed in crystals of the native protein and the protein with Ca-pTp bound at the active site [7]. The nuclease has two strong Ca^{2+}-binding sites at pH 7.5, in addition to several weaker metal-binding sites [30]. Under conditions of crystallization of the protein, only the "structural" metal-binding sites are revealed through x-ray analysis. The "catalytic" calcium site is not observed in the native crystals of DNase I, but was determined upon binding of Ca-pTp to the enzyme.

The first structural Ca^{2+}-binding site (assigned as site 1) of DNase I consists of residues of a peptide loop, Asp-198 to Thr-204, acting as ligands. These ligands are the main-chain carbonyl oxygens of Thr-200, Thr-202, and Thr-204, both oxygens of the carboxylate side-chain of Asp-198, the hydroxyl group of Thr-200, and two water molecules. The geometry of this peptide-metal complex is that of a distorted octahedron, resulting perhaps from the somewhat unusual array of the eight ligands involved in coordination. In this ligated arrangement, the two water molecules as well as the two carboxylate oxygens are symmetrically disposed relative to that of the expected octahedral positions. The second structural Ca^{2+} binding site (site 2) has six coordinating ligands, also arranged in a distorted octahedral fashion, with ligands contributed by the carboxylate oxygens of Asp-96, Asp-104, Glu-109, the carbonyl oxygen of Phe-106, and two water molecules. Both of these metal binding sites offer structural stability to the enzyme, protecting the metal-binding peptide loops from proteolytic digestion and preventing reduction of an essential disulfide bridge (Cys-170-Cys-206) [31], and also the presence of calcium ions has been reported to play a role in the refolding of the reduced protein [32].

A "catalytic" calcium-binding site can be visualized by x-ray analysis in the presence of thymidine-3',5'-diphosphate (Ca-pTp) diffused into the native crystals. Ca-pTp is located in a shallow groove at the surface of the enzyme. Studies involving the di-p-nitrophenylester of pTp have shown that it is hydrolyzed by DNase I [33] and, like the reaction of the enzyme with DNA, this reaction is catalyzed also by metal ions and is inhibited by carboxymethylation of a catalytically important residue, His-131. This finding suggests that the same substrate-binding site may be involved in both cases, thereby validating the structural depiction of the active site of DNase I determined by these crystallographic studies [27-29]. The role of the metal at the active site appears to be that of an electrophilic catalyst, donating positive charge at the 3'-phosphate, and serving also to position correctly the reactive phosphodiester bond. Among an intricate network of hydrogen bonding found at the active site, strong interactions are observed between the carboxylate group of Glu-75 and the N-1 position of His-131 and between Asp-209 and His-249, resulting in an increase in the pK_a values of the histidine residues involved in these salt bridges. Also from this description of the active site of DNase I, a strategically placed water molecule is suggested to be the probable candidate for the role of the attacking nucleophile in the hydrolysis of DNA.

Figure 2 illustrates the proposed mechanistic scheme and how the catalytic calcium ion may participate [28,29]. In a general acid-base catalytic scheme, the carboxylate anion of Glu-75 accepts a proton from its close neighbor, His-131, which in turn accepts a proton from an identified water molecule (water 390) at the active site. This activated water molecule then acts as a nucleophile, attacking the 3'-phosphate and severing the phosphodiester bond. A NMR investigation into the stereochemical course of reaction of DNase I, using substrates containing chiral phosphorus centers, has revealed that hydrolysis is accompanied by an inversion of configuration at the scissile phosphorus atom [34]. This result indicates that the reaction may involve a single displacement step and lends credence to the notion that this enzyme-catalyzed reaction may involve direct nucleophilic attack of a water molecule (hydroxide ion) at the phosphorus center. Moreover, there are other observations which are consistent with the proposed reaction mechanism: carboxymethylation of His-131 at the N-3 position [35] as well as protonation of histidine by decreasing the pH (pH optimum of this enzyme is between 7.5 and 8.5) serve to inactivate DNase I. The calcium ion is seen to be essential to this enzymatic scheme through donation of its positive charge to the phosphate oxygen atom, functioning in effect as a Lewis acid and facilitating nucleophilic attack of the hydroxide ion. The metal appears in addition to serve as template, holding the scissile phosphate in the correct orientation for attack by the nucleophile by directly coordinating the anion of the reactive phosphodiester group.

The geometry at the active site, along with the distribution of charged residues and van der Waals surfaces of the protein around this site, immediately suggested a structural model for the interaction of DNase I with double-stranded DNA, depicted in Fig. 2(b) which allows for the precise alignment of the protein with its DNA substrate [28,29]. A salient feature of this model is a rather tight fit of an exposed loop, consisting of residues Arg-70, Asn-71, Tyr-73, and Lys-74, into the minor groove of DNA. The electrostatic inter-

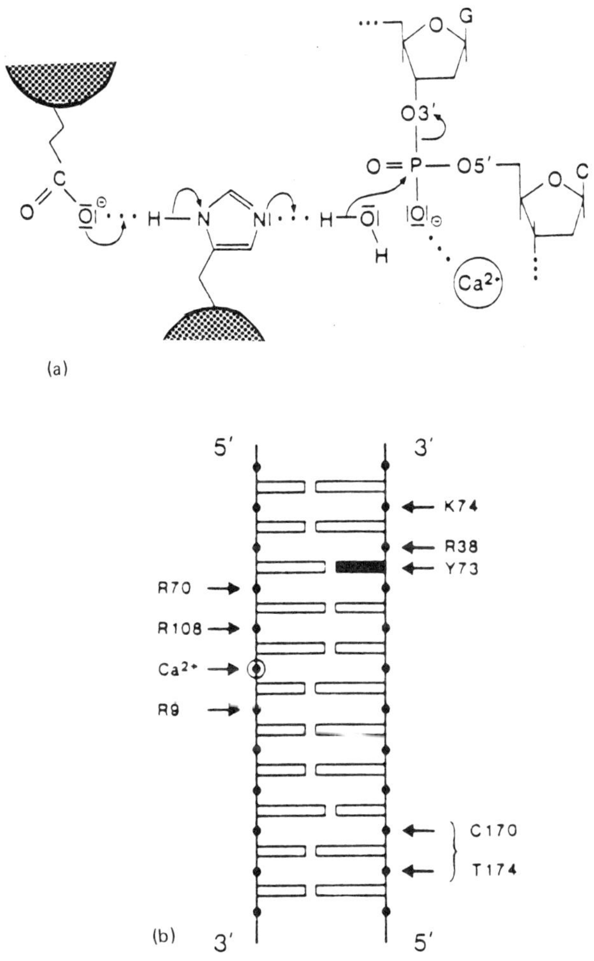

FIG. 2. Schematic illustrations of the proposed cleavage mechanism (a) and basis of DNA recognition (b) of DNase I. (Reproduced from Ref. 29 by permission of Macmillan Journals Limited.)

action of the positively charged residues contained in the loop with the negatively charged phosphate backbone of both DNA strands on either side of it are seen as important contributors to the binding interaction. Nearby glutamic and aspartic acid side chains are also suggested to interact with the DNA backbone through the binding of unidentified divalent metal ions. Van der Waals interactions, which occur five to six base pairs from the active site, are also significant to the correct alignment of the protein with DNA. Finally, the hydroxyl group of Tyr-73 is seen to penetrate into the minor groove and potentially hydrogen-bond to either the O-2 of pyrimidines or the N-3 of purines. Moreover, from steric considerations it can be envisioned that this tyrosine may have a modulating effect on the sequence selectivity of DNase I. Recently, cocrystals of DNase I and a self-complementary oligonucleotide have been obtained and confirm the basic features described for the interactive model of the enzyme with DNA [36].

Indirect evidence which also lends support to this interactive model of DNase I with the minor groove of DNA comes from a comparison of the enzyme cleavage patterns of synthetic DNAs of the same sequence in either a hairpin loop or duplex structure [17,37]. The results suggest that DNase I binds across two strands of DNA rather than one due to its inability to cleave within a loop of the hairpin structure. Moreover, the structural feature recognized by this nuclease did not appear to be the base pairs but rather the helical grooves. The strongest cleavage sites, found in the stems of the hairpin structures, lie directly opposite one another across the minor groove of DNA. Additionally, the cleavage rates are found to decrease dramatically where the groove is too wide, as one moves downward into an unpaired loop, or where phosphates no longer oppose one another. An implication of this model is that DNA segments displaying an A- or Z-like conformation would be resistant to digestion by DNase I. The size and shape of the minor groove is altered from that of B-DNA in these structural conformers, and it has been noted that brominated poly(dG-dC) in a Z-type conformation was cleaved by DNase I to only about 10% of the degradation of its B-DNA counterpart [38]. A comparison of nuclease rate of digestion along the duplex sequence d(CGCGAATTCGCG) also suggests that DNase I specificity may be purely structural, showing a correlation with the local helical twist of DNA. For example, steps T-C and G-C in this dodecamer sequence exhibit a high helical twist of 40-42° and are cleaved more rapidly than dinucleotide steps with lower values of helical twist [39], indicating that structural changes at the base pairs can transmit deformations out to the sugar-phosphate backbone causing a change in nuclease rate. Therefore, although DNase I does not recognize base pairs specifically, it may be sensitive to such local helical parameters as helix twist angle, base pair propeller twist, etc., which in turn may have an underlying sequence dependence [40,41]. It appears, then, that DNase I may sensitively probe the width of the minor groove of duplex DNA, a sequence-dependent structural feature, rather than recognizing hydrogen-bonding donor and acceptors on the base pairs.

2.1.3. Staphylococcal Nuclease

Staphylococcal or micrococcal nuclease, an extracellular nuclease of *Staphylococcus aureus*, can hydrolyze both DNA and RNA in the presence of Ca^{2+} ions (for reviews see Refs. 42-47). The enzyme, however, has a greater affinity for DNA than for RNA, and the activity is greater with heat-denatured DNA than native DNA, indicating that this nuclease has a preference for single-stranded nucleic acids [48]. It attacks at the 5' position (the 5' P-O bond is cleaved) of the phosphodiester linkage yielding a 5'-hydroxyl and a 3'-phosphate terminus. Simple phosphate diesters do not act as substrates, so this protein is indeed a nuclease and not a general phosphodiesterase. The globular enzyme (M_r = 16,800) contains a single polypeptide chain of 149 amino acid residues and no intrachain crosslinks. Because of its relatively simple structure and small size, ease of isolation and purification, and thermal stability, it has been well studied. Staphylococcal nuclease has been used widely in biochemistry, as a model system in studies of protein chemistry, and as a probe of nucleic acid structure.

Calcium ions, which are added as a cofactor, were strictly required for catalysis with all substrates, although Sr^{2+} can substitute for Ca^{2+} in the hydrolysis of DNA, but not of RNA [48]. The presence of Sr^{2+}, however, is inhibitory in the presence of Ca^{2+} as are many other metal ions, especially Zn^{2+}, Cd^{2+}, and Hg^{2+}. The pH range for activity is between 8.6 and 10.3, and varies inversely with calcium ion concentration. The optimal Ca^{2+} concentration at pH 8.8 is 10 mM, with higher metal ion concentrations inhibiting catalytic activity and producing an increase in the dissociation constant. In the presence of Ca^{2+} or other metal ions, pA, pdA, and pdT are good enzyme inhibitors, while pdTp is the best known inhibitor ($K_i \sim 10^{-7}$) [49]. The binding affinity appears to be greater for 5'-phosphoryl nucleotides than for their 3'-phosphoryl counterparts. Nucleosides themselves are not inhibitory. For the binding of a 5'-monophosphonucleoside, a single Ca^{2+} ion is involved, whereas the binding of pdTp and of DNA and RNA appears to require two calcium ions. In the absence of calcium, no detectable nucleotide binding is observed.

The structure of staphylococcal nuclease, crystallized in the presence of Ca^{2+} and the competitive inhibitor pdTp, has been solved at 1.5-Å resolution [50]. Presumably, the 5'-phosphate group of pdTp in this nonproductive enzyme-inhibitor complex mimics the binding of the phosphate of a diester substrate. This crystallographic information has led to a proposed mechanism of action for this nuclease, illustrated in Fig. 3, which elucidates a role for the divalent metal ions required for nuclease action. Here, as in the mechanism proposed for DNase I, the metal ion is seen to function as an electrophilic catalyst, stabilizing the negative charge of the nucleic acid substrate and polarizing the phosphate group through direct coordination of a phosphoryl oxygen. Once again the presence of and coordination by the metal ion appears to facilitate attack by an incoming nucleophile.

The metal ion is seen to be octahedrally coordinated, with the ligands donated by the carboxylate side chain of Asp-21 and of Asp-40, the carbonyl of a backbone peptide bond from Thr-41, two water molecules, and one of the anionic phosphoryl oxygens of the 5'-phosphate of the bound inhibitor. This phosphate is seen also to be within hydrogen-bonding distance of two guanidinium ions of Arg-35 and Arg-87. Recently, kinetic and EPR studies of active site mutants of staphylococcal nuclease indicate that Arg-87 may be interacting only

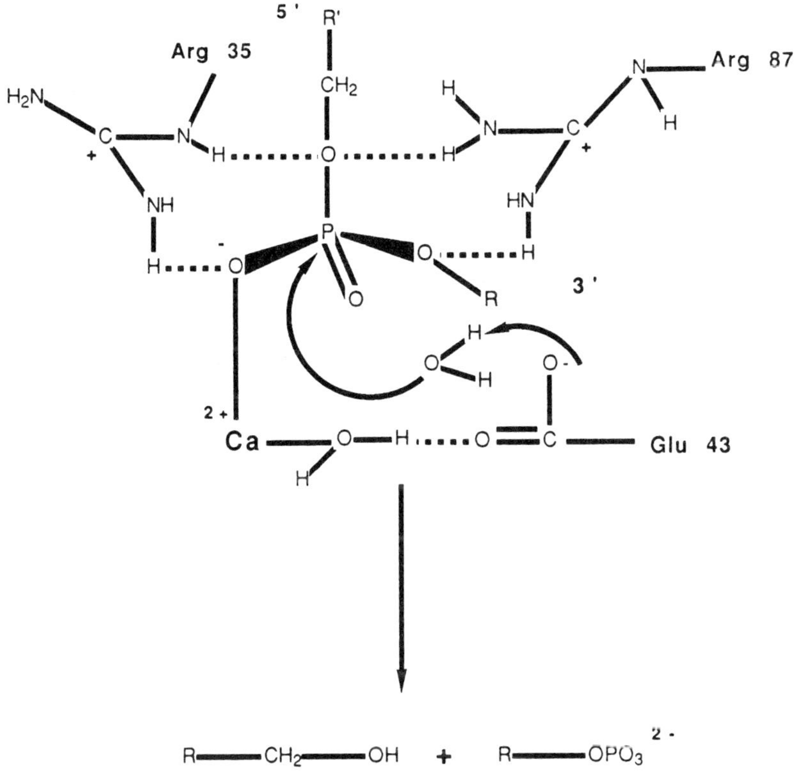

FIG. 3. Illustration of the proposed mechanism of action of staphylococcal nuclease.

with the trigonal bipyramidal transition state rather than the tetrahedral ground state of phosphorus [51]. The data that emerge from this crystal structure, however, call attention to the possible importance of arginine residues at the active site of these enzymes [52,53] and indicate that guanidinium ions of arginine residues may have the capacity to bind and catalytically activate a substrate [50,54]. However, it appears in this case that it is the sum of the effects of coordination of a 5'-phosphate oxygen to the metal plus the four hydrogen bonds from the Arg-35 and Arg-87 to the phosphate group which are important to the catalytic activation. These interactions result in both neutralizing the charge and enhancing the electrophilicity of the phosphorus, and thereby facilitate nucleophilic attack by hydroxide ion. Such interactions may also have the effect of lowering the energy of the transition state or intermediate [50]. Moreover, direct coordination of the phosphate group to the calcium ion may correctly position the phosphodiester bond, thereby activating it further to nucleophilic attack.

As can be seen in Fig. 3, the attack of water on the bound substrate is proposed to be catalyzed by the carboxylate group of Glu-43 acting as a general base. This "catalytic" residue is seen to be hydrogen-bonded to a water molecule from the inner coordination sphere of the calcium ion, and to a water molecule that is hydrogen-bonded to one of the oxygens of the 5'-phosphate. Consistent with the proposed scheme, cleavage of the phosphodiester proceeds with inversion of configuration at the phosphorus atom [55]. The crystal structure does not reveal the identity of a general-acid catalyst that may be required to facilitate the departure of the 5'-oxygen atom of the phosphate leaving group, although the phenolic hydroxyl group of Tyr-113 is offered as a possible candidate. Other amino acid residues that are seen to play a role in binding of the substrate are the ammonium group of Lys-84 and the hydroxyl group of Tyr-85, which are sufficiently close to hydrogen-bond to the 3'-phosphate of the bound inhibitor. A series of experiments involving the synthesis of enzyme segments with specific amino acid substitutions has confirmed the absolute requirement for Arg-35, and Asp-21 and Asp-40, which are coordinated to the calcium ion, and for Glu-43 [56,57]. Moreover, site-specific mutagenesis of Glu-43 has shown that V_{max}/K_m for the aspartate mutant is decreased 1,400-fold and the values for the charge-neutral mutations is decreased further (5,000-fold) relative to the wild-type enzyme [58]. The crystallographically derived depiction of the interactions involved in the binding of pdTp by the enzyme appears, then, to paint a vivid picture of a rigid and precisely positioned hydrolytic site containing the activating Ca^{2+} ion and the 5'-phosphate of the bound inhibitor. Our understanding of the role of the metals in this system is still somewhat incomplete [59], since 2 mol of calcium ions was found to bind to the enzyme-inhibitor complex, and also to the enzyme-substrate complex, but only 1 mol of metal ion is seen in the crystal structure [50].

The crystal structure of staphylococcal nuclease also reveals information concerning the overall shape of the active site, which may determine the cleavage specificities of the enzyme. The DNA binding surface is a long, narrow cleft, capable of containing one strand but not both; hydrophobic side chains residing just below the surface of the cleft are seen to create a binding site for the thymine base. The nuclease, although generally considered to be relatively nonspecific in its hydrolysis reaction with nucleic acids, does cleave pT and pA 30 times faster than either pC or pG [60,61], and this may suggest that thymine and adenine fit better in this hydrophobic cleft. Studies concerning the structural specificity of the cleavage reaction of this nuclease appear to confirm this notion [17]. Staphylococcal nuclease was found to cleave strongly at pT,pA bonds near the 5' ends or in a loop, but only moderately at pT,pA within a duplex structure. From this it was concluded that this enzyme appears to recognize both single-strand exposure and the identity of an unpaired base. Furthermore, rates of cleavage near 5' ends greatly exceed those near the 3' ends, and this may be attributed to the asymmetrical binding of the enzyme to its substrate, with a requirement for two phosphates to the 3' side of its cleavage site for optimal rate of attack.

TABLE 2

Features of Nonspecific DNA Cleavage by Metallonucleases

Nuclease	Ends generated	Structure attacked	Examples
S1	5'-Phosphate, 3'-hydroxyl	Exposed single strand	Tip of loop, ends of duplex
DNase I	5'-Phosphate, 3'-hydroxyl	Two sugar-phosphate strands, closely spaced	Minor groove of duplex
Staphylococcal	5'-Hydroxyl, 3'-phosphate	Exposed single strand, unpaired A or T	Ends of duplex, side of loop, weakly in duplex

Source: Adapted from Ref. 17.

Table 2 summarizes the features of the nonspecific DNA cleavage reactions of the metalloenzyme systems discussed (adapted from Ref. 17).

2.1.4. *Bacterial Alkaline Phosphatase*

Escherichia coli alkaline phosphatase is a zinc metalloenzyme which catalyzes the nonspecific hydrolysis of phosphate monoesters. Although this enzyme is not a nuclease and does not catalyze the hydrolysis of the phosphodiesters of DNA and RNA, understanding how zinc functions in this system can perhaps give some clues as to the possible roles played by metal cations in some nucleases. From this perspective, one is looking strictly at the reactivity of the metal ion, i.e., its role in phosphate binding and as a possible source of a coordinated nucleophile at neutral pH. In this system, however, the aspect of recognition of either base sequence and/or structural features of nucleic acids by the enzyme clearly does not apply. This phosphomonoesterase has, however, been well studied by a variety of physical and spectroscopic methods, and although not completely understood, much information has come forth regarding the role of zinc in the mechanism of action of this enzyme. The picture that emerges is complex, and reveals how multiple metals may function to facilitate the hydrolysis of phosphate anions.

Alkaline phosphatase isolated from *E. coli* (M_r = 86,000) is a dimeric protein consisting of two identical monomeric subunits [for reviews, see Refs. 62-64]. An absolute requirement for zinc in the enzymatic reaction scheme has been well established, although variable amounts of metal have been reported for the isolated enzyme. The metal content can range from 2 to 4 mol of zinc per mol of dimeric protein upon isolation, but readdition of 2 mol of zinc can completely restore activity to the apoenzyme. Interestingly, the dimeric structure of the enzyme is also affected by metal removal with an increase in the dissociation of the apoenzyme into monomer units. Data obtained from metal-binding studies of the apoenzyme revealed that alkaline phosphatase can potentially bind three pairs of metal ions over a free metal ion concentration range of 10^{-8} to 10^{-3} M. These three pairs of metal binding sites appear to have different functions and have been designated as "catalytic", "regulatory", and "structural" [65-71]. Both the catalytic and regulatory sites are occupied by zinc cations. Mg^{2+} ions are also usually found associated with alkaline phosphatase and are thought to bind at the designated structural site. Here the alkaline earth metal cations appear to function in the stabilization of the enzyme and the enhancement of activity; however, magnesium cannot substitute for the catalytically essential Zn^{2+} ions. ^{13}C NMR studies of the protein carrying ^{13}C-labeled histidyl residues indicate that the catalytic zinc binding site may contain three to four histidine residues as ligands, while the regulatory zinc ion may coordinate one histidine ligand [70,72]. Not surprisingly, the structural magnesium site does not appear to involve coordination by histidine residues. Chemical modification studies indicate that only three histidine residues per protein monomer are protected from ethoxyformylation [73]. With the Cu(II) enzyme, three magnetically equivalent nitrogen atoms were detected as ligands through ESR studies [74]. Consistent with these observations is the ability of zinc to protect selectively 2-3 of the total 16 histidine residues of the protein from photooxidation [75].

The minimal mechanistic scheme for the hydrolysis reaction of alkaline phosphatase with phosphate monoesters is given in the equation below:

$$\text{ROP} + \text{E} \rightleftharpoons \text{E}\cdot\text{ROP} \rightleftharpoons \text{E-P} + \text{ROH} \rightleftharpoons \text{E}\cdot\text{P} \rightleftharpoons \text{E} + \text{P}$$

In the course of its reaction with substrate, alkaline phosphatase forms two phosphoenzyme intermediates. These are a noncovalent complex E•P, in which the product phosphate is coordinated to one of the two zinc ions present at each active site, and a covalent intermediate E-P, which is formed by the phosphorylation of a serine residue [76,77]. Significant concentrations of both forms of the enzyme can be obtained under a variety of experimental conditions. The rate-limiting step for the hydrolysis reaction of alkaline phosphatase is pH-dependent [78,79]. At low pH, dephosphorylation of E-P becomes rate limiting, leading to an accumulation of the covalent intermediate which has been identified as the phosphate monester of Ser-102. At high pH, where maximal activity of the enzyme is exhibited, significant equilibrium concentrations of E-P disappear and dissociation of product from the E•P becomes rate limiting. The stereochemical course of the reaction catalyzed by this nonspecific phosphatase has been found to proceed with retention of configuration at a chiral phosphoryl group [80], and this is consistent with the formation of a phosphorylated enzyme intermediate.

Studies of several metallophosphatases made by the substitution of other metal cations for the native zinc ion have revealed the importance of the metal to various steps in the reaction pathway given above. First, the metal appears to be required for phosphate binding: Zn(II), Cd(II), Mn(II), and Co(II) all induce the binding of phosphate upon addition to the apoenzyme [81]. Ni(II), Cu(II), and Hg(II), however, which are capable of binding to the metal-free enzyme, do not induce significant phosphate binding [82]. Second, metal ion substitutions can markedly affect the relative rates of phosphorylation and dephosphorylation of the enzyme and thereby change the equilibrium concentration of the phosphoryl enzyme. Among the metal substitutions that produce a bound enzyme-phosphate complex, only Zn(II) and Co(II) show significant reactivity toward phosphate monoesters [65,83]. Co(II)-alkaline phosphatase, like its zinc counterpart, forms significant equilibrium concentrations of E-P at acid pH, but no detectable levels at alkaline pH. In contrast to this, the Cd(II) or Mn(II) derivative proteins form significant levels of E-P at alkaline pH, but the hydrolysis of phosphoserine intermediate is extremely slow. Therefore, the metal ions in this system are capable of inducing phosphate binding and can regulate the relative rates of phosphorylation and dephosphorylation.

The role of the metal ions in the catalytic mechanism of alkaline phosphatase, illustrated in Fig. 4, reveals numerous ways in which a metal ion may participate in the hydrolysis of phosphate anions. In the first step, coordination of the phosphate monoester

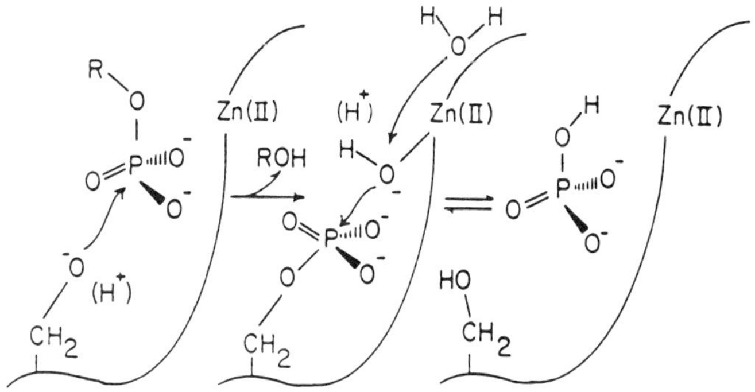

FIG. 4. Proposed molecular mechanism for bacterial alkaline phosphatase which illustrates a possible role for a metal-bound nucleophile in the dephosphorylation of the phosphoseryl ester. (Reproduced from Ref. 62 by permission of Elsevier Science Publishing Co., Inc.)

by the metal ion found at the catalytic site allows for the formation of the E•P complex (right, Fig. 4). Ligation by the metal to the phosphomonoester serves both to neutralize the negative charge of the substrate and to position strategically the phosphorus atom for attack by the hydroxyl oxygen of Ser-102. Upon formation of the seryl ester, the phosphate may move out of the coordination sphere of the metal [84]. At this point in the reaction scheme (middle, Fig. 4), a possible metal-bound hydroxide ion is postulated to participate in a second step leading to the hydrolysis of the phosphoseryl ester; moreover, the potency of a coordinated nucleophile has been demonstrated previously for small molecular complexes and enzyme systems [85-87]. For alkaline phosphatase, indirect evidence supports this notion. Variation in the pH value at which E-P/E•P = 1 from pH 5 for the zinc enzyme to pH 8 for the cadmium derivative, the induction of significant dephosphorylation of the cadmium enzyme at pH values above 9 rather than above pH 6 as for the zinc species, and the metal ion dependence of the dephosphorylation step all suggest that the mechanism may involve the dephosphorylation of the phosphoserine intermediate E-P by a metal-coordinated hydroxide [62]. Since the phosphate has moved out of the inner coordination sphere of the metal ion upon phosphorylation, this would apparently leave a position available for a coordinated solvent molecule, providing the nucleophile necessary for the finale of this mechanistic scheme. Although the precise function of the "regulatory" metal-binding site is less clear, having a metal ion bound here clearly accelerates phosphorylation, and may also serve to regulate the dissociation of the phosphate product, which is the controlling feature of the mechanism at neutral pH.

2.2. Metal Complexes that Cleave DNA Through Redox Chemistry

As in the case of the naturally occurring enzymes, for the synthetic complexes the primary role of the metal rests in strand cleavage reactivity. For the enzymes, as described above, however, the metal appears to aid in hydrolysis of the phosphate ester, either by providing Lewis acidity, or through the delivery of a coordinated nucleophile. In the case of synthetic molecules described below, the cleavage chemistry is not primarily hydrolytic. Instead, primary oxidative damage to the sugar leads eventually to scission of the sugar-phosphate backbone. In these examples, the metal provides neither acidity nor coordinated nucleophilicity but the redox source to effect the oxidative damage, either indirectly through the delivery of reactive hydroxyl radicals or singlet oxygen, or directly through hydrogen abstraction reactions or direct binding and activation of metal oxo species. We refer to these cleaving molecules as metallonucleases as a parallel to the true enzymatic systems. Actually, these complexes do catalyze scission of the DNA backbone, but only indirectly, through the catalytic generation of species which are themselves consumed in oxidatively cleaving the DNA. The small metallonucleases discussed are given in Fig. 5.

2.2.1. Methidiumpropyl-EDTA-Fe(II)

Methidiumpropyl-EDTA-Fe(II) (MPE-Fe(II)) may be considered to be the first example of a nonenzymatic metallonuclease to be widely used as

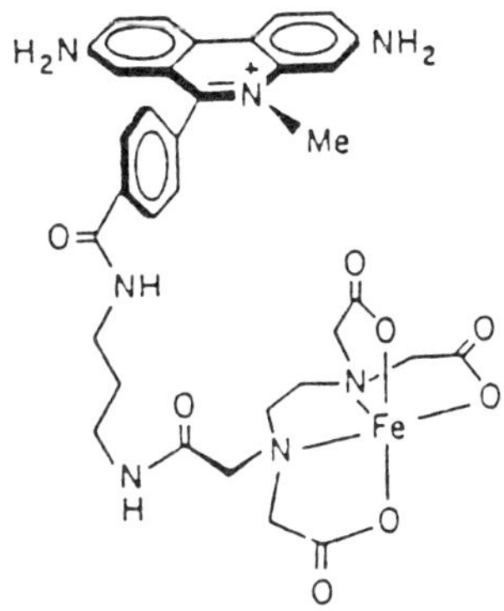

MethidiumpropylEDTA-Fe(II)

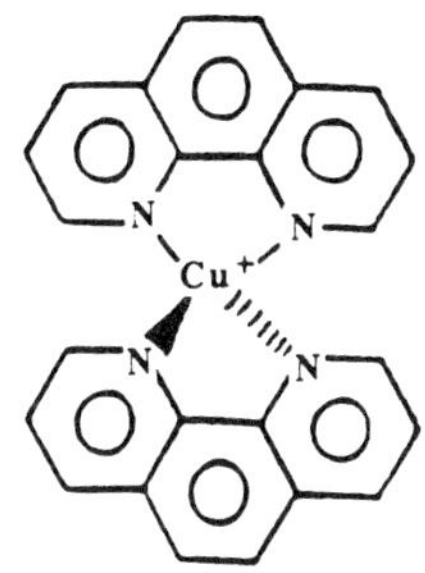

Bis(phen)Cu(I)

EDTA-Fe(II)$^{2-}$

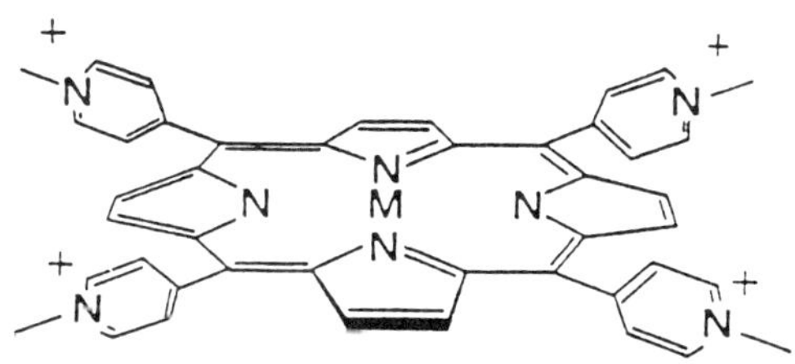

M-T4MPyP

M = Mn(III), Fe(III), Co(III)

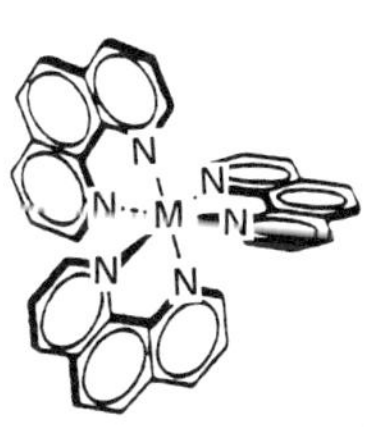

Λ

Δ

M(phen)$_3^{n+}$

M = Ru(II), Co(III), Rh(III)

FIG. 5. Redox-active artificial metallonucleases.

a probe for protein-nucleic acid interactions as well as for the interaction with DNA of smaller ligands, such as antiviral, antibiotic, and antitumor drugs (500-2,000 molecular weight) [88-91]. These interactions can be revealed through the use of footprinting techniques in which the bound molecule, i.e., the protein or drug, protects the DNA at its binding site(s) from the action of the nuclease. The result is a direct visualization of a "footprint" of the binding interaction on an autoradiogram of a sequencing gel. Previously the nuclease, DNase I, had been successfully used as an enzymatic probe of the sequence preferences of DNA-binding proteins in footprinting experiments [92]. In contrast to DNase I, however, MPE-Fe(II) offers the distinct advantage of yielding a higher resolution footprint of the interaction of small molecules, as well as proteins, with DNA. This increase in resolution is due to the smaller size of the synthetic footprinting tool compared with the enzymatic probe [29], and also because MPE-Fe(II) cleaves DNA with a lower sequence selectivity than does DNase I.

The synthetic nuclease MPE-Fe(II), elegantly designed by Dervan and coworkers, is the result of the synthetic coupling of a known DNA-binding moiety to a redox active metal center [88]. MPE-Fe(II) is composed of two "domainlike" functional regions, methidium, an aromatic intercalator which binds through the minor groove of DNA, and Fe(II)-EDTA, a metal-bound chelator, linked through a short hydrocarbon tether. DNA cleavage by MPE-Fe(II) is composed of two events: (1) the binding of the methidium group to DNA, delivering high local concentrations of reactive metal ions to the DNA-ribose backbone, and (2) the generation of a reactive species at the metal center which is responsible for strand scission. In the presence of ferrous ion and oxygen, this synthetic nuclease causes single-strand breaks in DNA in a reaction that is made more efficient upon the addition of reducing agents, such as dithiothreitol [88]. The reactive species responsible for the cleavage of DNA in this metal-mediated reaction is presumably a diffusible hydroxyl radical emanating from the iron center. For this nonenzymatic metallonuclease, then, binding is coupled to reactivity owing to the presence of the metal ion. The metal ion is essential to the activity of MPE, as it is for metalloenzyme systems. In this case, however, a redox active metal is required, whereas in the case of enzymatic nucleases, a nonredox metal, such as zinc, calcium, or magnesium, is required for the cleavage reaction. The metal chelator, EDTA, found to be arranged in a slightly distorted octahedral geometry about the central ferrous ion, is necessary to the reaction; many other strong chelators of ferrous ion, such as desferrioxamine, are not effective in producing strand scission [93]. The presence of an open coordination site (or one that is occupied by a readily dissociable ligand, such as H_2O) has been suggested as important to the reaction chemistry [94]. It is noted in this context that a seventh coordination site, occupied by H_2O, is seen in the crystal structure determination of Fe(III)-EDTA [95].

From a study of the reaction mechanism of MPE-Fe(II) with DNA, several key mechanistic features emerge [96]. From product analyses, the correlation of one DNA base released per cleavage event, and the nature of the DNA termini that result from the cleavage reaction, it was suggested that strand scission proceeds via oxidative degradation of the deoxyribose ring. Also the same cleavage products are obtained with EDTA-Fe(II) alone. The oxidizing species responsible for cleavage was deduced to be hydroxyl radicals generated near the deoxyribose in the minor groove of DNA, and several lines of evidence support this conclusion. Similar cleavage products are obtained using γ radiolysis, and hydroxyl radicals are known to be responsible, at least in part, for the DNA damage produced in the presence of such ionizing radiation. As stated above, the MPE cleavage reaction depends on ferrous ion and oxygen, and is enhanced in the presence of reducing agents, which allows the system to become catalytic by regeneration of the reduced form of the metal. This metal-mediated reaction is also inhibited by superoxide dismutase, which converts superoxide to hydrogen peroxide and oxygen, and catalase, which converts hydrogen peroxide to water and oxygen. Both superoxide and hydrogen peroxide are thought to be necessary for the generation of hydroxyl radicals in a metal-catalyzed Haber-Weiss reaction [97]:

$$O_2{\cdot}^- + M^{n+} \longrightarrow O_2 + M^{(n-1)+}$$
$$M^{(n-1)+} + H_2O_2 \longrightarrow M^{n+} + OH{\cdot} + OH^-$$

Therefore, taken together, the results are consistent with the iron-mediated reduction of oxygen to hydroxyl radical. Moreover, the binding to DNA of methidium, which binds in a sequence-neutral fashion, provides, in part, a basis for the low base selectivity of the cleavage reaction of MPE-Fe(II), as does the lack of specific reaction of hydroxyl radical with the sugar moiety of DNA.

2.2.2. *(Phenanthroline)$_2$Cu(I)*

The bis-1,10-phenanthroline-cuprous ion complex, $(phen)_2Cu^+$, acts as an efficient oxidative nuclease. The reaction requires binding of the coordination complex to its nucleic acid substrate and hydrogen peroxide as a coreactant [98-103]. In this reaction, $bis(phen)Cu^{2+}$ generated in situ is reduced to the cuprous form of the metal complex in the presence of a thiol. As with MPE-Fe(II), strand scission is seen to result from oxidative destruction of the deoxyribose moiety of DNA. Here again the role of the redox active metal is to deliver a reactive species to the nucleic acid. However, in this case, it is the redox active coordination complex itself that is intimately bound to duplex DNA. Importantly, cuprous ion is seen here also to play a structural role in this oxidative cleavage reaction by providing a geometry for the coordination complex which is uniquely suited for the binding interaction.

The reaction scheme proposed for this artificial metallonuclease involves the generation of a metal ion-associated radical species through a one-electron oxidation of the bound cuprous complex by hydrogen peroxide [103]. From analysis of the products of the scission reaction it was concluded that the initial site of attack by this hydroxyl radicallike species is the C-1' hydrogen of the deoxyribose ring, with a minor site of attack at the C-4' hydrogen [104]. Whether the reactive species may be considered to be diffusible, as it is in the case of MPE-Fe(II)(OH•), or whether it is better represented by a metal-oxo species is under some debate [105,106]. The phenanthroline ligand of this coordination complex can be seen to function both to provide a binding surface for interaction of the complex with nucleic acids and to modulate the redox potential of the Cu^{+}/Cu^{2+} couple.

Two possible binding modes have been proposed for the interaction of bis(phen)Cu^{+} with DNA: intercalation of the one of the planar phenanthroline ligands between adjacent base pairs, or insertion of the coordination complex into a groove of duplex DNA [107, 110]. A clear distinction between these two modes of interaction cannot be made at present, though it is clear that the binding of this complex takes place through the minor grove of DNA [103,104]. Interestingly, there appears to be some striking similarities in the sequence selectivities of bis(phen)Cu^{+} and DNase I. In a study concerned with probing the structural heterogeneity found locally along the DNA helix with various nucleases, it was observed that the global (overall as opposed to local) cleavage function of the nuclease DNase I resembles that of copper-phenanthroline. Both nucleases exhibit cleavage patterns that are displaced three bonds in the 3' direction for cuts on opposite strands, consistent with minor groove attack, and both exhibit minima at long runs of A/T or G/C [37], which may reflect differences in minor groove dimensions at these particular sequences; A/T sequences usually exhibit a narrow phosphate-phosphate distance across the minor groove (9-11 Å [108]), whereas sequences with high G/C content usually facilitate the formation of a wide minor groove, characteristic of A-DNA (15-16 Å [109]). Furthermore, both nucleases require double-stranded DNA as a substrate for cleavage. It was concluded from this that DNase I and the synthetic cuprous-nuclease may be recognizing a similar feature of DNA structure, more specifically, the width of the minor groove as it varies along the DNA helix in a sequence-dependent manner. It is also interesting to note in this context that for substrates where the minor groove of DNA is altered dramatically, as in A and Z forms of DNA, or lost entirely, as in single-stranded DNA, the cuprous-nuclease shows substantially diminished activity [110-112]. Both Z-DNA with its very narrow and deep minor groove, and single-strand DNA that is incapable of forming stable secondary structures, are not cleaved appreciably by $(phen)_2Cu^+$. A-form nucleic acid structures, however, containing a wide and somewhat shallow minor groove (~17 Å for A-DNA [113]) are cleaved, but with one-third the efficiency of B-DNA. It appears, therefore, that the synthetic cuprous-nuclease, like its natural analog DNase I, may sensitively discriminate among sequences of DNA with varying minor groove dimensions. Consequently, $Cu(phen)_2^+$ provides a quite useful probe of DNA structure [17,37,112,114] and protein-nucleic acid interactions [115-117].

2.2.3. *Metalloporphyrins*

Metal complexes of the water-soluble tetracationic porphyrin meso-tetrakis(N-methyl-4-pyridiniumyl)porphine (H_2T4MPyP) [118-123] are the most actively studied DNA-binding metalloporphyrins, and many of these complexes have been shown to act as metallonucleases through either chemical [122,123] or photochemical activation [119]. These metalloporphyrins are rigid structures containing a central metal ion and possessing four positively charged alkylated pyridine residues arranged about the macrocycle. Thus the binding of these inorganic nucleases to DNA appears to be governed largely by electrostatics.

The nature of the metal ion in this system is thought to regulate both the porphyrin-binding mechanism and its DNA base preference as evidenced by kinetic, optical, and physiochemical data [122,124-127]. At low binding ratios of the porphyrin/DNA, the Cu^{2+} and Ni^{2+} complexes of H_2T4MPyP, which are known to be four-coordinate structures in aqueous solution, bind by intercalation to GC-rich regions of double-helical DNA. The analogous Mn^{3+}, Fe^{3+}, Co^{3+}, and Zn^{2+} complexes, however, have been shown to bind through a nonintercalative process to AT regions of DNA; at neutral pH, these metalloporphyrins form five- and six-coordinate species containing axial ligands (e.g., OH^- and H_2O) above and below the porphyrin plane. Thus the metal ion is seen to modulate base preference and the mode of binding of these complexes largely through changes in coordination geometry.

Upon activation of Mn^{3+}-, Fe^{3+}-, Co^{3+}-T4MPyP with various chemical agents, it is possible to produce efficient metal-mediated DNA strand scission [122]. Here DNA cleavage is dependent on both the nature of the activator and the particular coordinated metal ion. Three chemical activators of the DNA scission reaction have been employed: iodosobenzene, superoxide, and ascorbate, with iodosobenzene yielding the greatest extent of activation in all cases. Details of the mechanism of reaction of these metalloporphyrins with DNA are not known. However, in studies of both iron and manganese porphyrin compounds, used as models for naturally occurring peroxidases, it was demonstrated that these metal complexes can be activated to form a metal-oxo species capable of oxygen insertion into alkanes and olefins [128,129]. Here again it appears that the role of the metal ion in this metalloporphyrin-mediated cleavage reaction may be to provide an active redox center in close proximity to the DNA helix.

These metallonucleases appear to cleave preferentially at AT sequences contained within a restriction fragment of pBR322 DNA, with the minimum binding/cleaving sequence requirement being variations of the trinucleotide sequence $(A{\cdot}T)_3$. Additionally, opposite strand DNase I footprinting analysis of these metalloporphyrins bound to this restriction fragment indicate interactions with AT-rich sequences in the minor groove of the helix. These metalloporphyrins are, however, capable of cleaving at all four nucleotide positions suggesting that perhaps it is the deoxyribose moiety which is the primary site of attack in the strand scission process. The preferential association in the minor groove with AT-rich sequences may be an indication that it is a groove-bound rather than intercalative interaction of the porphyrins with the DNA that leads to the cleavage chemistry. Although they have not yet been used as footprinting reagents, these metallonucleases may have potential as selective probes for DNA-binding ligands having A/T sequence specificity. A series of iron-porphyrins tethered to an acridine ligand, in contrast, have been shown to cleave DNA in a sequence-neutral manner [130], thus it is probably the binding interaction rather than the cleavage chemistry of a metalloporphyrin that is responsible for any selectivities observed. The very low sequence selectivity of the acridine-tethered complexes suggest their application to footprinting techniques. Here the lack of sequence preference is governed by the binding moiety, acridine, which intercalates into DNA in a sequence-neutral fashion.

2.2.4. Fe(II)-EDTA^{2-}

Fe(II)-EDTA^{2-} is perhaps the simplest inorganic coordination complex capable of nuclease activity. This complex, prepared in situ from shelf items available to any chemist or molecular biologist, provides a very simple yet powerful tool to probe the structure of DNA free in solution or bound to other molecules [131-133]. Its mode of action as well as the function served by the metal are presumably similar to that described for MPE-Fe(II), since in both cases it is the [Fe(II)-EDTA]$^{2-}$ moiety that is responsible for the DNA strand scission reaction. The role of the EDTA ligand here is to solubilize the metal ion and to modulate the redox potential of the complex, as it is for the tethered EDTA ligand of MPE-Fe(II). In the presence of oxygen or hydrogen peroxide, [Fe(II)-EDTA]$^{2-}$ generates a reduced oxygen species of high reactivity [96,134]. Here again, it is the hydroxyl radical that is thought to be the key species involved in the metal-mediated degradation of DNA [96,131]. In this system, however, since the ferrous complex is a dianion, radicals are generated away from the surface of the DNA polyanion and must diffuse toward the helix for reaction to occur. Secondary reactions of the resulting deoxyribose radicals lead eventually to a break in the DNA-phosphate backbone [135].

This inorganic nuclease offers some advantages over other chemical (and enzymatic) probes of DNA structure. A key feature in this system is that the negatively charged metal complex cannot electrostatically associate with the DNA molecule, thus eliminating the effect that a nucleic acid-binding ligand may have on the conformation of DNA or of a DNA-protein complex. (For example, molecules that bind by intercalation will cause the helix to unwind and to lose some of its conformational flexibility, yielding a DNA molecule that has been stiffened considerably.) As with MPE-Fe(II), this ferrous nuclease cleaves with no sequence specificity, breaking the backbone of DNA equally at every base position. In footprinting studies, however, this reagent yields even more highly resolved images of the protein-DNA contacts along the backbone, presumably because of the lack of DNA binding by the nuclease and the very small size of the reactive species. The binding interaction of both λ repressor and Cro protein with DNA were revealed using Fe(II)-$EDTA^{2-}$ [132]. In this study, additional protein contacts with the helix were observed that had not previously been resolved by other chemical footprinting methods.

Using Fe(II)-$EDTA^{2-}$, conformational bends in the adenine tracts of kinetoplast DNA, as well as in a restriction fragment of pBR322 DNA, were observed [133]. From the cleavage pattern, which progressively decreased in the 5' → 3' direction of each A tract, it was suggested that the structural bends at these purine tracts may stem from changes in the minor groove of DNA. The minor groove is believed to become narrower at runs of adenine residues, as revealed in the crystallographic analysis of the Dickerson-Drew dodecamer with a narrow minor groove width seen for the central AATT sequence (~9 Å) [108a]. If the minor groove were rather narrow, the sugar hydrogen atom(s) that is attacked preferentially by hydroxyl radical might be less accessible to solvent, and consequently attacked less often. This result leads to the conclusion that accessibility to and reactivity of the sugar ring may depend on the local structure of the helix and, as such, may have an inherent sequence dependence. Thus this simple inorganic nuclease is capable of probing the structural details of DNA conformation in solution.

2.2.5. *Tris(phenanthroline)metal Complexes*

A family of tris(1,10-phenanthroline)metal complexes has been developed in our laboratory which displays stereoselective binding to duplex DNA [136-140] and can be activated in the presence of light to produce strand scission reactions [140-142]. These cationic metallonucleases display octahedral coordination about a central metal ion, yielding complexes which are rigid in structure and exist in one of two enantiomeric configurations: a right-handed spiral, the Δ enantiomer, or a left-handed spiral, the Λ enantiomer. Here the chiral metal center provides a structural handle, affording a unique geometry and ligand disposition to each enantiomer of the inorganic racemate.

Spectroscopic studies using ruthenium complexes were extremely valuable in elucidating binding characteristics of the complexes to DNA [137-139]. Tris(phenanthroline)ruthenium(II) complexes are intensely colored, highly luminescent, and these spectroscopic features are perturbed upon binding to DNA. At least two binding modes of the complexes with DNA are apparent: intercalation and groove (surface) binding. In the case of intercalative binding, it is thought that one of the phenanthroline ligands partially inserts and stacks between the base pairs, anchoring the complex, while leaving the two nonintercalated ligands to lie along the helical groove. For groove binding, the chiral metal complex may now be oriented in the opposite fashion with one ligand out toward the solution, while the other two are aligned along the surface of the duplex. Moreover, in forming diastereomeric complexes with DNA, which, after all, is also chiral, enantiomeric selectivities are evident. These stereoselective binding interactions are illustrated in Fig. 6 and, as can be seen in the figure, for each binding mode, there is an associated and distinct enantiomeric selectivity. In binding studies of $Ru(phen)_3^{2+}$ to a right-handed B-DNA helix, the Δ enantiomer is favored for intercalation, whereas for surface binding, it is the Λ enantiomer that is found to be preferred.

Tris(phenanthroline)metal complexes furthermore can serve to cleave the DNA strand. $Ru(phen)_3^{2+}$, $Co(phen)_3^{3+}$, and $Rh(phen)_3^{3+}$ have been shown to induce strand scission upon photoactivation [142].

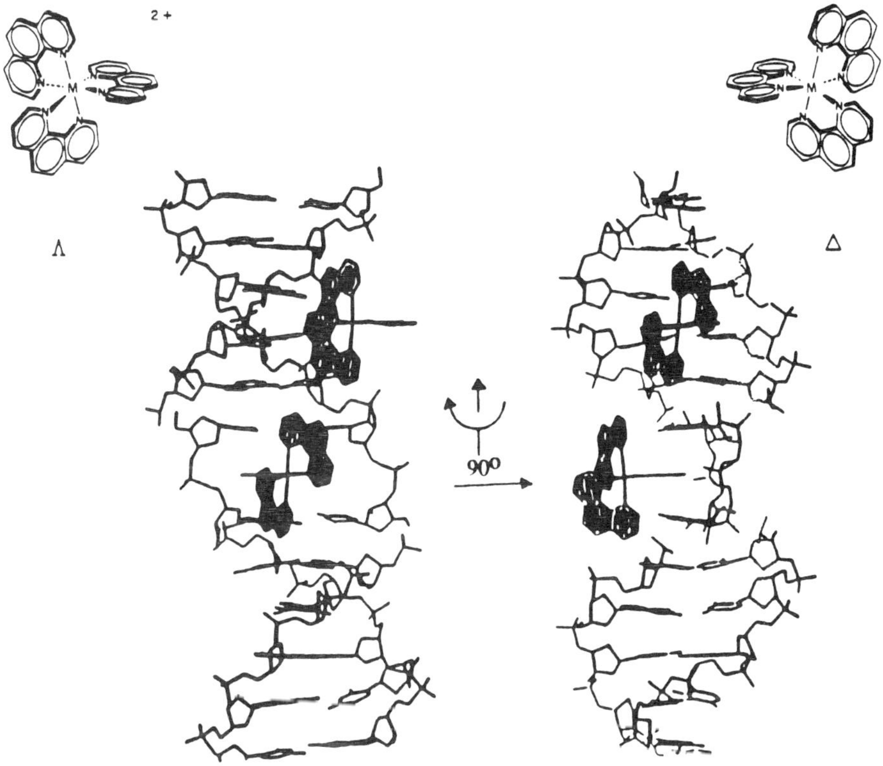

FIG. 6. Enantiomers of $Ru(phen)_3^{2+}$ bound to B-DNA either intercalatively or in a surface-bound fashion. Shown on the left is the Λ isomer bound through surface binding (above) and the Δ isomer bound intercalatively (below). The preferential intercalation of Δ-$Ru(phen)_3^{2+}$ into a right-handed helix can be seen in the figure where one ligand of the metal complex is inserted into the helix, while the two ligands remaining point outward and align along the right-handed helical groove. In contrast, the model for the Λ isomer, intercalated into the helix (not shown), reveals nonintercalated ligands disposed contrary to the right-handed groove. Also, from the view on the right, which is the result of a 90° rotation about the helix axis, the enantioselectivity associated with the surface-binding interaction is made apparent. In this surface- or groove-binding mode for the Λ isomer, one ligand is pointed outward, toward the solvent, while the other two ligands are aligned along the surface of the right-handed helical groove.

These complexes, therefore, represent examples of synthetic light-activated metallonucleases (noncatalytic, however), which, as a result of their binding characteristics, also have the remarkable capacity to discriminate helical handedness.

DNA strand scission reactions by these complexes occur as a result of coupling the DNA binding to a range of inorganic photochemistry; photoredox reactions result from the initial excitation of ligand field, intraligand, and charge transfer states. These transition metal complexes exist in a low-spin d^6 electronic configuration, are substitutionally inert in the absence of light, and are similarly inert to racemization. The electronic properties of the central metal ion, including both the ground and excited states, determine reactivity. The relative quantum yields for the ruthenium, cobalt, and rhodium polypyridyl complexes are found to be 0.017:1:1.5, respectively [142a]. The excited states involved depend on the metal and the wavelength of irradiation: a metal-to-ligand charge transfer state for the ruthenium complex irradiated at 442 nm [143], and intraligand states for both the cobalt and rhodium complexes irradiated at 325 nm [144,145]. The common feature to the cleavage reactions of $Ru(phen)_3^{2+}$, $Co(phen)_3^{3+}$, and $Rh(phen)_3^{3+}$ is the photoreactive role played by the metal.

The different photochemical profiles also lead to different kinds of cleavage chemistry by each metal complex with DNA. For example, cobalt and rhodium complexes are photooxidizing agents, undergoing photoreduction upon primary excitation of the ligand bands. This photoreduction may yield direct oxidative DNA damage, and in this case the reactive species may not be diffusible. The lack of dependence of these cleavage reactions on oxygen supports this notion [146]. Product analysis suggests that the sugar ring may be the initial point of attack with subsequent backbone breakage; product identification is in progress. Here the cleavage reaction is expected to be inherently nonspecific, assuming equal reactivity of the sugar moieties along the phosphate backbone. This sequence neutrality may not always be found, however, independent of binding preferences, if the reactivity of the sugar ring depends on DNA conformation. To a first approximation, however, since the metal complexes bind to DNA in a nonspecific fashion, it is expected that a uniform cleavage pattern should be obtained. This uniformity in cleavage has been confirmed through high-resolution DNA-sequencing experiments for both the cobalt and rhodium complexes [146b].

The excited state chemistry of ruthenium polypyridyl complexes is known to be quite different from that of its cobalt and rhodium analogs, involving a long-lived triplet metal-to-ligand charge transfer state which is known to sensitize singlet oxygen formation [147]. Not surprisingly, the photoreaction of $Ru(phen)_3^{2+}$ with DNA has been shown to be oxygen-dependent [146]. Here sensitization of oxygen through excitation of the tris(phenanthroline)ruthenium(II) complex results in DNA damage mediated by a diffusible $^1O_2$ species [142,148, 149]. Not only is the generation of a diffusible species a source of contrast to the cobalt and rhodium chemistry, but singlet oxygen reacts at a DNA base rather than with the sugar ring, the primary target for other synthetic metallocleaving agents. Targeting of the purine or pyrimidine bases produces, then, an inherent variation in cleavage chemistry dependent on sequence. Consistent with this notion, and in contrast to the DNA cleavage chemistry generated by the cobalt and rhodium nucleases, the singlet oxygen-mediated cleavage reaction of ruthenium complexes shows a clear base selectivity, i.e., a preferential reactivity at guanine sites. Singlet oxygen reacts preferentially with guanine residues (the rate constants for reactions with nucleotides vary G >> T > C > A) [150]. Table 3 summarizes the different kinds of cleavage chemistry exhibited by the small molecular DNA metallonucleases.

As expected, then, the DNA cleavage pattern obtained on a restriction fragment of pBR322 is seen to be somewhat different depending on the metallonuclease employed in the strand scission reaction, even though the structures of the coordinatively saturated complexes are quite similar [146b]. As the binding characteristics of these metal complexes do not appear to vary substantially with

TABLE 3

DNA Cleavage Properties of Artificial Metallonucleases

	Target	O_2 dependence	Photoactivated	Diffusible species
MPE-Fe(II)	Sugar	Yes	No	Yes
$Cu(phen)_2^+$	Sugar	Yes	No	?
Fe(II)-EDTA	Sugar	Yes	No	Yes
$Ru(phen)_3^{2+}$	Base	Yes	Yes	Yes
$Co(phen)_3^{3+}$	Sugar	No	Yes	No
$Rh(phen)_3^{3+}$	Sugar	No	Yes	No

sequence, the variance in cleavage patterns may be explained by the different cleavage mechanisms exhibited by each metal complex. The singlet oxygen-mediated cleavage reaction of the ruthenium nuclease leads to a strong guanine preference in cleavage pattern and, interestingly, resembling the pattern obtained by the Maxam-Gilbert guanine-specific sequencing reaction. The cobalt and rhodium cleaving molecules, however, behave as nonspecific photonucleases and yield a cleavage pattern with nearly equal reactivity with each base position along the DNA strand. Clearly, with these two complexes, the coupling of the nonspecific binding of the tris(phenanthroline) structure with the nonspecific photochemistry of the metal yields a reaction with DNA of low sequence specificity ideally suited for footprinting experiments. Furthermore the light dependence of these reactions makes them potentially useful tools in following time-dependent processes, such as resolving the movements of a protein as it traverses the DNA helix.

2.3. Metal Ions and Complexes that React Hydrolytically at Phosphate Esters

Several simple metal complexes, which strive to mimic the mode of action of metalloenzymes involved in nucleophilic displacements reactions on phosphate esters, have been investigated. These model complexes serve to exemplify the possible role for metal ions in metalloenzyme systems, such as alkaline phosphatase and in the phosphate ester hydrolysis reactions carried out by nucleases. These complexes have not been shown to cleave DNA, like their redox-cleaving counterparts, but rather attack simple phosphate esters. Yet a cursory examination of some of these systems might provide an understanding of how enzyme-bound metal ions of metallonucleases exert control over the phosphate hydrolysis reactions which they perform. Figure 7 shows some of the complexes described.

In the model system with zinc ion coordinated to phenanthroline-2-carbinol [see Fig. 7(b)], for example, it has been demonstrated that the zinc ion catalyzes the phosphorylation of 1,10-phenanthroline-2-carbinol by ATP [151]. Here the reaction was seen to have an absolute dependence on the presence of the metal cation and proceed through the formation of a reactive ternary complex composed of 1,10-phenanthroline-2-carbinol, ATP, and zinc. Interestingly, the rate is enhanced further with the addition of stoichiometric calcium ion. The zinc dication is proposed to play three crucial roles: (1) to serve as a template in appropriately positioning the metal-bound ATP for transfer of its terminal phosphoryl residue; (2) to enhance the nucleophilicity of the carbinol group by reducing its pK_a; and (3) to neutralize the negative charge density associated with the phosphate moieties of ATP, thereby lowering the electrostatic barrier and facilitating attack by the anionic nucleophile. Here the added source of charge neutralization by calcium ions probably accounts for the additional rate enhancement (from 26% for the zinc-phen-2-carbinol system to 42% for the same system in the presence of stoichiometric calcium).

Coordinated metal ions also promote the hydrolysis of 8-quinolinyl phosphate, and in this system Cu(II) appears particularly effective. It is interesting that the metal-promoted hydrolysis is increasingly effective with increasing pH, indicating involvement of OH^- as an external or metal-bound nucleophile [152,153]. Since 3-pyridyl phosphate cannot be metal-activated, an essential requirement for catalysis appears to be at least bidentate coordination to the metal

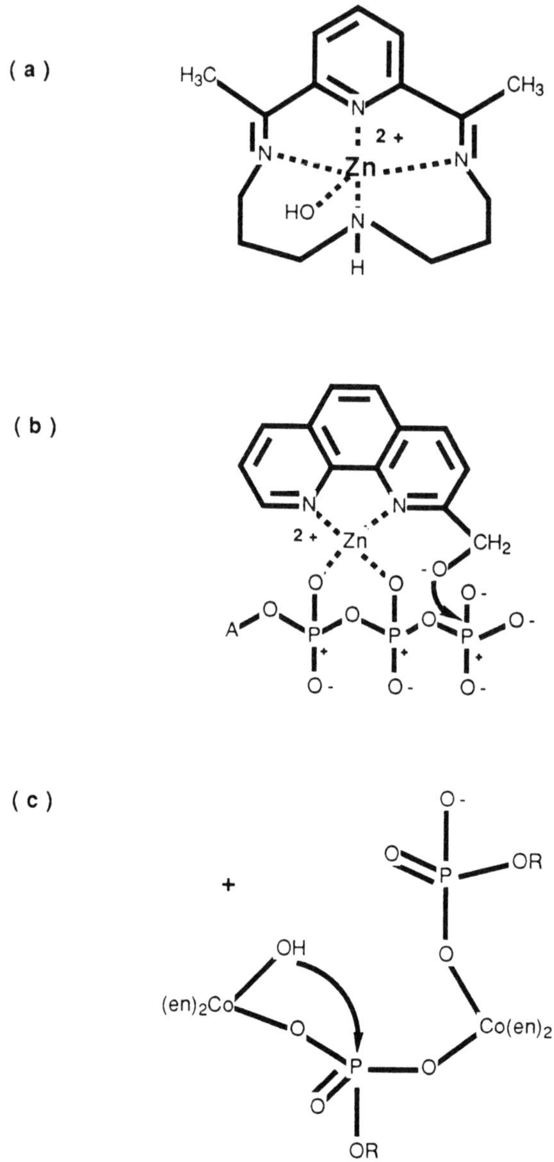

FIG. 7. Model complexes which promote the hydrolysis of phosphate esters.

by phosphate and the heterocyclic nitrogen atom. It is argued that were the role of the metal simply to neutralize charge, the second coordination site would not be required (although larger formation constants do result from bidentate coordination). An additional role of the metal in this system may be also to stabilize the leaving quinolinol anion in the transition state, through coordination to the ester oxygen atom.

Metal ions, even lacking additional coordinated ligands, themselves facilitate the dephosphorylation of nucleoside triphosphates [154,155]. When the rates at the pH of maximum promotion of ATP dephosphorylation are compared, it appears that the effectiveness of the metal ion decreases in the order $Cu^{2+} > Cd^{2+} > Zn^{2+} > Ni^{2+} > Mn^{2+} > Mg^{2+}$. The maximal pH for promoting dephosphorylation, however, is found to increase in the order $Cu^{2+} < Zn^{2+} < Ni^{2+} < Cd^{2+} < Mn^{2+} << Mg^{2+}$. Interestingly, this latter order parallels that for the pK_a for formation of the metal-bound hydroxo complexes and suggests that nucleophilic attack at the phosphorus atom may occur in an intramolecular fashion through a coordinated hydroxide ion.

The hydrolysis of the dimeric cation bis(m-4-nitrophenylphosphato)[bis(1,2-ethylenediamine)cobalt(III)]$^{2+}$ has also been investigated [86]. It was demonstrated in this case that ester hydrolysis occurs in this dimeric system exclusively via an intramolecular path [see Fig. 7(c)] and that intermolecular attack of hydroxide ion on the phosphorus atom of the dimer pair is not a competitive process. The isolation of this cobalt dimer permitted an assessment of the effect of simultaneous coordination of two metal centers on the rate of phosphate ester hydrolysis. A comparison of the rate enhancements obtained for the dimeric cobalt species with that of a monomeric cobalt complex, cis-$[Co(en)_2(OH)O_3POC_6H_4NO_2]$, indicated that the presence of a second cobalt center results in a 26-fold increase in the rate over that of the mononuclear complex. From this result an estimate of the relative contributions of "phosphate charge neutralization" and "intramolecular attack" was made for the rate enhancements of several coordinated nitrophenyl phosphate metal complexes relative to that of the uncoordinated ester. It was concluded from these analyses that monodentate coordination of a phosphate ester to a metal ion will not by itself lead to large rate enhancements; furthermore, the principal function of the metal in these systems appears to be to afford a template on which the nucleophile and phosphate ester are held in close proximity, thereby leading to an intramolecular hydrolysis reaction. It was further argued, therefore, that in enzymatic systems, monodentate coordination of a phosphate group alone may be inadequate to account for the great acceleration of hydrolysis rates obtained.

Catalytic hydrolysis of a phosphate triester, diphenyl p-nitrophenyl phosphate, by a zinc complex of a tetraaza macrocycle has also been observed [see Fig. 7(a)] [156]. The variation of the observed rate constant with pH implicated a coordinated zinc-bound hydroxide or the zinc complex with free hydroxide ion as active species in the mechanistic scheme for hydrolysis. Here two limiting possibilities are given for the catalytic species: a nucleophilic process involving attack on the phosphorus center by a zinc-associated hydroxide, or an electrophilic activation of the P-O bond to attack by free hydroxide through direct coordination of the phosphoryl oxygen to the zinc center. Additionally, a "hybrid" mechanism was postulated where the zinc center serves to deliver a coordinated nucleophile (OH^-) to the phosphorus center while simultaneously interacting with the phosphoryl oxygen, withdrawing electron density here possibly through direct coordination. The hybrid mechanism appeared to be most appealing.

In summary, from just a brief review of some synthetic metal-activated systems reactive in phosphate ester hydrolysis, several roles for the metal ion become evident, as illustrated in Fig. 8, and these resemble suggestions made earlier for the functioning of the metal ions in enzymatic systems. The metal ions may serve to group reactants, acting as a template, and perhaps provide a coordinated nucleophile in an intramolecular or intermolecular reaction. They can polarize the phosphorus center to enhance nucleophilic

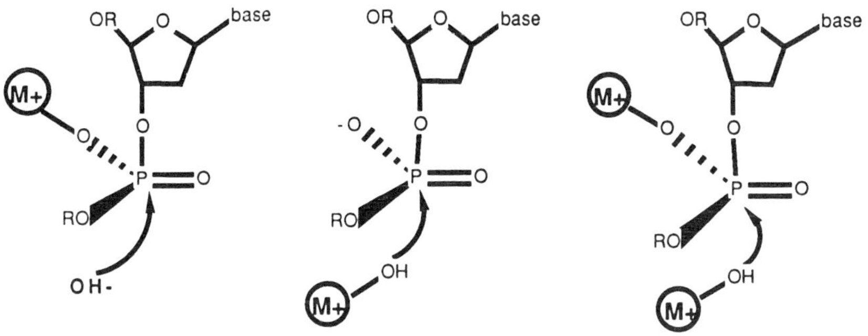

FIG. 8. Roles for metal ions in promoting hydrolysis of DNA: (left) as a Lewis acid; (center) in delivering a coordinated nucleophile; (right) serving both functions.

attack either by bound residues or a general base, and simultaneously the metal can serve to neutralize the negative charge on the phosphate group. Metal cations bound to phosphoryl oxygen atoms might furthermore facilitate P-O bond rupture by coordination to the leaving group, thus increasing its acid strength. Metal ions may therefore be uniquely suited to accelerate hydrolytic reactions of phosphate esters, and below we examine some systems where in specific reactions with DNA, a complex phosphodiester, this chemistry may be applied.

3. SITE-SPECIFIC DNA CLEAVAGE

Thus far we have focused on the requirements to achieve nonspecific cleavage of DNA. Here we encounter reactions at specific sites along the strand and the requirements to achieve such reactions actually parallel those necessary in achieving sequence neutrality. The enzymatic system examined is that of a restriction endonuclease, an enzyme able to recognize a particular six-base pair sequence and, in the presence of divalent metal ion, to cleave each of two DNA strands to produce 5'-phosphoryl and 3'-hydroxyl termini. With respect to site-specific cleavage by small molecules, the tactic is largely to couple nonspecific reactivity, usually metal-promoted, to a site-specific DNA-binding moiety, thereby delivering the metal reactivity, locally at high concentration, to sites along the strand. Site-specific DNA-cleaving molecules described include known DNA-binding peptides coupled to simple redox-cleaving moieties, conformation-specific metal complexes armed with photoactivated cleaving functionality, and, as a switch, oligonucleotides coupled to a natural nuclease. Through this variety of schemes, we examine and illustrate some of the principles necessary for specific targeting of a DNA site.

3.1. Restriction Endonuclease, EcoR1

The prototypical system in which to examine site-specific recognition and reaction at a DNA site is that of a restriction enzyme. The discovery of restriction enzymes in many respects marked the beginning of biotechnology since they serve as the essential enzymatic tools for the specific splicing of gene segments. Endonucleases form a class of DNA site-specific enzymes capable of recognizing and cleaving a unique sequence of base pairs along the DNA strand, usually four to eight base pairs in length. Minimal recognition requires a double-stranded DNA sequence with twofold rotational symmetry, and nuclease activity, hydrolytic cleavage of each of the two strands, is dependent on the presence of divalent metal cations.

EcoR1, a restriction endonuclease from *Escherichia coli*, binds specifically to the duplex palindromic sequence 5'-GAATT C-3' and cleaves both strands by hydrolyzing the phosphodiester bond between symmetrically disposed guanine and adenine residues [for reviews, see Refs. 157-159]. The strand scission reaction results in the production of 5'-phosphoryl and 3'-hydroxyl termini. This reaction is seen to proceed with inversion of configuration at the reactive phosphorus, implying that there may be an odd number of chemical steps in the hydrolysis mechanism [160]. This observation is also consistent with a direct, enzyme-catalyzed nucleophilic attack by H_2O (or hydroxide ion) at the phosphorus center, without involvement of a covalent enzyme intermediate. The metallonuclease EcoRl is active as a dimer [161,162] (M_r = 31,000 for the monomer unit) and has been shown to require both divalent magnesium [157] and zinc for activity [163]. Specific DNA binding by this enzyme occurs in the absence of magnesium with a dissociation constant of 10^{-11} M^{-1}; however, under these conditions the cleavage reaction does not proceed [157,164].

Recently, cocrystals of EcoRl and its recognition fragment, d(TCGCGAATTCGCG), have been obtained in the absence of magnesium, providing structural data concerning the specific recognition and binding of DNA by the endonuclease [165]. Since one of the catalytically essential cations, magnesium ion, is not present in the crystal structure, no information regarding the nucleolytic conformer of this protein is evident. The crystallographic complex is seen to be a twofold symmetrical globular structure with the protein embedded in the major groove of the DNA, leaving the minor groove completely exposed. Minor groove contacts made by the protein at the central adenine of the recognition sequence were inferred, however, from methylation protection studies of the protein-DNA complex in solution [166]. An important feature of the binding interaction, and perhaps also with respect to the strand scission reaction, appears to be the presence of several structural perturbations in the DNA, termed neokinks. These kinks, observed in the crystallographic structure, unwind the DNA by 25° per binding site and are induced by association of the endonuclease. An identical unwinding value has also been obtained from solution studies of the specific binding interaction of EcoRl; no unwinding and, perhaps consequently, no cleavage, is observed for the nonspecific interaction of this protein with DNA [167]. The key structural features of the protein include a five-stranded β-sheet, which may provide elements responsible for both recognition and strand scission, α-helices, and a disordered extension, called the "arm", which wraps around the DNA. The first 14 amino acid residues of the amino terminus of the protein, which form part of the armlike segment, have been reported to be essential to the catalytic cleavage reaction of this metallonuclease, but are not required for specific binding and recognition [168]. Thus it appears that the specific binding and cleavage functions of the enzyme may be somewhat distinct entities, and that the amino terminal ends of the arms of the dimer may contain critical features necessary for catalysis.

Both magnesium and zinc ions appear essential to the strand scission reaction of EcoR1. Magnesium is required as a cofactor, added in situ. Zinc ion is found to be a tightly bound, intrinsic component to the wild-type enzyme [163]. Removal of the zinc dication from each monomer subunit of the endonuclease results in a concomitant loss of enzyme activity, while readdition of the metal can be seen to restore its cleavage ability [169]. Cobalt can be exchanged for the native zinc ion, yielding an enzyme which displays the same substrate specificity [170]. The cobalt derivative protein, however, exhibits a higher rate of reaction, as compared with the native zinc protein, and an increase in product dissociation rate [171]. Since dissociation of the doubly cleaved product is rate limiting in the reaction mechanism of EcoR1 [172], the increase in the maximal velocity observed for the cobalt-substituted enzyme may reflect the lower affinity of cobalt rather than zinc for the phosphate product. It is notable that an increase in the reaction rate of the cobalt analogs of several zinc metalloenzymes has been observed [173,174]. The binding site for the zinc ion is not evident in the crystal structure. Manganese has been reported to substitute for the essential magnesium cofactor in this metal-mediated reaction; however, here the recognition capacity of the endonuclease is relaxed, yielding an enzyme of much lower sequence specificity [175]. This finding suggests that the metal cofactor may play a role in the site-specific recognition as well as in the subsequent hydrolytic cleavage step. Specific details of the metal-mediated reaction of EcoR1, however, have not been elucidated. From the above discussion of the metal-activated chemistry of other phosphoesterases and synthetic model complexes, a picture emerges of how metal ions may function to assist in the cleavage of a phosphate ester bond. For this site-specific endonuclease, similar roles for the essential metal ions in this system may be envisioned. Possibly metal ions serve to aid both in the nucleophilic attack and in the electrophilic polarization of the diester. The bimetallic nature of this protein-DNA complex, consisting of both an alkaline earth and transition metal ion, suggests that perhaps a more intricate picture for the metal-mediated chemistry may eventually emerge here. From the crystallographic data, the active site for cleavage, which may include both zinc and magnesium dications in the native state, was not fully assembled, thus limiting any interpretation of the role of these essential metal ions in the enzymatic mechanism.

3.2. Complexes that Cleave DNA Through Coupled Redox Chemistry

In the case of redox-activated DNA site-specific cleaving molecules, similar chemistry to that seen above for nonspecific reactions is employed. The major difference is that coupling of the nonspecific cleaving functionality is made now to a site-specific DNA-binding moiety. The site specificity is derived from this binding functionality. If the reactive characteristics of the cleaving element are innocent or neutral to sequence, as in the case of the small and uncharged hydroxyl radical, then the characteristics of recognition by the DNA-binding element will be preserved upon its conversion to a DNA-cleaving element. Many examples of this notion, often termed *affinity cleaving*, are given below.

3.2.1. Fe(II)-EDTA Coupled to Sequence-Specific DNA-Binding Functions

As previously discussed, Fe(II)-EDTA is a general, nonspecific DNA-cleaving function capable of inducing strand scission in the presence of oxygen and a reducing agent. The coupling of the nonspecific nucleolytic chemistry of Fe(II)-EDTA to a sequence-specific DNA-binding molecule, therefore, creates a sequence-specific nuclease. The new coordination complexes that result are now capable of both specific recognition and cleavage along the DNA strand based on the site affinity of the tethered binding moiety. Numerous examples of such converted DNA-binding molecules have been synthesized by Dervan and coworkers [176-181].

One of the first examples is that of ferrous EDTA tethered to the antibiotic distamycin. This natural tripeptide contains three N-methylpyrrolecarboxamides which interact with the minor groove of B-DNA, preferentially at A-T-rich sequences [182-184]. Here the EDTA moiety is tethered to either the amino or carboxyl terminus of tris-N-methylpyrrolecarboxamide yielding distamycin-EDTA (DE) or EDTA-distamycin (ED) [176-178]. From sequencing gel analyses, both DE•FE(II) and ED•FE(II) are found to cleave adjacent to five base pair sites consisting of A-T-rich DNA sequences. Furthermore, the cleavage patterns on opposite strands, produced from either of two binding orientations of the tripeptide (DE or ED), are skewed in the 3' direction, indicating that the metal complex of distamycin binds also in the minor groove of DNA. Because of its high DNA affinity, the pentapeptide penta-N-methylpyrrolecarboxamide-EDTA•Fe(II) (P5E•Fe(II)) was seen to achieve double-stranded scission of DNA adjacent to a six-seven base pair A + T recognition site in a catalytic reaction [177].

Another very interesting and novel application of a peptide-EDTA•Fe(II) complex is seen with bis(netropsin)-3,6,9,12,15-pentaoxa-heptadecanediamide, shown in Fig. 9 and synthesized to illustrate metalloregulation in the sequence-specific binding of a small molecule to DNA [179]. Here a dimer of netropsin, a dipeptide containing two N-methylpyrrolecarboxamides, is linked together by a tetraethylene glycol tether which provides a multidentate acyclic neutral ligand for metal cations. Also tethered to the molecule is Fe(II)-EDTA, lending affinity-cleaving capabilities to the molecule. In the absence of metal ions the complex binds poorly to DNA. A metal ion, however, may serve to orient structurally the pyrrolecarboxamide

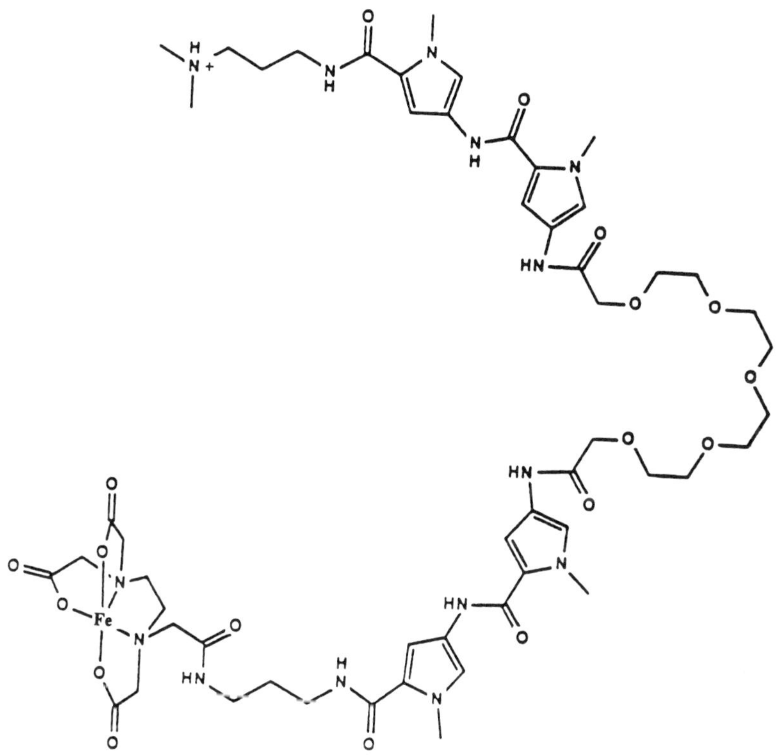

FIG. 9. Bis(netropsin)-3,6,9,12,15-pentaoxaheptadecanediamide-EDTA•Fe(II), a synthetic nuclease whose site-specific binding is regulated by metal ions. (Reproduced from Ref. 179 by permission of the American Chemical Society.)

units along the DNA strand when bound to the central crown ether portion of the molecule. DNA cleavage by bis(netropsin)-3,6,9,12,15-pentaoxaheptadecanediamide-EDTA•Fe(II) was analyzed on a restriction fragment of pBR322, and it was found that in the absence of any metal cation, or in the presence of several metal cations, including magnesium and calcium, little DNA strand scission results. Addition of Sr^{2+} or Ba^{2+}, however, produces strong patterns of sequence-specific DNA cleavage at the sites 5'-TATAGGTTAA-3' and 5'-AATA-3'. This result indicates that simultaneous binding of the two dipeptide units is induced in the presence of certain metal cations, thus affording a metalloregulatory function to this netropsin analog. In this molecule, then, metal ions serve both a function in cleavage, through the redox chemistry of Fe(II)-EDTA, and a structural function in recognition, the binding of the strontium or barium ions to the polyether.

A synthetic 52-residue peptide, forming the DNA-binding domain of Hin recombinase which recognizes a 13 base pair sequence, has also been equipped with the Fe(II)-EDTA-cleaving function, tethered, in this case, to the amino terminus of the peptide fragment [180]. Once again the attachment of Fe(II)-EDTA provides a route to transform a specific DNA-binding molecule into a site-specific nuclease because now this synthetic-EDTA•Fe(II) peptide is capable of cleaving rather than simply binding DNA at Hin recombination sites. Because this hybrid peptide can now recognize and cleave at specific DNA sequences of approximately 13 base pairs in length, it may also prove very useful in the isolation of long DNA fragments for genome mapping, surpassing the utility of restriction enzymes in this regard.

Affinity cleaving is illustrated also in the attachment of Fe(II)-EDTA to a strand of homopyrimidine oligonucleotides containing 11-15 base residues. In this case, it is shown that this homopyrimidine "nuclease" specifically binds and cleaves at the precisely corresponding homopyrimidine-homopurine tracts of DNA [181]. What proves to be extremely interesting, moreover, is that the affinity cleavage is achieved through triple-helix formation at the site where the homopyrimidine strand binds. From the location and asymmetry in the resultant cleavage pattern, it appears that the homopyrimidine oligonucleotide binds in the major groove parallel to the homopurine strand of double-stranded DNA. The cleavage data additionally indicate that the binding interaction results from triple-helix formation rather than from displacement of the homopyrimidine strand of the DNA. These oligonucleotides armed with a redox-cleaving functionality provide a site specificity (a unique 15-base pair homopurine-homopyrimidine site with no mismatch) that is unrivaled, and these complexes may be extremely important as precise cutting tools for the mapping of human chromosomes.

3.2.2. (Phenanthroline)copper(I) Coupled to the Trp Repressor

The attachment of a 1,10-phenanthroline-copper complex, a nonspecific DNA-cleaving function, to the tryptophan gene (trp) repressor of *E. coli* transforms this DNA-binding protein into a site-specific nuclease [185]. This chemical conversion is accomplished by the covalent attachment of 5-iodoacetamido-1,10-phenanthroline to certain (unidentified) modified lysyl residues of the repressor protein. This procedure results in the incorporation of 4 mol of phenanthroline per protein subunit. As previously discussed, the artificial nuclease $(phen)_2Cu^+$ cleaves DNA through oxidative degradation of the deoxyribose moiety in a reaction that requires H_2O_2 as a coreactant [103]. Presumably, when linked to the trp repressor, the resulting (phenanthroline)copper(I)-mediated strand scission reaction proceeds in a similar fashion. However, for this reaction only one phenanthroline ligand appears to be necessary to induce strand scission, whereas two ligands are necessary for the $(phen)_2Cu^+$ reaction (suggesting that the presence of two phenanthroline ligands may be necessary only for effective binding).

Both the binding and nucleolytic activity of the phenanthroline-derivatized trp repressor (OP-trp repressor) were tested on a restriction fragment derived from the control region of the *aroH* transcription unit. In the presence of L-tryptophan, a corepressor, both the native and modified proteins yield the same DNase I footprint and, as expected, in the absence of the corepressor, both do not bind to the operator sequence. The nucleolytic action of the OP-trp repressor was clearly revealed on sequencing gels upon addition of cupric ion and a thiol to the corepressed protein. What is apparent from inspection of these gels is the presence of a "reverse footprint" at the binding site of the trp repressor. Moreover, no scission is observed outside of the binding site. The specificity of the OP-trp repressor was also tested by using an operator containing a different sequence, *trpEDCBA*, as a substrate. Again the OP-trp-modified protein cleaved this operator sequence. The presence of a reverse footprint at the binding site of the trp repressor was observed here, but a somewhat different cutting pattern was also revealed. Interestingly, this difference observed in the cleavage patterns produced by the OP-trp repressor may suggest that the binding orientation of the protein relative to the DNA helix is different when *aroH* as compared to *trpEDCBA* is used as a recognition sequence; thus some of the details of the binding interaction are revealed through the coupled reactivity of the copper-phenanthroline complex.

This coupling of a nonspecific DNA-cleaving function to a specific DNA-binding function yields a molecule which retains its specific binding capacity, but now has the acquired ability to cleave DNA at its binding site. Moreover, binding is coupled to nucleolytic activity owing to the metal ion. Numerous uses can be envisaged for such artificially constructed metallonucleases. Since the binding site for the trp repressor is approximately 20 base pairs, OP-trp protein may be used to isolate long DNA fragments for sequencing and chromosomal mapping. Additionally, this copper-dependent nuclease may also provide a new reagent for analysis of DNA structure and the location of unidentified repressor-binding sites, and potentially for the therapeutic damaging of such repressor-binding sites.

3.2.3. *Tris(4,7-diphenylphenanthroline)cobalt(III)*

Site-specific reactions with tris(4,7-diphenylphenanthroline)cobalt(III) ($Co(DIP)_3^{3+}$), a bulky dissymmetric derivative of $Co(phen)_3^{3+}$ shown in Fig. 10, illustrate that, by simply changing the size and shape of the ligand, a complex can be created that retains the nonspecific, photoactivated cleavage chemistry associated with the central metal ion, but which now displays altered DNA-binding properties. Given that the electronic properties of $Co(DIP)_3^{3+}$ and $Co(phen)_3^{3+}$ are quite similar, it is reasonable to consider that their photochemical reactions with DNA would also proceed in a similar manner. The

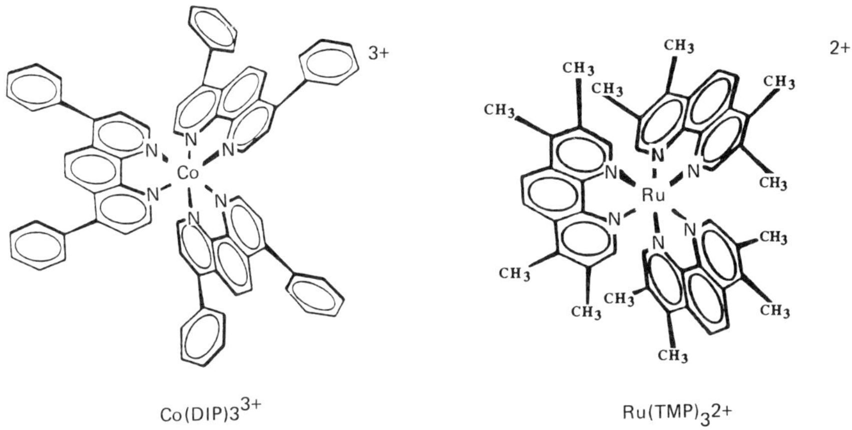

FIG. 10. (left) Tris(4,7-diphenylphenanthroline)Co(III): a probe for Z-DNA. (right) Tris(3,4,7,8-tetramethylphenanthroline)Ru(II): a probe for A-DNA.

deoxyribose moiety is the most likely site of attack by these photoexcited metal complexes, and analyses of the cleavage products after reaction of both cobaltic complexes with DNA are consistent with this notion [186]. However, what is quite different about these two inorganic photonucleases is their reaction specificities toward DNA. $Co(DIP)_3^{3+}$ is a site-specific enantioselective DNA-cleaving molecule, whereas $Co(phen)_3^{3+}$ is not. Illustrative of this point are the results obtained for the cleavage reaction of $Co(DIP)_3^{3+}$ enantiomers with pCol El DNA. DNA strand scission, in this case, can be observed only with the Δ enantiomer of $Co(DIP)_3^{3+}$, while the Λ enantiomer displays little cleavage [141]. Because Λ-$Co(DIP)_3^{3+}$ can bind to the B-form DNA only through a surface (or groove) interaction, these results seem to indicate that only close association of the metal complex with DNA, presumably by intercalation, will lead to efficient cleavage. Λ-$Co(DIP)_3^{3+}$, sterically precluded from binding to B-DNA by intercalation, may, however, bind to left-handed Z-DNA [21,187], or in fact other DNA sites which exhibit a conformation that is sufficiently unwound to accommodate the bulky trication. Here, then, the metal serves both as the reactive element and as the structural core of the complex to specify sites, really shapes, for recognition along the DNA strand.

To determine if Λ-Co(DIP)$_3^{3+}$ could recognize local regions of Z-DNA, and hence provide a conformation-specific DNA-cleaving moiety, a plasmid containing a 16 base pair GC insert (pLP32, a derivative of the plasmid pBR322), which is known to adopt a Z conformation under certain conditions [188], was used as a substrate in the cleavage reaction by Co(DIP)$_3^{3+}$ [21]. In this case, both enantiomers show significant cleavage of the plasmid DNA. The approximate sites of cleavage by Λ-Co(DIP)$_3^{3+}$ along the DNA strand was then located by a coarse mapping procedure with the major cleavage site obtained at the GC insert. Thus Λ-Co(DIP)$_3^{3+}$ may bind and cleave a Z segment located within a predominantly B-DNA helix. In addition to the cleavage site found at the GC insert of pLP32, Λ-Co(DIP)$_3^{3+}$ was seen to cleave the plasmid at four additional sites, and these same four sites were cleaved also in the parent DNA molecule, pBR322. The sites correspond to regions of long alternating purine/pyrimidine sequences known to favor the formation of Z sites under conditions of high superhelical density and relatively low salt (pBR322, superhelical density = -0.058, [Na^+] = 40 mM). Another intriguing observation was that these four sites cleaved by Λ-Co(DIP)$_3^{3+}$ all appeared to border the gene-coding regions of the plasmid. Surely, given the steric constraints on Co(DIP)$_3^{3+}$ in binding to DNA, these sites are necessarily conformationally distinct. Λ-Co(DIP)$_3^{3+}$ becomes, then, a sensitive, site-specific structural probe of conformationally distinct DNA sites that exist along the strand.

Recently, specific cleavage sites by Λ-Co(DIP)$_3^{3+}$ along the SV40 viral DNA were determined using high-resolution sequencing techniques [187]. Among the cleavage sites obtained for this plasmid was an eight base pair segment of alternating purine/pyrimidine residues in the enhancer region, suggested earlier to adopt an altered Z conformation [189]. Furthermore the 5' skewing in the cleavage pattern obtained indicates that the bulky trication may be accessing its DNA-binding sites through association with the major groove, consistent with earlier binding data for the ruthenium analogs [139]. The specificity of cleavage at this site is remarkable and presents a graphic illustration of how shape or conformation selection coupled to a specific cleaving reaction, one without a diffusible species, can generate high specificity in reaction along the strand. Interestingly, cleavage sites for the cobalt complex were observed also in the promoter region where the transcription factor Spl binds. The cleavage pattern, differing from that seen in the enhancer segments, suggests a different distinct conformation at these sites. Perhaps the distinct conformations at these sites recognized by the simple coordination complex also provide features for recognition by the DNA regulatory protein. Interestingly, Spl has been suggested quite recently to be a zinc finger protein, and thus perhaps in this protein we see another example of a metal as a coordinatively saturated structural core for a complex to recognize sites along the strand.

3.2.4. Tris(3,4,7,8-tetramethylphenanthroline)ruthenium(II)

Tris(3,4,7,8-tetramethylphenanthroline)ruthenium(II) ($Ru(TMP)_3^{2+}$, shown also in Fig. 10, is another member of the family of chiral "metallophotonucleases". This complex binds preferentially to DNA conformations which are A-like. The complex maintains the DNA photocleavage chemistry characteristic of ruthenium polypyridyl complexes, but the binding properties of the complex, in comparison to $Ru(phen)_3^{2+}$, are altered with the addition of four methyl groups to the phenanthroline ligand. The presence of these methyl groups surrounding each ligand of $Ru(TMP)_3^{2+}$ precludes intercalative binding and augments the hydrophobicity of the metal complex. Thus the binding of this ruthenium photonuclease to nucleic acids takes place at the surface of the helix. Furthermore, the increase in the size of the complex compared with $Ru(phen)_3^{2+}$ serves to limit its interactions in helical grooves, such as those in B-DNA, instead facilitating binding to the more shallow minor groove surface of A-form helices. The photochemistry and excited state properties of the tris(phenanthroline)-ruthenium(II) core is preserved, however.

Binding studies reveal that this metal complex binds avidly to synthetic polynucleotides that can adopt an A-form helix. An A-like conformation, found in double-stranded RNA, DNA-RNA hybrids, and DNA of certain sequences, contains a deep but narrow major groove, and a shallow but wide minor groove. This relatively flat and wide surface of A-form polynucleotides appears to enhance association of the methylated ruthenium photonuclease. $Ru(TMP)_3^{2+}$ is seen to bind preferentially to poly(rI)•poly(rC) and poly(rG)•poly(dC), while little or no binding is observed for both B-form and Z-form DNA helices for which groove shapes and dimensions are considerably altered from that of the A structure [148]. Moreover, consistent with a surface-binding interaction proposed for the bulky ruthenium complex, is the observation that the Λ enantiomer is found to be preferentially bound to the helix at equilibrium.

$Ru(TMP)_3^{2+}$ has recently been used as a structural probe of the conformational heterogeneity that may exist along the DNA strand visualized through the use of high-resolution mapping techniques [149]. Along a restriction fragment of pBR322, several specific sites of binding and therefore reaction by $Ru(TMP)_3^{2+}$ were observed, and these suggest that some sequences may exert a propensity to adopt A-like features locally along the strand. The major cleavage sites for $Ru(TMP)_3^{2+}$ are seen to correspond to homopurine-homopyrimidine stretches. The pattern of cleavage at the metal-bound sites contrasts the specific scissions seen with $Co(DIP)_3^{3+}$ and reflects instead the diffusion of singlet oxygen emanating out from the DNA-bound ruthenium complex. Consistent with a singlet oxygen-mediated cleavage mechanism is the enhancement of the cleavage and increase in bandwidth in D_2O, which lengthens the lifetime of this activated oxygen species. This structurally specific metallonuclease is now being used to map sequences in other DNAs that may also exist in altered non-B conformation. Interestingly, in the promotor region of SV40 viral DNA, $Ru(TMP)_3^{2+}$ is found to cleave specifically at sequences that border the binding sites for the gene-activating protein, Sp1 [190]. $Co(DIP)_3^{3+}$, with a different characteristic structure, recognized central features of the distinct transcription factor binding site, and $Ru(TMP)_3^{2+}$, binding to DNA in a different fashion and with differing steric constraints, appears to detect different structural features at the borders of the transcription factor-binding sites. Again the prevalence of altered DNA conformations at positions where they could act as recognition signals for biologically important DNA regulatory proteins appears evident.

3.3. Complexes that Cleave DNA Through Coupled Phosphodiester Hydrolysis

Just as it has become clear that metal ions aid in DNA cleavage by efficient delivery of redox chemistry and aid in specific recognition by acting as a central template for the structure of a DNA-binding moiety, it is also apparent from that described above that metal ions are useful in promoting hydrolytic chemistry. The development of true synthetic restriction enzymes as useful tools for molecular biology will require the application of such hydrolytic chemistry. Hydrolytic cleavage of the DNA phosphodiester backbone is not an easily accomplished task, but some schemes to accomplish this result are now being examined.

3.3.1. Hybrid Staphylococcal Nuclease

The relatively nonspecific, single-stranded DNA cleavage chemistry associated with the staphylococcal metallonuclease was used as a starting point for the creation of a semisynthetic nuclease, armed with the capacity to perform instead a site-specific strand scission reaction with DNA [191]. The construction of a site-specific staphylococcal nuclease was accomplished by selectively fusing an oligonucleotide of a defined sequence to the protein. This approach involves redesigning an existing enzyme, maintaining, in this case, the catalytic domain of the nuclease, but adding to it a new binding "domain" in the form of an oligonucleotide sequence. This single-stranded binding domain can then form a duplex structure at a specific site dictated by the base sequence covalently attached to the protein, hence providing a scheme to recognize and cleave at a specified site along a single-stranded polymer.

As previously discussed, crystallographic information obtained for the complex of staphylococcal nuclease-diphosphothymidine-Ca^{2+} indicates that the pyrimidine ring of the bound inhibitor, Ca-pTp, fits into a hydrophobic pocket at the enzyme surface and that the 5'-phosphate is positioned at the active site by a calcium ion as well as certain arginine residues [50]. From inspection of the geometry at the site of the bound inhibitor, Corey and Schultz [191] devised a synthetic plan for the redesigning of this calcium-dependent nuclease. The molecular geometry suggested that the coupling of the 3' terminus of an oligonucleotide-binding domain to lysine-116 on the enzyme surface would allow for the correct alignment of a hybridized DNA substrate to the active site of the nuclease. An oligomer, 15 nucleotides in length, was then coupled to staphylococcal nuclease by a disulfide exchange reaction, which required the introduction of a free thiol at both the 3' terminus of the oligonucleotide and at residue 116 of the protein; a free cysteine was substituted at position 116 of the enzyme by oligonucleotide-directed mutagenesis. The subsequent disulfide exchange between the mutant protein, containing cysteine-116, and the 3' terminus of the oligomer proceeded with high selectivity due to the lack of free thiol in both the DNA and staphylococcal nuclease.

This staphylococcal hybrid metalloenzyme is found to site-selectively cleave a single-stranded DNA substrate aligned with its oligonucleotide-binding site, 5'-CCCGCACAAGCCGCT-3', in a reaction also dependent on the presence of calcium ions. In contrast, the underivatized cysteine-116 enzyme cleaved nonspecifically at T-rich sites in the single-stranded substrate, as does the native protein. As expected, this hybrid DNA/nuclease retained the metal-mediated hydrolytic chemistry associated with staphylococcal nuclease, cleaving at the phosphodiester backbone of DNA by nucleophilic attack at the activated phosphorus atom. This cleavage chemistry may be compared to that of the site-specific cleaving agents discussed above.

For most other synthetic site-specific nucleases, strand scission is instead accomplished by a metal-mediated redox chemistry resulting in the degradation of the deoxyribose moiety of DNA. This approach, utilizing the hydrolytic chemistry of a true nuclease, results in the recreation of a biological catalyst but with newly designed specificities.

3.3.2. $Ru(DIP)_2Macro^{n+}$

$Ru(DIP)_2Macro^{n+}$, where DIP = 4,7-diphenyl-1,10-phenanthroline and Macro = 4,7-[$(NH_2CH_2CH_2)_2NCH_2CH_2NHSO_2C_6H_4$]-1,10-phenanthroline, shown schematically in Fig. 11, is a synthetic metallonuclease resembling in its central structure its analog $Co(DIP)_3^{3+}$, shown earlier to cleave DNA upon photoactivation at structurally distinct sites along the strand [21,187]. The nucleolytic ability of this ruthenium metal complex, however, is conferred instead by the presence of two polyamine arm-like segments which can bind various metal cations and, in so doing, deliver metal-activated chemistry to each strand of duplex DNA. In the presence of copper and a thiol, $Ru(DIP)_2Macro^{n+}$ cleaves DNA efficiently to form a mixture of both single-stranded and double-stranded products [192]. The double-stranded/single-stranded product ratio depends on the copper content in the polyamine arms; when both arms are filled with metal, maximal double-stranded scission ensues. Thus in the presence of a reducing agent, this coupled ruthenium-copper nuclease cleaves by delivering redox-active metal ions to the DNA, much like other synthetic nucleases previously mentioned.

The general utility of redox nucleases as molecular cutting tools is limited, however, because oxidative DNA strand scission results in the production of DNA terminal ends which cannot be religated enzymatically. Natural nucleolytic systems do not cleave DNA through metal-mediated redox chemistry but instead utilize hydrolytic chemistry, so as to cleave at the phosphodiester backbone of DNA, and yielding products, in most cases, that can be religated enzymatically. As described extensively in this review, metal ions are useful also in promoting this chemistry. Based on this premise, $Ru(DIP)_2Macro^{n+}$

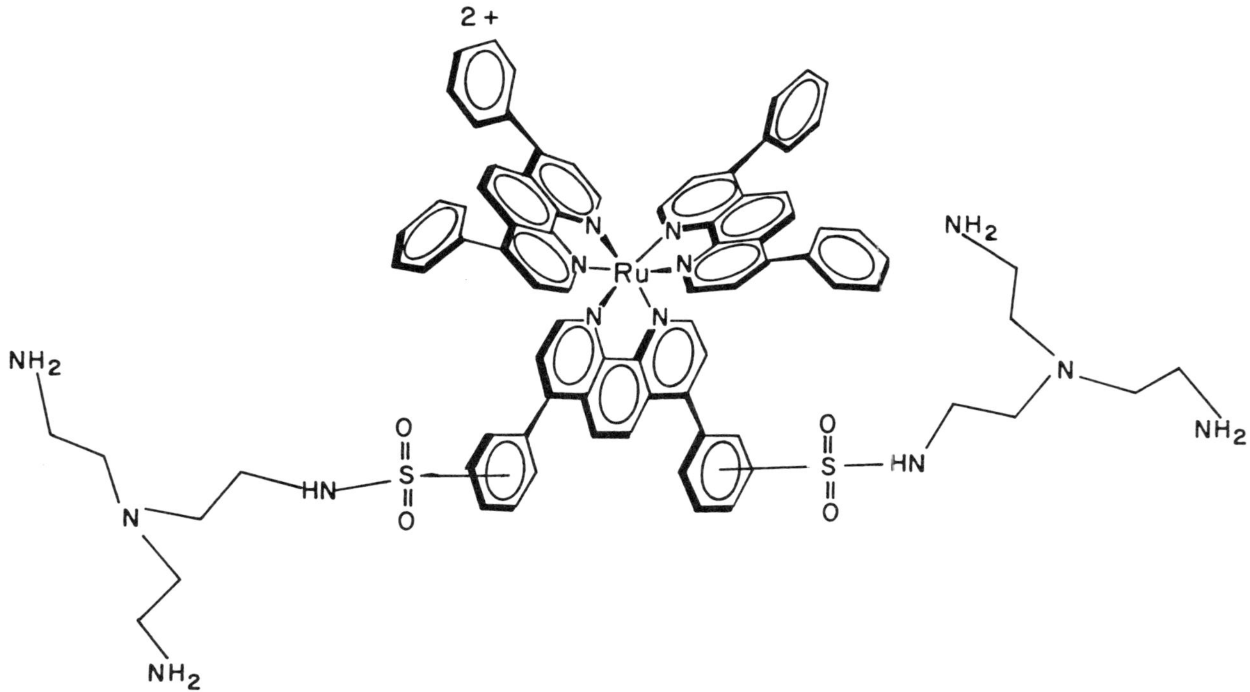

FIG. 11. $Ru(DIP)_2Macro^{n+}$, a synthetic metallonuclease. (Reproduced from Ref. 192 by permission of the American Chemical Society.)

was developed as a first approximation to a true synthetic nuclease, one capable of cleaving DNA through phosphodiester hydrolysis, yielding products which may be religated enzymatically and, of course, utilizing non-redox-active metal ions to accomplish this result [193].

Cleavage of pBR322 DNA by the ruthenium nuclease was monitored in the presence of various divalent metal ions, Cu(II), Co(II), Zn(II), Cd(II), and Pb(II), and without the addition of a reducing agent. Strand scission, though quite inefficient, is apparent in the presence of both the redox-active metals, Cu(II) and Co(II), and the non-redox-active metals, Zn(II), Cd(II), and Pb(II). This result suggests that a nonredox mechanism may be needed to account, at least in part, for these strand scission reactions. To test this notion, an enzymatic assay was performed, utilizing T4 DNA ligase, to determine whether the products of the strand scission reaction were consistent with hydrolysis of the phosphodiester linkages. T4 DNA ligase, in the presence of ATP, will ligate a severed DNA product only when the appropriate ends are presented to the enzyme for religation; these termini must contain both 5'-phosphoryl and 3'-hydroxyl groups [194]. The efficiency of religation after cleavage by these differing metal ions in the presence of $Ru(DIP)_2Macro^{n+}$ is shown in Table 4, along

TABLE 4

Metal Ion-Promoted Cleavage by $Ru(DIP)_2Macro^{n+}$ and Religation of pBr322 DNA

Added metal ion	% Cleavage	% Religated/% cleaved
Cu(II)	73	0.14
Co(II)	64	0.09
Zn(II)	30	0.68
Cd(II)	29	0.53
Pb(II)	29	0.57

Source: Adapted from Ref. 193.

with the cleavage efficiencies for the different added metals. As may be seen in the table, the greatest extent of religation is apparent with the non-redox-active metal ions: Zn(II), Cd(II), and Pb(II). In fact, despite their high cleavage levels, only low levels of religation were observed with cobalt and copper; with these metal ions, therefore, DNA cleavage must be operating, at least in part, through a redox-promoted, oxidative path.

The efficiency of religation of the cleavage products produced by $Ru(DIP)_2Macro^{n+}$, in the presence of non-redox-active metals, is frequently seen to be greater than the 50% yield expected if hydrolysis were to give an equal distribution of both 3'-phosphate and 5'-phosphate products. This result suggests that perhaps the DNA binding moiety of this metallonuclease, $Ru(DIP)_3^{2+}$, presents a stereochemical bias through binding, so that the hydrolysis reaction leads to the generation of the appropriate ends for religation. If displacement at the phosphate ester proceeds by attack of an incoming nucleophile from the major groove of DNA, this stereochemical approach might explain the lack of an equal distribution of phosphate products; furthermore, binding to DNA through the major groove has been observed for both structural analogs of $Ru(DIP)_2Macro^{n+}$, $Co(DIP)_3^{3+}$ [187], and $Ru(DIP)_3^{2+}$ [139].

This cleavage reaction performed by the ruthenium nuclease, in the presence of non-redox-active metal ions, represents the first example of hydrolysis of the phosphodiester backbone of DNA by a fully synthetic cleaving agent, and furthermore is illustrative of how DNA strand scission reactions, performed by synthetic nucleases, may be accomplished nonoxidatively. The coupling of the nonredox chemistry associated with this trinuclear metallonuclease to other synthetic molecules may be useful to the development and application of artificial restriction enzymes.

Interestingly, many comparisons can be made between the synthetic nuclease, $Ru(DIP)_2Macro^{n+}$, and some of the other enzymatic nucleases previously discussed. In a very simple way, this analog nuclease contains the salient structural features found in the restriction enzyme EcoR1, i.e., a central binding domain and armlike segments that are essential for reaction. Here in a similar fashion, the central core of the ruthenium nuclease is seen to provide the binding and recognition domain of the molecule, while the metal-binding arms confer reactivity. Additionally, both the bimetallic restriction endonuclease and trinuclear metal complex are capable of producing double-stranded scissions in duplex DNA, and both are associated with DNA through interactions with the major groove. Although the recognition capacity of the synthetic metallonuclease has not yet been assessed, the potential to recognize certain structural features of DNA may be provided by its DNA-binding moiety. Moreover, as seen in numerous enzyme systems, multiple metals are also needed for reaction by this synthetic nuclease: both "structural" and "catalytic" metals are present in $Ru(DIP)_2Macro^{n+}$.

4. CONCLUSIONS

In this chapter we have seen the essentiality of metal ions in performing reactions by nucleases. The metal ions are clearly useful in promoting cleavage of nucleic acids. In enzymatic systems they function in activating the phosphate ester for hydrolysis, either electrophilically through direct coordination or by delivery of coordinated nucleophiles. In synthetic systems, the metal has functioned primarily to deliver instead redox chemistry to the DNA helix. The metalloenzyme systems described above represent only a small portion of the metallonucleases active in phosphate ester hydrolysis and consequently may not give a complete picture of the role of the metal ions involved. As more metallonucleases are discovered, and the role of the metal ions in these systems elucidated, perhaps a more vivid picture will emerge. Furthermore, as we begin to better describe and understand the metal-mediated chemistry in natural nucleases, we can also move away from redox chemistry in synthetic reactions, and instead mimic more closely the hydrolytic chemistry at phosphate esters associated with true metallonucleases. Indeed, current efforts in the construction of artificial nucleases, though still in their infancy, suggest that such synthetic hydrolytic chemistry promoted by metal ions may be achieved.

Metal ions serve structural roles as well in nucleases, those real and artificial. Metal ions are seen to orient substrates for reactivity and also in synthetic systems to orient ligand structures for nucleic acid binding and even specific site recognition. While we know that transition metal ions can bind both to phosphates and to the nucleic acid bases, studies on natural DNA-binding enzymes have focused on metal-phosphate interactions; perhaps there may be a role for certain metal ions, maybe zinc, in the recognition of DNA through interactions with the nucleic acid bases. The roles for metals in nucleic acid recognition in the biological arena are still very unclear, but there are some indications of structural roles for metal ions. Certainly studies with synthetic site-specific metallonucleases containing a metal ion as the structural core highlight the utility of metal ions in this context.

Metallonucleases, both real and artificial, can be powerful tools for biotechnology and critical elements in dissecting and describing nucleic acid biochemistry. While some features and clues as to roles of metal ions in nucleases have been obtained, much remains to be done. The elucidation and application of transition metal chemistry in the functioning of metallonucleases remains an exciting challenge for the future.

ACKNOWLEDGMENTS

We are grateful to the National Institutes of Health (GM33309) and the National Science Foundation (Alan T. Waterman Award CHE-85-17354 to J.K.B.) and the National Foundation for Cancer Research for their financial support.

REFERENCES

1. W. Saenger, in "Principles of Nucleic Acid Structure" (C. R. Cantor, ed.), Springer-Verlag, New York, 1984, p. 253 ff.
2. V. M. Vogt, *Methods Enzymol.*, *65*, 248 (1980).
3. M. Laskowski, Sr., *Methods Enzymol.*, *65*, 263 (1980).
4. S. Linn, *Methods Enzymol.*, *12A*, 247 (1967).
5. J. H. Waterborg and C. M. Kuyper, *Biochim. Biophys. Acta*, *571*, 359 (1979).
6. J. H. Waterborg and C. M. Kuyper, *J. Biochemistry*, *92*, 1655 (1982).
7. S. A. Martin, R. C. Ullrich, and W. L. Meyer, *Biochim. Biophys. Acta*, *867*, 67 (1986).
8. S. A. Martin, R. C. Ullrich, and W. L. Meyer, *Biochim. Biophys. Acta*, *867*, 76 (1986).
9. T. Y. Chow and M. J. Fraser, *J. Biol. Chem.*, *258*, 12010 (1983).
10. D. M. Hanson and J. L. Fairley, *J. Biol. Chem.*, *244*, 2440 (1969).
11. M. J. Fraser and H. Cohen, *J. Bacteriol.*, *154*, 460 (1983).
12. W. K. Holloman and R. Holliday, *J. Biol. Chem.*, *248*, 8114 (1973).
13. M. Fujimoto, A. Kuninaka, and H. Yoshino, *Agric. Biol. Chem. (Japan)*, *38*, 777 (1974).
14. M. Fujimoto, A. Kuninaka, and H. Yoshino, *Agric. Biol. Chem. (Japan)*, *38*, 1555 (1974).
15. A. Ahmad, W. K. Holloman, and R. Holliday, *Nature*, *258*, 541 (1975).
16. Y. A. Shin and G. L. Eichhorn, *Biochemistry*, *7*, 1026 (1968).
17. H. R. Drew, *J. Mol. Biol.*, *176*, 535 (1984).
18. D. M. J. Lilley, *Proc. Natl. Acad. Sci. USA*, *77*, 6468 (1980).
19. C. K. Singleton, J. Klysik, S. M. Stirdivant, and R. D. Wells, *Nature*, *299*, 312 (1982).
20. L. G. Sheflin and D. Kowalski, *Nucleic Acids Res.*, *13*, 6137 (1985).
21. J. K. Barton and A. L. Raphael, *Proc. Natl. Acad. Sci. USA*, *82*, 6460 (1985).
22. M. Laskowski, Sr., in "The Enzymes", 3rd ed., Vol. 4 (P. D. Boyer, ed.), Academic Press, New York, 1971, p. 289 ff.
23. S. Moore in "The Enzymes" 3rd ed., Vol. 14 (P. D. Boyer, ed.), Academic Press, New York, 1981, p. 281 ff.
24. P. A. Price, *J. Biol. Chem.*, *250*, 1981 (1975).
25. E. Junowicz and J. H. Spencer, *Biochim. Biophys. Acta*, *312*, 72 (1973).

26. V. W. Campbell and D. A. Jackson, *J. Biol. Chem., 255,* 3726 (1980).

27. D. Suck, C. Oefner, and W. Kabsch, *EMBO J., 3,* 2423 (1984).

28. C. Oefner and D. Suck, *J. Mol. Biol., 192,* 605 (1986).

29. D. Suck and C. Oefner, *Nature, 321,* 620 (1986).

30. P. A. Price, *J. Biol. Chem., 247,* 2895 (1972).

31. P. A. Price, T.-Y. Lui, W. H. Stein, and S. Moore, *J. Biol. Chem., 244,* 917 (1969).

32. P. A. Price, W. H. Stein, and S. Moore, *J. Biol. Chem., 244,* 929 (1969).

33. T.-H. Liao, *J. Biol. Chem., 250,* 3721 (1975).

34. S. Mehdi and J. A. Gerlt, *Biochemistry, 23,* 4844 (1984).

35. P. A. Price, S. Moore, and W. H. Stein, *J. Biol. Chem., 244,* 924 (1969).

36. C. Oefner and D. Suck, *Nature, 332,* 464 (1988).

37. H. R. Drew and A. A. Travers, *Cell, 37,* 491 (1984).

38. A. Moller, A. Nordheim, S. A. Kozlowski, D. J. Patel, and A. Rich, *Biochemistry, 23,* 54 (1984).

39. G. P. Lomonossoff, P. J. G. Butler, and A. Klug, *J. Mol. Biol., 149,* 745 (1981).

40. A. J. H. Wang, G. J. Quigley, F. J. Kolpak, J. L. Crawford, J. H. van Boom, G. van der Marcel, and A. Rich, *Nature, 282,* 680 (1979).

41. R. E. Dickerson and H. R. Drew, *J. Mol. Biol., 149,* 761 (1981).

42. P. W. Tucker, E. E. Hazen, Jr., and F. A. Cotton, *Mol. Cell. Biochem., 22,* 67 (1978).

43. P. W. Tucker, E. E. Hazen, Jr., and F. A. Cotton, *Mol. Cell. Biochem., 23,* 3 (1979).

44. P. W. Tucker, E. E. Hazen, Jr., and F. A. Cotton, *Mol. Cell. Biochem., 23,* 67 (1979).

45. P. W. Tucker, E. E. Hazen, Jr., and F. A. Cotton, *Mol. Cell. Biochem., 23,* 131 (1979).

46. C. B. Anfinsen, *Science, 181,* 223 (1973).

47. C. B. Anfinsen and H. A. Scheraga, *Adv. Protein Chem., 29,* 205 (1975).

48. P. Cuatrecasas, S. Fuchs, and C. B. Anfinsen, *J. Biol. Chem., 242,* 1541 (1967).

49. P. Cuatrecasas, S. Fuchs, and C. B. Anfinsen, *J. Biol. Chem., 242,* 3063 (1967).

50. F. A. Cotton, E. E. Hazen, Jr., and M. J. Legg, *Proc. Natl. Acad. Sci. USA, 76,* 2551 (1979).

51. E. H. Serpersu, D. Shortle, and A. S. Mildvan, *Biochemistry, 26,* 1289 (1987).

52. F. A. Cotton, V. W. Day, E. E. Hazen, Jr., S. Larsen, and S. T. Wong, *J. Am. Chem. Soc., 96,* 4471 (1974).

53. J. F. Riordan, K. D. McElvany, and C. L. Borders, Jr., *Science, 195,* 884 (1977).

54. B. Springs and P. Haake, *Tetrahedron Lett., 37,* 3223 (1977).

55. S. Mehdi and J. A. Gerlt, *J. Am. Chem. Soc., 103,* 3223 (1982).

56. I. M. Chaiken and C. B. Anfinsen, *J. Biol. Chem., 246,* 2285 (1971).

57. G. R. Sanchez, I. M. Chaiken, and C. B. Anfinsen, *J. Biol. Chem., 248,* 3653 (1973).

58. D. W. Hibler, N. J. Stolowich, M. A. Reynolds, and J. A. Gerlt, *Biochemistry, 26,* 6278 (1987).

59. A. Fersht, in "Enzyme Structure and Mechanism", W. H. Freeman, New York, 1985, p. 432ff.

60. C. Dingwall, G. P. Lomonossoff, and R. A. Laskey, *Nucl. Acids Res., 9,* 2659 (1981).

61. W. Horz and W. Altenburger, *Nucl. Acids Res., 9,* 2643 (1981).

62. J. E. Coleman and J. F. Chlebowski, in "Advances in Inorganic Biochemistry", Vol. 1 (G. L. Eichhorn and L. G. Marzilli, eds.), Elsevier/North-Holland, New York, 1979, p. 2ff.

63. J. E. Coleman, in "Zinc Enzymes", Vol. 5 (Metal Ions in Biology Series) (T. G. Spiro, ed.), John Wiley and Sons, New York, 1983, p. 219ff.

64. J. F. Chlebowski and J. E. Coleman, in "Metal Ions in Biological Systems", Vol. 6 (H. Sigel, ed.), Marcel Dekker, New York, 1976, p. 2ff.

65. R. T. Simpson and B. L. Vallee, *Biochemistry, 7,* 4343 (1968).

66. J. M. Sowadski, B. A. Foster, and H. W. Wyckoff, *J. Mol. Biol., 150,* 245 (1981).

67. W. F. Bosron, F. S. Kennedy, and B. L. Vallee, *Biochemistry, 14,* 2275 (1975).

68. R. A. Anderson, F. S. Kennedy, and B. L. Vallee, *Biochemistry, 15,* 3710 (1976).

69. W. F. Bosron, R. A. Anderson, M. C. Falk, F. S. Kennedy, and B. L. Vallee, *Biochemistry, 16,* 610 (1977).

70. J. D. Otvos and I. M. Armitage, *Biochemistry, 19,* 4021 (1980).

71. J. E. Coleman, K. Nakamura, and J. F. Chlebowski, *J. Biol. Chem., 258,* 386 (1983).

72. J. D. Otvos and D. T. Browne, *Biochemistry, 19,* 4011 (1980).

73. S. McCracken and E. A. Meighen, *J. Biol. Chem., 256,* 3945 (1981).

74. J. Taylor and J. E. Coleman, *Proc. Natl. Acad. Sci. USA, 69,* 859 (1972).

75. G. H. Tait and B. L. Vallee, *Proc. Natl. Acad. Sci, USA, 56,* 1247 (1966).

76. J. E. Coleman and P. Gettins, *Adv. Enzymol., 55,* 381 (1983).

77. P. Gettins, M. Metzler, and J. E. Coleman, *J. Biol. Chem., 260,* 2875 (1985).

78. T. W. Reid and I. B. Wilson, in "The Enzymes", Vol. 4 (P. D. Boyer, ed.), Academic Press, New York, 1971, p. 373.

79. W. E. Hull, S. E. Halford, H. Gutfreund, and B. D. Sykes, *Biochemistry, 15,* 1547 (1976).

80. S. R. Jones, L. A. Kindman, and J. R. Knowles, *Nature, 275,* 564 (1978).

81. M. L. Applebury, B. P. Johnson, and J. E. Coleman, *J. Biol. Chem., 245,* 4968 (1970).

82. (a) M. I. Harris and J. E. Coleman, *J. Biol. Chem., 243,* 5063 (1968). (b) M. Gottesman, R. T. Simpson, and B. L. Vallee, *Biochemistry, 8,* 3776 (1969).

83. J. F. Chlebowski and J. E. Coleman, *J. Biol. Chem., 249,* 7192 (1974).

84. J. D. Otvos, J. R. Alger, J. E. Coleman, and I. M. Armitage, *J. Biol. Chem., 254,* 1778 (1979).

85. P. Wooley, *Nature, 258,* 677 (1975).

86. D. R. Jones, L. F. Lindoy, and A. M. Sargeson, *J. Am. Chem. Soc., 106,* 7807 (1984).

87. S. Lindskog and J. E. Coleman, *Proc. Natl. Acad. Sci. USA, 70,* 2505 (1973).

88. R. P. Hertzberg and P. B. Dervan, *J. Am. Chem. Soc., 104,* 313 (1982).

89. M. W. Van Dyke, R. P. Hertzberg, and P. B. Dervan, *Proc. Natl. Acad. Sci. USA, 79,* 5470 (1982).

90. M. W. Van Dyke and P. B. Dervan, *Science, 225,* 1122 (1984).

91. P. B. Dervan, *Science, 232,* 464 (1986).

92. D. J. Galas and A. Schmitz, *Nucleic Acids Res., 5,* 3157 (1978).

93. (a) H. Rosen and S. J. Klebanoff, *Arch. Biochem. Biophys.*, *208*, 512 (1981). (b) J. Gutteridge, R. Richmond, and B. Halliwell, *Biochem. J.*, *184*, 469 (1979).

94. E. Graf, J. R. Mahoney, R. G. Bryant, and J. W. Eaton, *J. Biol. Chem.*, *259*, 3620 (1984).

95. M. D. Lind, M. J. Hamor, T. A. Hamor, and J. L. Hoard, *Inorg. Chem.*, *3*, 34 (1964).

96. R. P. Hertzberg and P. B. Dervan, *Biochemistry*, *23*, 3934 (1984).

97. A. Samuni, M. Chevion, and G. Czapski, *J. Biol. Chem.*, *256*, 12632 (1981).

98. D. S. Sigman, D. R. Graham, V. D'Aurora, and A. M. Stern, *J. Biol. Chem.*, *254*, 12269 (1979).

99. B. G. Que, K. M. Downey, and A. G. So, *Biochemistry*, *19*, 5987 (1980).

100. K. A. Reich, L. E. Marshall, D. R. Graham, and D. S. Sigman, *J. Am. Chem. Soc.*, *103*, 3582 (1981).

101. S. Goldstein and G. Czapski, *J. Am. Chem. Soc.*, *105*, 7276 (1983).

102. S. Goldstein and G. Czapski, *Inorg. Chem.*, *24*, 1087 (1985).

103. D. S. Sigman, *Acc. Chem. Res.*, *19*, 180 (1986).

104. M. Kuwabara, C. Yoon, T. Goyne, T. Thederahn, and D. S. Sigman, *Biochemistry*, *25*, 7401 (1986).

105. M. L. Kremer, *Israel J. Chem.*, *9*, 321 (1971).

106. C. Walling, *Acc. Chem. Res.*, *8*, 125 (1975).

107. D. R. Graham and D. S. Sigman, *Inorg. Chem.*, *23*, 4188 (1984).

108. (a) A. V. Fratini, M. L. Kopka, H. R. Drew, and R. E. Dickerson, *J. Biol. Chem.*, *257*, 14686 (1982). (b) S. Arnott, R. Chandrasekaran, I. H. Hall, and L. C. Puigjaner, *Nucleic Acids Res.*, *11*, 4141 (1983).

109. A. H. J. Wang, S. Fujii, J. H. van Boom, and A. Rich, *Proc. Natl. Acad. Sci. USA*, *79*, 3968 (1982).

110. L. E. Marshall, D. R. Graham, K. A. Reich, and D. S. Sigman, *Biochemistry*, *20*, 244 (1981).

111. D. S. Sigman, A. Spassky, S. Rimsky, and H. Buc, *Biopolymers*, *24*, 183 (1985).

112. L. Pope and D. S. Sigman, *Proc. Natl. Acad. Sci. USA*, *81*, 3 (1984).

113. S. Arnott and D. W. L. Hukins, *Biochem. Biophys. Res. Commun.*, *47*, 1504 (1982).

114. B. Jessee, G. Garguilo, F. Razvi, and A. Worcel, *Nucleic Acids Res., 10,* 5823 (1982).

115. A. Spassky and D. S. Sigman, *Biochemistry, 24,* 8050 (1985).

116. A. Spassky, *J. Mol. Biol., 188,* 99 (1986).

117. M. D. Kuwabara and D. S. Sigman, *Biochemistry, 26,* 7234 (1987).

118. R. J. Fiel, M. J. Carvlin, R. W. Byrnes, and E. H. Mark, in "Molecular Basis of Cancer, (Part B)" (R. Rein, ed.), Alan R. Liss, New York, 1985, p. 215ff.

119. J. M. Kelly, M. J. Murphy, D. J. McConnell and C. OhUigin, *Nucleic Acids Res., 13,* 167 (1985).

120. R. F. Pasternack, A. Autebi, B. Ehrlich, and D. Sidney, *J. Mol. Catal., 23,* 235 (1984).

121. B. Ward, A. Skorobogaty, and J. C. Dabrowiak, *Biochemistry, 25,* 7827 (1986).

122. B. Ward, A. Skorobogaty, and J. C. Dabrowiak, *Biochemistry, 25,* 6875 (1986).

123. R. J. Fiel, T. A. Beerman, E. H. Mark, and N. Datta-Gupta, *Biochem. Biophys. Res. Commun., 107,* 1067 (1982).

124. R. F. Pasternack, E. J. Gibbs, and J. J. Villafranca, *Biochemistry, 22,* 5409 (1983).

125. R. F. Pasternack, E. J. Gibbs, and J. J. Villafranca, *Biochemistry, 22,* 2406 (1983).

126. M. J. Carvlin and R. J. Fiel, *Nucleic Acids Res., 11,* 6121 (1983).

127. G. Dougherty, J. R. Pilbrow, A. Skorobogaty, and T. Smith, *J. Chem. Soc. Faraday Trans. 1, 81,* 1739 (1985).

128. J. T. Groves and T. E. Nemo, *J. Am. Chem. Soc., 105,* 5786 (1983).

129. J. T. Groves and T. E. Nemo, *J. Am. Chem. Soc., 105,* 6243 (1983).

130. J. W. Lown, S. M. Sondhi, and C.-W. Ong, *Biochemistry, 25,* 5111 (1986).

131. T. D. Tullius and B. A. Dombroski, *Science, 230,* 679 (1985).

132. T. D. Tullius and B. A. Dombroski, *Proc. Natl. Acad. Sci. USA, 83,* 5469 (1986).

133. A. M. Burkhoff and T. D. Tullius, *Cell, 48,* 935 (1987).

134. C. Bull, G. J. McClune, and J. A. Fee, *J. Am. Chem. Soc., 105,* 5290 (1983).

135. J. C. Wu, J. Kozarich, and J. Stubbe, *J. Biol. Chem., 258,* 4694 (1983).

136. J. K. Barton, J. J. Dannenberg, and A. L. Raphael, *J. Am. Chem. Soc., 104,* 4967 (1982).

137. J. K. Barton, A. T. Danishefsky, and J. M. Goldberg, *J. Am. Chem. Soc., 106,* 2172 (1984).

138. J. K. Barton, L. A. Basile, A. Danishefsky, and A. Alexandrescu, *Proc. Natl. Acad. Sci. USA, 81,* 1961 (1984).

139. J. K. Barton, J. M. Goldberg, C. V. Kumar, and N. J. Turro, *J. Am. Chem. Soc., 108,* 2081 (1986).

140. J. K. Barton, *Science, 233,* 727 (1986).

141. J. K. Barton and A. L. Raphael, *J. Am. Chem. Soc., 106,* 2466 (1984).

142. (a) M. B. Fleisher, K. C. Waterman, N. J. Turro, and J. K. Barton, *Inorg. Chem., 25,* 3549 (1986). (b) J. M. Kelly, A. B. Tossi, D. J. McConnell, C. OhUigin, *Nucleic Acids Res., 13,* 6017 (1985).

143. T. J. Meyer, *Pure Appl. Chem., 58,* 1193 (1986).

144. V. Balzani, L. Moggi, F. Scandola, and V. Çarassiti, *Inorg. Chem. Acta Rev., 1,* 7 (1967).

145. R. Ballardini, G. Varani, and V. Balzani, *J. Am. Chem. Soc., 102,* 1719 (1980).

146. (a) M. B. Fleisher, Ph.D. thesis, Columbia Univ., 1988. (b) M. B. Fleisher, H.-Y. Mei, and J. K. Barton, in "Nucleic Acids and Molecular Biology", Vol. 2, 65 (1988).

147. S. K. Chatopadhyay, C. V. Kumar, and P. K. Das, *J. Photochem., 24,* 1 (1984).

148. H.-Y. Mei and J. K. Barton, *J. Am. Chem. Soc., 108,* 7414 (1986).

149. H.-Y. Mei and J. K. Barton, *Proc. Natl. Acad. Sci. USA, 85,* 1339 (1988).

150. I. Rosenthal and J. N. Pitts, Jr., *Biophys. J., 11,* 963 (1971).

151. D. S. Sigman, G. M. Wahl, and D. J. Creighton, *Biochemistry, 11,* 2236 (1972).

152. R. S. Brown, J. Huguet, and J. Curtis, in "Metal Ions in Biological Systems", Vol. 15 (H. Sigel, ed.), Marcel Dekker, New York, 1983, p. 55ff.

153. Y. Murakami and J. Sunamoto, *Bull. Chem. Soc. Japan, 44,* 1827 (1971).

154. R. M. Milburn, M. Gautam-Basak, R. Tribolet, and H. Sigel, *J. Am. Chem. Soc., 107,* 3315 (1985).

155. H. Sigel, F. Hofstetter, R. B. Martin, R. M. Milburn, V. Scheller-Krattiger, and K. H. Scheller, *J. Am. Chem. Soc., 106,* 7935 (1984).

156. S. H. Gellman, R. Petter, and R. Breslow, *J. Am. Chem. Soc., 108,* 2388 (1986).

157. P. Modrich, *Q. Rev. Biophys., 12,* 315 (1979).

158. P. Modrich, *Crit. Rev. Biochem., 13,* 287 (1982).

159. S. E. Halford, *Trends Biochem. Sci., 8,* 455 (1983).

160. B. A. Connolly, F. Eckstein, and A. Pingoud, *J. Biol. Chem., 259,* 10760 (1984).

161. P. Modrich and D. Zabel, *J. Biol. Chem., 251,* 5866 (1976).

162. L. Jen-Jacobson, M. Kurpiewski, D. Lesser, J. Grable, H. W. Boyer, J. M. Rosenberg, and P. J. Greene, *J. Biol. Chem., 258,* 14638 (1983).

163. J. K. Barton, L. A. Basile, and S. R. Paranawithana, *J. Biol. Chem., 257,* 7911 (1982).

164. J. M. Rosenberg, H. W. Boyer, and P. J. Greene, in "Gene Amplification and Analysis: Volume 1, Restriction Endonucleases" (J. G. Chirikjian, ed.), Elsevier/North-Holland, Amsterdam, 1981, p. 131ff.

165. J. A. McClarin, C. A. Frederick, B.-C. Wang, P. Greene, H. W. Boyer, J. Grable, and J. M. Rosenberg, *Science, 234,* 1526 (1986).

166. A.-L. Lu, W. E. Jack, and P. Modrich, *J. Biol. Chem., 256,* 13200 (1981).

167. R. Kim, P. Modrich, and S.-H. Kim, *Nucleic Acids Res., 12,* 7285 (1984).

168. L. Jen-Jacobson, D. Lesser, and M. Kurpiewski, *Cell, 45,* 619 (1986).

169. L. A. Basile, S. R. Paranawithana, and J. K. Barton, unpublished results in our laboratory.

170. J. K. Barton and L. A. Basile, *Inorg. Chim. Acta, 79,* 152 (1983).

171. S. R. Paranawithana, Ph.D. thesis, City University of New York (Hunter College), 1986.

172. C. Walsh, in "Enzymatic Reaction Mechanisms", W. H. Freeman, New York, 1979, p. 208.

173. M. F. Dunn, *Structure Bond., 23,* 1 (1975).

174. P. M. Harrison and R. J. Hoare, in "Metals in Biochemistry", Chapman and Hall, London, 1980, p. 17.

175. M. Hsu and P. Berg, *Biochemistry, 17,* 131 (1978).

176. P. G. Schultz and P. B. Dervan, *J. Am. Chem. Soc., 105,* 7748 (1983).

177. P. G. Schultz and P. B. Dervan, *Proc. Natl. Acad. Sci. USA, 80,* 6834 (1983).

178. J. S. Taylor, P. G. Schultz, and P. B. Dervan, *Tetrahedron, 40,* 457 (1984).

179. J. H. Griffin and P. B. Dervan, *J. Am. Chem. Soc., 109,* 6840 (1987).

180. J. P. Sluka, S. J. Horvath, M. F. Bruist, M. I. Simon, and P. B. Dervan, *Science, 238,* 1129 (1987).

181. H. E. Moser and P. B. Dervan, *Science, 238,* 645 (1987).

182. J. D. McGhee, *Biopolymers, 15,* 1345 (1976).

183. D. J. Patel and L. L. Canuel, *Proc. Natl. Acad. Sci. USA, 74,* 5207 (1977).

184. K. Zakrzewska, R. Lavery, and B. Pullman, *Nucleic Acids Res., 11,* 8825 (1983).

185. C.-H. B. Chen and D. S. Sigman, *Science, 237,* 1197 (1987).

186. A. L. Raphael, Ph.D. thesis, Columbia Univ., 1987.

187. B. C. Muller, A. L. Raphael, and J. K. Barton, *Proc. Natl. Acad. Sci. USA, 84,* 1764 (1987).

188. L. J. Peck, A. Nordheim, A. Rich, and J. C. Wang, *Proc. Natl. Acad. Sci. USA, 79,* 4560 (1982).

189. A. Nordheim and A. Rich, *Nature, 303,* 674 (1983).

190. H.-Y. Mei and J. K. Barton, unpublished results in our laboratory.

191. D. R. Corey and P. G. Schultz, *Science, 238,* 1401 (1987).

192. L. A. Basile and J. K. Barton, *J. Am. Chem. Soc., 109,* 7548 (1987).

193. L. A. Basile, A. L. Raphael, and J. K. Barton, *J. Am. Chem. Soc., 109,* 7550 (1987).

194. N. E. Murray, S. A. Bruce, and K. J. Murray, *J. Mol. Biol., 132,* 493 (1979).

3

Metalloregulatory Proteins: Metal-Responsive Molecular Switches Governing Gene Expression

Thomas V. O'Halloran
Department of Chemistry and
Department of Biochemistry,
Molecular Biology, and Cell Biology
Northwestern University
2145 Sheridan Road
Evanston, IL 60201

1. INTRODUCTION

1.1. Molecular Basis of Metal-Responsive Gene Expression

1.1.1. Scope of Review

Critical changes in the intracellular or extracellular concentrations of heavy metals can be transduced by metal-responsive regulatory proteins into alterations in the expression of specific genes. The mechanisms by which these inorganic signals produce their biological response are being delineated in increasing detail. After defining the types of inorganic signals and their general roles in cellular metabolism, this chapter will focus on the only two systems to date in which a majority of the components of the metal-responsive gene regulatory apparatus have been isolated. The mercuric ion-responsive protein MerR, which controls bacterial mercury resistance genes, and the ferrous ion-responsive protein Fur, which controls microbial iron uptake genes, have both been overproduced and purified to homogeneity. The metal-responsive genetic switching events which these proteins mediate in the cell have been reproduced in vitro. Both of these regulatory proteins appear to play two roles: heavy-metal receptor and transcriptional effector. The Fur and MerR proteins are the first recognized examples of heavy-metal receptors that are involved in signal transduction. They provide an opportunity to explore the molecular determinants of heavy-metal recognition in biological systems. An understanding of the molecular basis of the recognition process will ultimately require precise definition of the coordination environments in these heavy-metal receptors.

The first step in the process of inorganic signal transduction is the binding of a metal ion to a receptor protein. Although the receptor must have the appropriate sensitivity to different concentrations of the target metal, they are not simply preformed binding pockets. The metal-protein interaction must lead to a physical change in the receptor which propagates the signal to other biopolymers in the circuit. Ultimately we would like to understand the nature of the conformational changes that are induced by the metal-protein interactions. Both metal ion specificity and sensitivity will have a molecular basis, and the elements of the macromolecular structure which determine the specificity and the sensitivity will vary depending on the coordination chemistry of the metal.

Following a review of two systems in which a purified protein has been shown to mediate metal-responsive gene expression and a description of what is known about the molecular basis of their action, a brief survey of other examples of metal-responsive gene expression will be presented. All of these cases are likely to involve metalloregulatory proteins of one type or another. Metal-responsive regulation of the metallothionein genes is reviewed separately in Chap. 4 of this volume and will not be covered in this chapter.

1.1.2. Metalloregulatory Proteins

Metalloregulatory proteins are a subclass of metal-activated proteins involved in gene expression and can be defined as multifunctional heavy-metal receptors, where one function is to directly *regulate* gene expression in response to changes in concentrations of heavy-metal ions. These gene regulatory proteins can thus be thought of as metal-responsive genetic switches. Metalloregulatory proteins can be distinguished from metalloproteins and other metal-activated proteins involved in gene expression by virtue of their role in signal transduction. The gene 32 protein described in Chap. 5 of this volume is a DNA-binding zinc metalloprotein. The zinc-binding site involves coordination of three cysteine sulfurs and a histidine nitrogen. Zinc binds very tightly to the gene 32 protein and can only be displaced by organomercurials, which have a higher affinity for the thiolate side chains. There is no evidence that the metal-free form of the protein acts as a signal transducer for changes in zinc ion concentration. Since the gene 32 protein is isolated with the zinc tightly bound, it is a metalloprotein rather than a metal-activated protein [1,2]. In addition, it does not have a zinc-responsive receptor role in any known biological process and so is not a metalloregulatory protein. There are a variety of similar zinc metalloproteins (see Sec. 3.2.3 and Chaps. 5 and 6 of this volume) that play key roles in gene expression which, however, do not act as metal-responsive genetic switches.

Cells frequently encounter environmental conditions that require rapid triggers of cell functions. Cellular response to heavy-metal stress, whether it originates in the sudden increase in the concentration of a toxic metal or the sudden decrease in the concentration of an essential metal, does not always involve alterations in the patterns of gene expression. For instance, heavy metals such as Fe(II) and Mn(II) can regulate various biochemical pathways by substitution into inactive apoenzymes thus increasing the concentrations of catalytically competent metal-activated enzymes [3]. Metalloregulatory proteins, in contrast, provide the cell with the means to articulate rapid expression of specific proteins involved in the response to a particular heavy-metal stress.

As demonstrated repeatedly in Sec. 3 of this chapter, genes for metabolic pathways that require heavy metals (e.g., copper in plastocyanin) are often regulated by changes in the concentration of these cofactors. Metalloregulatory proteins regulating many different processes in cellular metabolism have been implicated in several of these systems. They may mediate decisions between metabolic pathways based on the availability of inorganic resources, as well as adjusting the expression of proteins involved in metal uptake and the expression of detoxification systems.

1.2. Transduction of Inorganic Signals in Living Systems

An inorganic signal is simply a change in the concentration of an inorganic solute, within a critical range of values, which can affect the metabolic state of the cell. If the inorganic solute is a metal ion that plays an essential role, such as iron, the cell must respond to limitation stress by more actively accumulating iron, or it will not survive. Conversely, if the inorganic signal is the sudden increase in the concentration of a nonessential metal ion which can exert toxic effects, the cell must respond by decreasing the intracellular concentration of that substance until it no longer presents a threat to the survival of the cell. Inorganic signals can be changes in the concentrations of metals outside the cell (extracellular signals) or they can be changes in the concentrations of metal ions within the cell (intracellular signals).

All types of living organisms must struggle with the basic dilemma presented by heavy metals. On one hand, they must elaborate specific systems and expend energy to sequester sufficient quantities of essential metals, which may be present at low concentrations in the extracellular environment. On the other hand, concentrations of these same metal ions can disrupt normal metabolism when the capacity of the cell to store or utilize the metal ion is exceeded [4]. Maintenance of essential metal ion concentrations within a narrow range is achieved through the interaction of many different proteins in energy-requiring processes. These multiprotein systems allow cells to maintain heavy-metal concentrations within a narrow optimum range and provide the basis for *heavy-metal homeostasis*.

1.3. Proteins that Regulate Gene Expression: Background and Terminology

Cells can regulate the synthesis of a specific protein by controlling gene expression at one of several levels. Signal-responsive gene regulation can occur at a transcription level, where the quantities of messenger RNA (mRNA) are controlled by the rate at which a gene is transcribed by the enzyme RNA polymerase. Alternatively, the cell can regulate the levels of protein encoded by a specific gene by adjusting the rate at which the mRNA molecules are translated into protein. While both transcriptional and translational metal-responsive gene regulation are known, there are many more examples of the former at present. A brief description of transcriptional regulation is provided below [5].

Signal-responsive transcription of prokaryotic and eukaryotic genes by RNA polymerases can involve both cis-acting elements and trans-acting factors [5,6]. The cis-acting elements are sequences of DNA which play an important role in facilitating the transcription of the adjacent genes. When a cis-acting DNA sequence serves as a binding site for trans-acting regulatory factors, it is referred to as an operator. Cis-acting sequences which specify the binding site for RNA polymerase and include a start site for transcription are referred to as promoters. Some trans-acting factors are DNA-binding proteins which either interact with other proteins, such as RNA polymerase, or change the local structure of the DNA to turn on or off the expression of adjacent genes.

1.3.1. Negative Control

Negative control characterizes promotors which are transcribed by RNA polymerase at moderate levels in the absence of a regulatory protein and/or signal, and which are transcribed at low levels in the presence of the regulatory protein. Basal level transcription, or constitutive transcription, is measured by the amount of RNA polymerase activity at a promoter sequence in the absence of a regulatory protein and/or signal. Negative control, or repression, thus involves a gene regulatory protein, or a repressor, which lowers the amount of transcription from a specific promoter.

1.3.1.1. Signal-Responsive Repression. The Fur metalloregulatory protein described below acts as a signal-responsive repressor. In the presence of some divalent transition metals, such as ferrous ion, the protein assumes a conformation which allows it to bind tightly to DNA sequences in the promoter region where RNA polymerase normally interacts. Apparently, the interaction of the Fur protein at this site blocks the entrance of RNA polymerase and/or prevents the formation of a transcriptionally active complex (vide infra).

1.3.1.2. Signal-Responsive Derepression. Signal-responsive derepression is another mechanism involving negative control. The gene regulatory protein blocks the transcription of a specific gene until a signal molecule binds and effects the dissociation of the DNA-binding protein from the promoter region, thus allowing RNA polymerase to initiate transcription. The Lac repressor of *E. coli* and numerous other microbial gene regulatory proteins exert negative control over gene expression in this manner [5].

1.3.2. Positive Control

The term *positive control of transcription* implies that factors other than RNA polymerase are required for full induction of transcription from a given promoter. Positively controlled promoters typically exhibit low levels of expression in the absence of a regulatory protein or signal. When the activated regulatory protein is present, transcription is greatly stimulated. The MerR metalloregulatory protein exhibits metal-responsive positive activation of a set of genes responsible for heavy-metal detoxification. The molecular basis of positive control in both prokaryotes and eukaryotes is not well understood at present [5-7].

Positive activator proteins may not have any DNA-binding properties and may act by binding to or modifying other proteins in transcription complexes [6]. Signal transduction often involves activation of a target protein by phosphorylation. Heavy-metal receptors may be involved in these indirect paths; however, there is no evidence at present for such activities in any known metalloregulatory systems.

2. CHARACTERIZATION OF PURIFIED METALLOREGULATORY PROTEINS

Initial work on the mechanisms of the Fur and MerR proteins suggests that they both act as heavy-metal receptors and both have specific DNA-binding activities. However, the mechanism by which they transduce this signal into a change in gene expression is very different. While the Fur protein is a metal-responsive repressor of gene expression, it apparently has no function in regulating gene expression in the absence of metal ions. On the other hand, the MerR protein is a *repressor* of gene expression in the absence of metals and is converted to a positive *activator* of gene expression in the presence of inducing metal ions. The molecular properties of these two systems are described below.

2.1. The MerR Protein: Heavy-Metal Responsive Positive Activator of Mercury Resistance Genes

2.1.1. Background

The microbial mercury resistance operons (*mer*) are the best understood heavy-metal detoxification systems [8]. The core of the resistance system is the NADPH-dependent flavoenzyme, mercuric ion reductase, which reduces mercuric ion to elemental mercury, a volatile substance that diffuses out of the bacteria. The properties of many of the proteins that process mercuric ion have recently been reviewed [9] and the molecular genetics of the mercury-resistant operons have also recently been reviewed [10,11].

A variety of different types of experiments, including transposon mutagenesis [12,13] and translational and transcriptional fusion experiments [14,15], have shown that the trans-acting *merR* gene product regulates the expression of structural genes for mercury resistance in a negative manner in the absence of mercuric ion and in a positive manner in the presence of mercuric ion. These results suggested that the *merR* gene product was required for positive activation of the mercury resistance operons but could not establish whether other cellular factors might also be required for positive activation. These studies also indicated that the *merR* gene product negatively regulated the expression of its own gene (i.e., the *merR* gene) in the presence and absence of mercuric ion.

Elucidation of the mechanism of positive regulation in this divergent dual-promoter system required isolation of the regulatory protein. This facilitated studies of its interaction with the inducer, any other regulatory factors present in the cell, and with the operator(s) of the *merT* and *merR* promoters. Since the *merR* gene product was negatively autoregulated, a successful overproduction strategy necessitated the disruption of the signals at the 5' end of the *merR* gene responsible for the negative autoregulation. This strategy resulted in MerR expression at levels which constitute approximately 20% of total cellular protein. Initial DNase I footprinting studies on crude extracts from these overproducers (where MerR constituted approximately >80% of the total protein) revealed that the MerR protein bound tightly to a region directly between the -10 and -35 regions of the *merT* promoter and directly downstream of the -10 region for the *merR* promoter. The protection was the same in the presence and absence of mercuric ion [16]. These studies demonstrated that the MerR protein was an essential component of the regulatory apparatus for the *mer* operon but failed to show any evidence for mercury-responsive switching. These studies also left open the possibility that other cellular factors, which were not present in the crude extract from the overproducer, were required for positive activation.

The amino acid sequence of the MerR protein has been determined by peptide sequencing for the first 10 amino acid residues [16]. The rest of the sequence was deduced from the DNA sequence of the *mer* operon [17]. The reading frame encoding the MerR protein was incorrectly assigned in initial studies and was later shown to correspond to ORF3 in the TN501 *merR* gene [18]. Key promoter elements for the divergent *merR* and *merT* genes are shown in Fig. 1.

Recently, the *merR* gene from the highly homologous TN21 *mer* operon has been cloned into an expression of vector that allowed the visualization of a 16-kD protein via ^{35}S labeling [19]. Cell extracts containing this protein exhibit DNA-binding activities with DNA fragments containing the *mer* promoter region. The location of the binding site with respect to the -10 and the -35 regions of the promoter is apparently similar to that of the MerR protein from TN501 described above. In this system, as with TN501, the occupation of the operator by the MerR protein was the same in the presence and absence of mercury. Detailed binding studies with the purified TN501 MerR protein, however, show that the affinity of MerR protein for its operator sequence is decreased by threefold in the presence of mercuric ion [20].

Recent studies of constructs where the *merT* promoter is fused to the *lacZ* gene and the *merR* gene is provided in trans, show that mercuric ion-induced in vivo transcription from the *mer* promoter can occur in strains that lack all of the other *mer* operon gene products. Thus none of the factors encoded in the *mer* operon, including the periplasmic protein (MerP) and the transport protein (MerT), are required for metal-responsive induction. However, these in vivo results do not exclude the possibility that other types of cellular factors may play a role in mediating metal-responsive induction [21].

Another question raised by these in vivo results is how the mercuric ion enters the cell to cause the initial induction of the

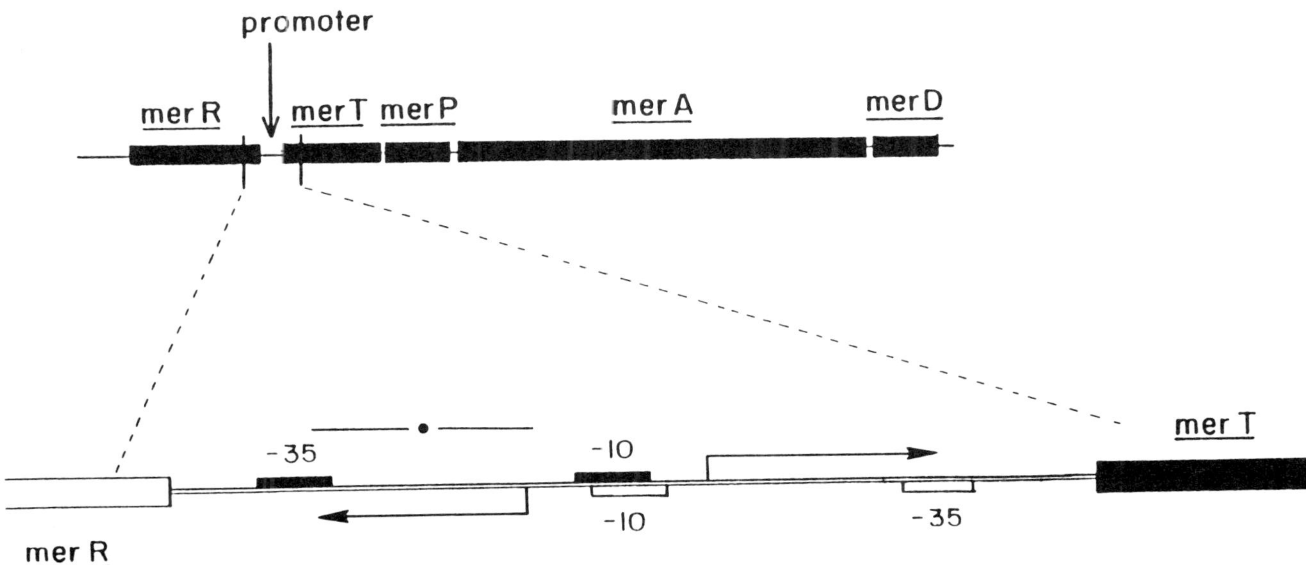

FIG. 1. The organization of the genes in the Mer operon is shown above (top). Key promoter elements (bottom) for the *merT* and the divergent *merR* genes are shown as filled and open boxes, respectively. The lower strand codes for *merR* and the upper strand codes for *merT*. Arrows mark the in vitro transcription start sites, which are separated by 14 base pairs [20]. The start codons for the *merR* and *merT* genes are separated by 77 base pairs [14]. The binding site for the MerR protein is shown as an interrupted line above the DNA [16].

mer genes. One possibility involves a general cellular response to toxic metals (see Sec. 3.1.1.2). In this scenario, a chromosomally encoded heavy-metal receptor could signal the presence of a toxic metal. MerR could then function as an intermediate signal transducer (e.g., phosphorylation), mediating communication between the receptor and the transcriptional apparatus. There are many precedents for this type of two-component signal transduction system in the response of wild-type *E. coli* to inorganic solutes such as phosphate [22]. Another possibility is that sufficient mercuric ion can enter the cytoplasm even in the absence of the transport system to bind to MerR, which in turn would induce expression of the *mer* genes.

2.1.2. *Transcriptional Activation*

The MerR protein was recently purified to homogeneity in large quantities in order to resolve some of the above questions and to establish whether cellular factors other than mercuric ion and MerR protein were required to stimulate *E. coli* RNA polymerase transcription from the *mer* promoter in vitro. The results of in vitro transcript "runoff" experiments are shown in Fig. 2. In these experiments RNA polymerase, promoter DNA, and ^{32}P-labeled ribonucleotides are incubated in the presence and absence of MerR protein and Hg^{2+}. When transcription occurs, ^{32}P-labeled mRNA chains of specific lengths are formed. These mRNAs are then separated by gel electrophoresis and an autoradiograph is obtained. In the absence of the MerR protein (lane 1), the principal mRNA transcripts produced correspond to the *merR* promoter and activity is not detected from the *merT* promoter. When the purified MerR protein is added to the transcription assay, the concentration of *merR* mRNA drops dramatically, consistent with the role of the MerR protein as a repressor. Note that concentration of the *merT* transcript remains below a detectable level in this assay (lane 2). However, upon addition of mercuric ion (lane 4), the concentrations of the *merT* message increase substantially, indicating a stimulation of transcription from the *merT* promoter. The level of induction is dependent both on the concen-

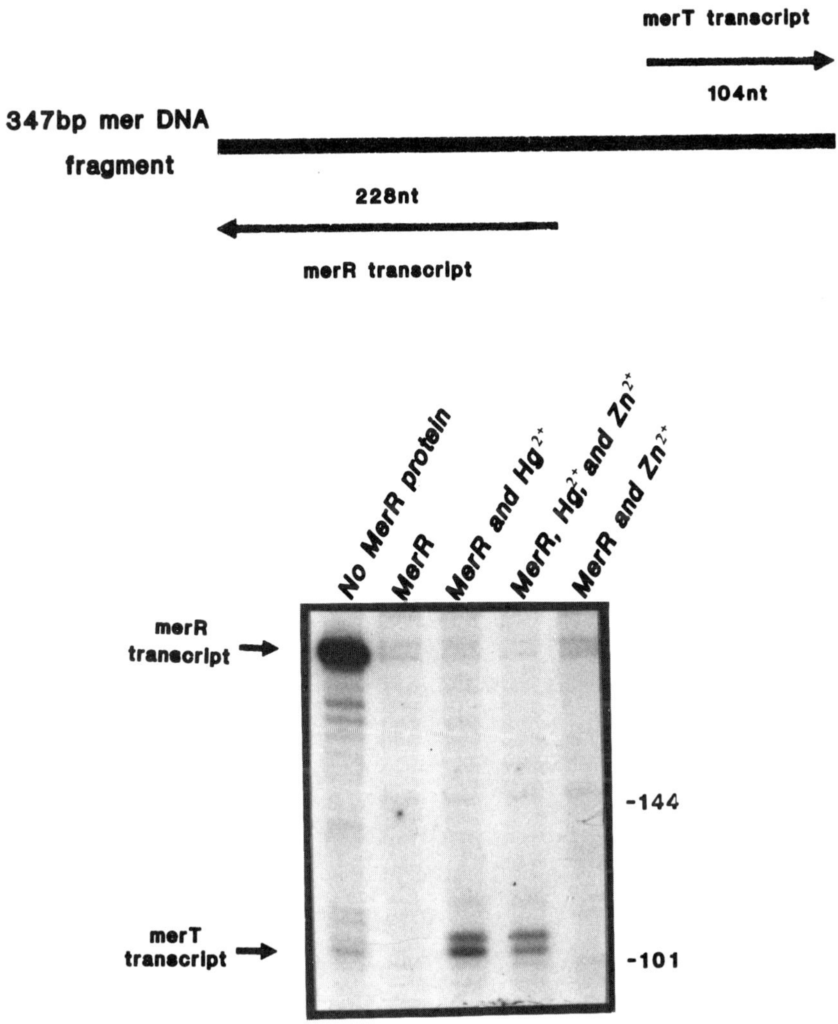

FIG. 2. In vitro transcription runoff experiments showing mercuric ion and MerR protein are necessary for production of a 104 base pair nucleotide *merT* transcript by RNA polymerase in the presence of nucleotides. In the absence of purified MerR protein, *merT* promoter activity is not detectable whereas *merR* promoter activity is high (note the abundance of the 228 nucleotide message). In the absence of mercuric ion, the MerR protein represses both promoters. The last lane demonstrates that Zn^{2+} cannot activate *merT* transcription.

tration of mercuric ion and the concentration of MerR protein in the transcription assay. Concentrations of mercuric ion as low as 1 x 10^{-8} M result in a detectable increase in *merT* transcription. These in vitro results with the purified components of the metal-responsive switch unequivocally establish that cellular proteins, other than the purified MerR protein and *E. coli* RNA polymerase holoenzyme, are not required for mercury-dependent activation of transcription at the *merT* promoter [20].

2.1.3. Mercuric Ion/Receptor Interactions

The MerR protein binds mercuric ion very tightly even in the presence of excess thiol competitors [23]. The exact Hg/MerR stoichiometry depends on the initial concentration of thiol competitor and the initial concentration of mercuric ion. Stable complexes with a stoichiometry of 1 mol of mercury per mol of MerR dimer have been isolated. Complexes of this Hg/MerR stoichiometry can be separated from unbound mercury through dialysis. These complexes can then be utilized in the above transcription assays. Complexes with an Hg/MerR stoichiometry of 1:1 are capable of stimulating transcription from the *merT* promoter without added excess mercuric ion. It is interesting to note that the transcription assays are conducted using nanomolar quantities of the Hg/MerR complex, and in the presence of 1 mM DTT. Thus, 10^6-fold excess of competing thiol is not sufficient to inactivate the Hg/MerR complex in the transcription assay, which implies that it is not sufficient to remove mercuric ion from the protein.

Mercuric ion has an extremely high affinity for thiolate ligands and can adopt several geometries in these complexes. The formation constant for $Hg(cys)_2$ complexes is 10^{45} [24]. The preferred coordination geometry in many monodentate, mercury-thiolate complexes is linear, as exemplified in the structure of a mercury-cysteine complex (**1** in Fig. 3). Secondary bonding interactions with adjacent sulfurs are frequently observed in the solid state. Distorted tetrahedral and trigonal Hg^{2+}-thiolate complexes are

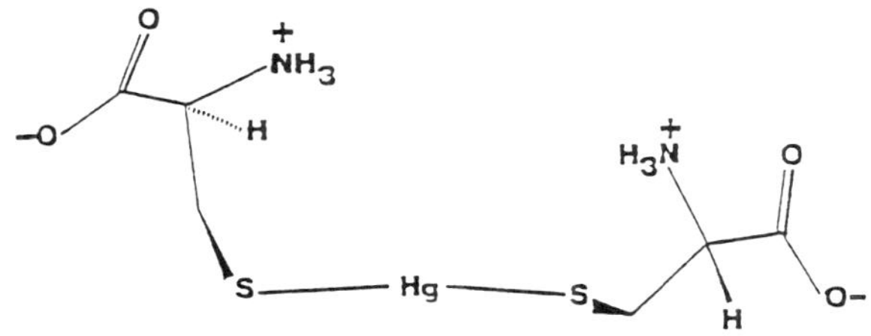

1. $\bar{r}_{Hg-S} = 2.34$ Å

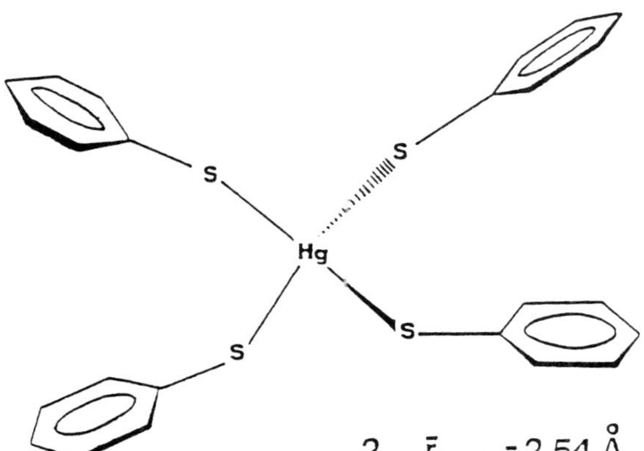

2. $\bar{r}_{Hg-S} = 2.54$ Å

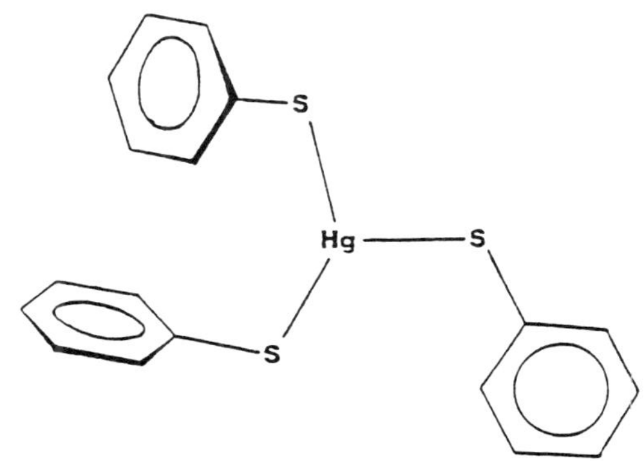

3. $\bar{r}_{Hg-S} = 2.448$ Å

FIG. 3. These three mononuclear mercury-thiolate complexes demonstrate the diversity in Hg-SR stoichiometry and geometry. Linear RS-Hg-SR coordination, such as that in compound 1, $Hg(cys)_2Cl_2 \cdot 0.5H_2O$ [92], is a common type of mercury-thiolate geometry obtained in the solid state. The only known crystal structures of monomeric mercury compounds with tetrahedral (2) [93] and trigonal (3) [94] mercury-sulfur geometries are anionic complexes of aromatic thiolates.

rarer, and the only mononuclear representatives of this class whose structures are known are complexes **2** and **3**, respectively, shown in Fig. 3. The average Hg-S bond distance increases as the number of coordinated thiolates increase.

The resistance of the Hg/MerR complex to thiolate exchange with excess competing thiol suggested an unusual thermodynamic stability (even for mercury-thiol complexes) in this metal-protein interaction. In order to understand the molecular basis of this stability and the preferred coordination geometry, EXAFS experiments have recently been conducted on Hg-MerR complexes that were isolated by equilibrium dialysis [25]. Preliminary results reveal mercury-sulfur bonding in Hg/MerR complexes of defined stoichiometry. Since the EXAFS method cannot distinguish between mercury centers in a heterogeneous sample, we do not know at present whether there is more than one type of coordination environment in the MerR protein. ^{199}Hg experiments are currently under way to resolve the number of different types of mercuric ion coordination environments in this mercury-protein complex.

2.1.4. Hg/MerR Interaction with DNA and RNA Polymerase

Other studies currently under way are aimed at resolving the nature of the interactions between the Hg/MerR complexes with DNA and RNA polymerase. Hydroxy-radical footprinting techniques coupled with native gel electrophoresis methodologies allow the characterization of specific protein DNA interactions at high resolution [26]. Such experiments provide a detailed picture of the interaction of the Hg-activated MerR protein with RNA polymerase in transcriptionally competent complexes [20]. These studies and other chemical probes of protein-DNA interactions have confirmed earlier observations that show the MerR protein binds to the same operator in the presence and absence of mercuric ion.

Recent high-resolution footprinting results suggest that the MerR protein binds predominantly to one face of the duplex DNA with some extension of the peptide wrapping around the DNA in the center of the operator sequence [20]. The MerR-DNA contracts are apparently unchanged in the presence of mercury. The question then arises, how can a protein bound to a single site on the DNA first block RNA polymerase transcription of a specific promoter, and then after modification by Hg^{2+}, remain bound at the same site and *stimulate* the entry of RNA polymerase at the same promoter. To resolve these questions it was necessary to directly examine the interaction of MerR and RNA polymerase at this promoter.

Gel retardation studies demonstrate that *E. coli* RNA polymerase holoenzyme (RNAP) does not bind tightly to the mercury-responsive *merT* promoter in the presence of the MerR protein, unless micromolar mercuric ion is added. Under these conditions, the binding of Hg/MerR and RNA polymerase to the *merT* promoter is synergistic. Chemical probes of the complex formed between Hg-MerR and RNAP at the *merT* promoter demonstrate that the DNA in the -10 region of this promoter is unpaired and that the complex is transcriptionally competent when NTPs are added. Analysis of the protein content and footprinting data for the activated complex demonstrate that MerR and RNAP are bound to the same region of DNA on different faces of the helix and the backside contacts that arose from the putative MerR peptide arms are no longer present. Thus these results suggest that the Hg/MerR complex with DNA undergoes a conformational change that is induced by the interaction with polymerase at the promoter site. This cooperativity of metal-protein and protein-protein interaction explains why the metal-responsive switching event could not be observed in footprinting of the MerR/DNA complexes.

2.1.5. Regulatory Logic

Based on the results outlined above, the following events are envisioned to occur in the activation of this metal-responsive switch. In the absence of mercury, MerR protein extends around the backside of the helix into a region that precludes the binding and/or formation of RNA polymerase open complexes. As mercuric ion concentrations in the cell rise, the metal ion binds to cysteine sulfurs in the mercuric ion receptor domain of MerR, while this protein is bound to the DNA. This initial event allows conversion of the MerR protein to an intermediate form which retains many of the same contacts with DNA that are found in the absence of mercuric ion. However, RNA polymerase can now recognize this activated form of MerR and binds tightly to the promoter on an alternate face of the DNA. RNA polymerase is now in a position to melt out the DNA in the -10 region, and transcriptional activation is achieved. We have characterized several regulatory states of the transcription complexes in the *mer* promoter region and these are shown in Fig. 4.

Many details about the molecular basis of transcriptional activation within the ternary complex remain unresolved. The MerR protein may function by altering the DNA conformation in a manner which facilitates RNA polymerase binding and/or the unwinding of the DNA in the -10 region. On the other hand, the MerR protein may simply provide a complementary face for favorable interactions with RNA polymerase only when mercury is bound to MerR. Detailed kinetic studies of the activation of transcription at this promoter may help resolve these two possibilities.

2.1.6. Functional Components of the MerR Protein

Now that several elements of the regulatory logic and biological activity of the mercuric ion-responsive switch have been elucidated, the molecular determinants in the MerR protein responsible for the high selectivity of this protein for mercuric ion, operator DNA, and RNAP are being probed. The unusual thermodynamic stability of the MerR-Hg complex relative to other Hg-thiolate complexes may arise

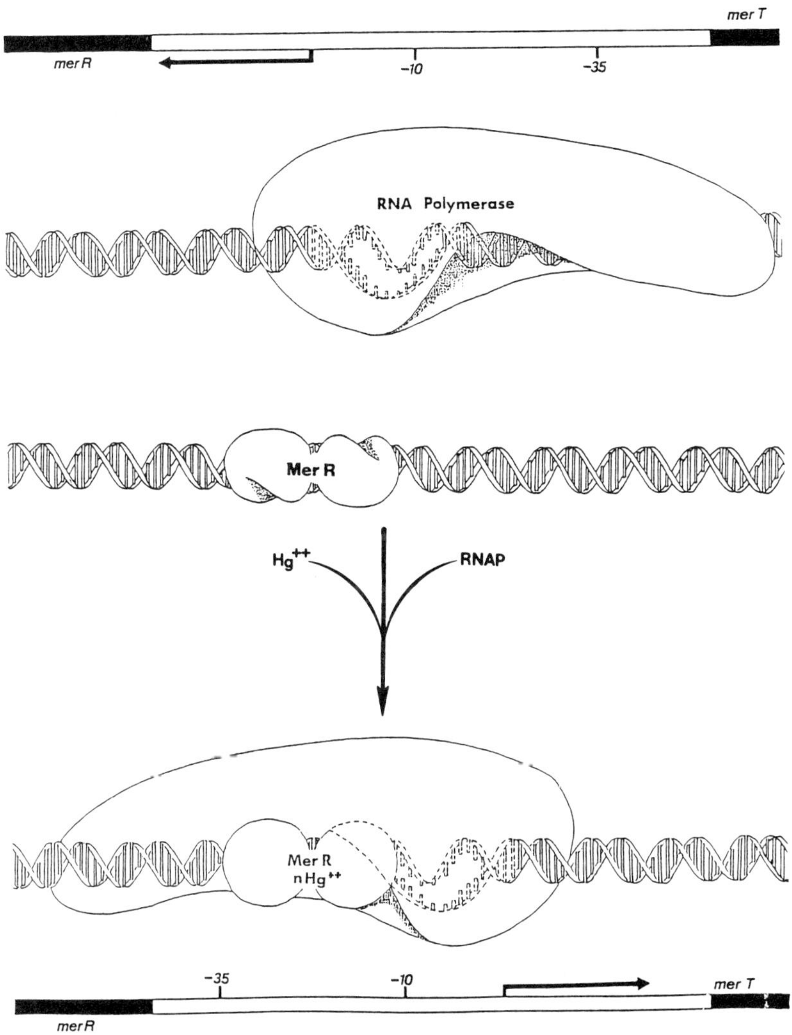

FIG. 4. Models for three regulatory states of the *mer* operon based on chemical protection data, protein analysis, and transcription runoff assays [20]. In the absence of the MerR protein, RNA polymerase preferentially transcribes the *merR* promoter. As the MerR concentration rises, it binds to the operator and prevents RNA polymerase from transcribing *merR* or *merT* promoters. When Hg^{2+} is added, it binds to cysteine side chains and converts the MerR protein to a conformation which facilitates the binding of RNA polymerase and the formation of an open complex with the *merT* but not the *merR* promoter. This latter complex is transcriptionally competent and the circuit is complete.

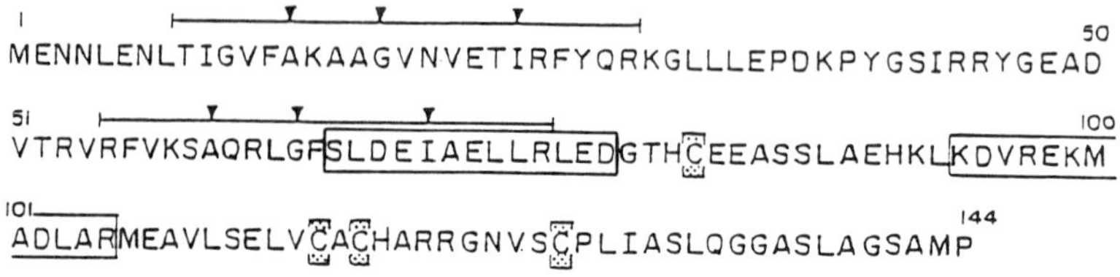

FIG. 5. Predicted functional domains in the MerR protein. Single boxed residues indicate most probable Hg^{2+}-binding sites. Lines over the sequence indicate regions similar to sequences in DNA-binding domains of structurally characterized DNA-binding proteins. The boxed sequences may correspond to RNA polymerase interaction domains based on similarities to other positive activator proteins.

from the utilization of four donor ligands and the chelate effect. Several residues including histidines and methionines have been shown to interact with mercuric ion in crystal structures of other mercury-protein complexes. Based on the thermodynamics of mercury thiolate complexation and on the basis of the recent EXAFS results, the thiolates are the likely sites of mercuric ion interaction. The sensitivity of the MerR protein to mercuric ion may arise from the ability of the protein to restrain the thiolate donors in a distorted coordination geometry that cannot accommodate a metal preferring a regular tetrahedral environment. The four cysteine residues in the 144 amino acid MerR protein are blocked out in Fig. 5 and are found in the carboxy terminal half of the peptide.

There are two regions of the MerR protein that exhibit similarity to domains of structurally characterized DNA-binding proteins which have been shown to mediate specific protein-DNA interactions. These helix-turn-helix motifs are both located in the N-terminal half of the MerR protein [11,16]. One or both of these domains may be involved in recognition of the operator by the MerR protein.

Finally, the synergistic binding of Hg-MerR and RNA polymerase to the promoter suggest that the MerR protein may facilitate the formation of transcriptionally active complexes with RNA polymerase through specific protein-protein interactions. Recently, several positive activator proteins known to interact with RNA polymerase through specific protein-protein interactions have been shown to possess an amphipathic α helix which has a high density of negatively charged residues on one helical face, and it is suggested that this serves as a molecular determinant which can interact with a complementary site on the RNA polymerase enzyme [27]. The MerR protein possesses several regions which have a high probability of forming an α helix based on Chou-Fassman predictions. Model construction suggests that two of these α-helical regions would be amphipathic with a high density of negatively charged residues on one surface. Such a surface could be exposed in the mercury-induced conformation of the MerR protein. However, a role for these putative domains in the MerR function is purely conjectural at this point.

2.1.7. Intracellular Receptor for an Environmental Signal

One striking characteristic of the mer system is that the MerR protein apparently acts as an intracellular receptor for an environmental signal. This phenomenon is even more intriguing in view of the fact that induction of the mercury resistance operons results in the production of membrane-spanning transport functions which initially pump mercuric ion into the cell for delivery to the reductase enzyme. This system may have evolved to protect functions in the periplasmic membrane which are sensitive to mercuric ion. It is curious that the sensor/receptor part of this transduction apparatus is not a transmembrane protein. Apparently, mercuric ion or, more likely, small-molecule complexes, e.g., $Hg(cys)_2$, can cross biological membranes relatively easily [28]. In this sense the MerR protein is analogous to the DNA-binding steroid receptors found in eukaryotic cells. Steroid hormones diffuse into the cell, where they are bound by receptor proteins. The steroid-receptor complex then activates the expression of specific genes [29].

The Fur protein also acts as an intracellular receptor which mediates both intracellular and extracellular signals. This situation is understandable when the metal is an essential nutrient and its concentrations must be balanced by a precisely poised homeostatic mechanism. Such an intracellular receptor must allow the cell to gauge the specific intracellular iron levels at a specific point in time, and facilitate the production or repress the transcription of additional uptake proteins.

An intracellular receptor for the extracellular toxic metal signal may allow the cell to utilize other lines of defense before activating the transcription of an energy-intensive pump and NADPH-consuming detoxification system. Thus if the concentration of low molecular weight thiols which can sequester Hg^{2+} (such as glutathione or cysteine) is relatively high, the cell can protect itself against low concentrations of mercuric ion. Once initial defenses have been overwhelmed and the mercuric ion concentration inside the cell begins to exceed a toxic threshold, then an intracellular receptor can allow a very rapid response of the transcriptional apparatus. Efficient response is achieved in the MerR system by having the intracellular receptor bound to the DNA directly in front of the genes that are to be activated by the heavy-metal signal. This efficiency may be the hallmark of plasmid-mediated resistance systems or it may be a general mechanism utilized by chromosomal heavy-metal homeostasis systems and plasmid-based resistance systems alike. Resolution of this point will require the investigation of additional plasmid and chromosomal detoxification loci and their respective metalloregulatory proteins.

2.2. The Fur Protein: Heavy-Metal Responsive Repressor of Iron Uptake Genes

Under conditions of iron starvation, the transcription of a group of genes scattered through the *E. coli* chromosome is coordinatively activated in what is best described as an iron-controlled regulon. The function of genes characterized as iron-responsive to date include all of the high-affinity iron uptake systems and also a toxin production system. Fifteen years ago Neilands postulated the existence of a single protein which acts in a metal-responsive manner to regulate the transcription of these genes [30]. In a recent review he summarizes the properties of an iron-responsive repressor which is now known as the Fur protein [31]. Fur was the first metalloregulatory protein to be isolated.

All of the high affinity iron uptake systems in *E. coli* are regulated by the *fur* (ferric uptake regulation) gene [32]. The *fur* gene maps at 15.5-15.7 min in the *E. coli* chromosome [33,34]. Cloning and sequencing of the *fur* gene showed that it encodes a 17-kD polypeptide containing 12 histidines [35]. Mutations in the *fur* locus in either *S. typhimurium* or in *E. coli* constitutively expressed proteins involved in iron transport [36]. The fur protein has recently been purified and shown to be a dimer [31] that apparently uses iron(II) or a variety of other divalent metals as a cofactor in the regulation of *lacZ*-aerobactin fusion genes in vivo [37]. The aerobactin operon of pColV-K30 encodes a high-affinity, siderophore-mediated iron uptake system. The aerobactin promoter has been studied by quantitative S1 nuclease mapping, which demonstrated that iron regulates the levels of messenger RNA production and identified start sites for transcription [38]. Based on the properties of *fur* mutants and the S1 analysis, the Fur protein is thought to play the role of a classical repressor that binds iron as a corepressor. DNase I footprinting of the interaction of the Fur protein with the iron-responsive aerobactin promoter in the presence and absence of various divalent metals support some aspects of the above model. The purified Fur protein binds to two contiguous binding sites downstream of the -35 region of the aerobactin promoter sequence only in the presence of divalent metals such as Mn^{2+}, Fe^{2+}, Co^{2+}, Cu^{2+}, and Cd^{2+}. Zn^{2+}, however, did not produce any detectable effect and apparently acted as a competitor for activation by Mn^{2+} [39]. In the absence of these metals or in the presence of excess 2,2'-bipyridine, no binding of the Fur protein to either of the operator sites was observed.

Thus at low Fe(II) concentrations the aerobactin promoter region is not occupied by the Fur protein and is apparently available to interact with RNA polymerase molecules inside the cell. As the Fe(II) concentration inside the cell increases, the Fur protein acquires the ability to bind to this promoter and in doing so, inhibits RNA polymerase transcription. It is important to note, however, that gene regulatory proteins binding in this region of a promoter may also facilitate RNA polymerase binding and thus activate transcription (see Sec. 2.1). This mechanism implies that the divalent metal ion interacts directly with the Fur protein, activating a specific DNA binding activity. The nature of metal ion interactions with the Fur protein interactions have not yet been described.

Analysis of the Fur-binding sites in the aerobactin promoter in comparison with other iron-regulated promoters, such as the *fhuA* and the *fepA* promoters, allows the establishment of a consensus DNA sequence for the *fur* operator. The homology has allowed the establishment of a 19 base pair dyad consensus sequence 5'-GATAATGATAATCATTATC-3' [31]. The presence of this DNA sequence, dubbed an "iron box", suggests that an adjacent gene may be responsive to Fur-mediated, iron-induced repression. This short consensus sequence has shown to be sufficient to function as an operator site for the binding of the Fur protein in vivo [40,41].

Several bacterial toxins, including *shiga* toxin, are repressed in the presence of high concentrations of iron in the growth medium and derepressed when the concentration of iron is lowered. Gene fusion experiments have established that repression of a shiga-like toxin operon in *E. coli* (SLT-1) is mediated by the *fur* locus. A 21 base pair dyad symmetry element identical to the *fur* consensus sequence discussed above was isolated [41]. Thus the Fur protein also regulates toxin production in bacteria as well as biosynthesis of siderophores. Low concentrations of extracellular iron may be an important environmental signal to bacteria that they have entered a mammalian host. This signal then leads to the coordinated expression of several virulence determinants.

Several other genes in *E. coli* may be responsive to the concentrations of iron in the media. Now that a consensus *fur* sequence has been established, iron-regulated genes may be identified by searches for the consensus sequence. Other genes known to be regulated by the fur protein, such as the colicin (I) resistance gene (*cir*), are strongly induced by iron starvation [42]. However, their function is not known.

Iron control in all the above cases is a result of negative regulation of transcription by the Fur protein in the presence of iron. The absence of iron triggers the derepression of the genes and allows constitutive expression. While the above genes may be classified as a stress response of one type or another, Fur may also control genes involved in metabolism of succinate and other dicarboxylic acids [43]. The effects of Fur on metabolic genes has not yet been elucidated.

One interesting aspect of the Fur metalloregulatory protein is that it is apparently functioning as an intracellular receptor which gauges the intracellular iron status of the cell. This is in contrast to the mechanisms for homeostasis of other essential inorganic solutes, such as phosphate, where a periplasmic domain of a transmembrane protein acts as a sensor and independent gene regulatory proteins act in concert to transduce an extracellular signal [22].

3. ADDITIONAL SYSTEMS EXHIBITING METAL-RESPONSIVE GENE EXPRESSION

3.1. Prokaryotes

3.1.1. Metal-Inducible Homeostasis, Detoxification, and Metabolic Pathways

Microbes respond to heavy-metal stress by translocating the metal or by carrying out a variety of chemical transformations. Proteins responsible for efflux and chemical transformations are frequently encoded on plasmids (or extrachromosomal DNA molecules) which often carry drug resistance determinants as well [8,44].

Some of these resistance mechanisms are always in place. The proteins responsible for mediating such resistance are always expressed and are present in the cell, regardless of the concentration of the toxic metal in the media. Such resistance systems are said to be constitutive and, therefore, no factor is required for metalloregulation. On the other hand, many of the proteins responsible for carrying out specific chemical transformations are not present in the cell at detectable levels until a subtoxic amount of metal ion is present in the media. At this point the biosynthesis of the resistance proteins is induced, and the cells exhibit a heavy-metal resistance phenotype. Such detoxification systems are said to be inducible and therefore will require the existence of an inorganic signal transduction system, perhaps including a metalloregulatory protein. Several examples of heavy-metal inducible detoxification systems in bacteria are described below.

Although there are numerous reports of bacteria that are resistant to wide varieties of metals, there are few cases where the specific organization of genes or the resistance proteins have been identified.

3.1.1.1. Arsenic. The *ars* operon confers resistance to arsenate, arsenite, and antimony(III). This operon is inducible and encodes three peptides, one of which, ArsA, has recently been shown to be an arsenite-stimulated ATPase [45,46]. Resistance in this system is thus achieved by an energy-dependent efflux mechanism. Detoxification results from the active efflux of the oxy anions (i.e., AsO_4^{3-}), preventing their concentration from reaching toxic levels inside the cell. The nucleotide sequence of the arsenical resistance operon reveals three genes—*arsA*, *arsB*, and *arsC*—in addition to the unsequenced *arsR* regulatory gene. No protein corresponding to the *arsR* gene has yet been identified, and genetic evidence concerning the nature of the regulation is not yet available.

3.1.1.2. Cadmium. Plasmid-determined cadmium resistance has been found in *S. aureus* [43]. Two separate genes are responsible for cadmium resistance in *S. aureus* strains. However, there is no evidence that these genes are inducible [47,48].

The presence of heavy metals, such as cadmium, in the media induces the de novo synthesis of at least 35 different proteins in *E. coli*. Seventeen of these uncharacterized proteins induced by cadmium chloride are not induced by other types of environmental stress, such as oxidizing agents or heat shock [49]. In addition, cadmium chloride induces the de novo protein synthesis of a variety of other proteins that are known to be induced by the above environmental stresses. Thus, different types of environmental stress result in the coordinated expression of a group of unlinked and individually controlled genes [45]. In several cases one common regulatory gene is known to control the expression of these unlinked genes which are grouped together as functional units called regulons. Two such stress systems are the heat shock regulons [50,51] and the oxidation stress regulon [52]. A subset of the genes involved in the heat stress response and a set of genes involved in the oxidation stress regulon are induced by the presence of $CdCl_2$. Since $CdCl_2$ is one of the most potent inducers of many oxidative stress proteins, it has been classified as an oxidant. In view of the redox properties of Cd^{2+} and the range of biological redox potentials, it is unlikely that cadmium is directly involved in any kind of chemical oxidation. It may well affect the activity of enzymes which mediate biological oxidations, but is itself not an oxidant. This heavy metal induces several global responses in bacteria as well as a set of unique responses, and it is very possible that further investigation will reveal the existence of a heavy-metal regulon analogous to the heat shock and oxidant stress in *E. coli*. Characterization of a metalloregulatory gene that controls a variety of unlinked genes that respond to the presence of cadmium in the media are required before the existence of heavy-metal regulon in *E. coli* can be established.

3.1.1.3. Copper. Copper is an essential trace element yet it is also highly toxic. Intracellular levels must therefore be tightly regulated, and several genes involved in copper homeostasis in *E. coli* K-12 have been identified.

Plasmid free strains of *E. coli* grow normally in common media (where the copper concentration is 5 μM) and can tolerate no more than 4 mM copper sulfate. The presence of the plasmid pRJ1004 in standard test strains of *E. coli* allow this bacteria to tolerate up to 20 mM copper sulfate [53,54]. The plasmid-mediated copper resistance was shown to be inducible by studying the effect of a sudden increase in copper concentration on the growth rate. A number of copper-dependent and copper-sensitive mutants have been characterized. Using these mutants distinct uptake, efflux, and storage components have been defined.

A copper resistance determinant *pCo* has been localized on pRJ1004 and been shown to code for an inducible mechanism which results in decreased intracellular copper accumulation. The *pCo* determinant can be cloned and contains at least four genes (pCoA, B, C, and R) [55]. The *pCoC* gene product is an inducible 25-kD Cu-binding protein, while the *pCoR* gene is reported to have trans-acting regulatory role and the roles of *pCoA* and *pCoB* are not yet known. Expression of the *pCo* genes is apparently controlled at the transcriptional level, and it is known that at least two chromosomal gene products are essential for the function of *pCo*-encoded copper resistance. Kinetic results show that reduced Cu accumulation is due to an energy-dependent efflux of copper. It has been proposed that the mechanism of copper resistance conferred by *pCo* involves intracellular copper binding as well as exclusion by an enhanced efflux and chemical modification system in a manner that prevents the exported copper from being recognized by chromosomally encoded copper uptake systems [55].

Plasmid-encoded copper resistance has also been observed in *Pseudomonas* species and the copper resistance determinant has recently been subcloned [56]. Other plasmid-encoded copper resistance mechanisms, one of which is based on the precipitation of copper sulfide, have been observed but none of the genes responsible for metal resistance have been identified nor has the inducibility of the resistances been investigated [44].

3.1.1.4. Iron: Superoxide Dismutase. Superoxide dismutases (SOD) are a ubiquitous class of metalloenzymes which confer some resistance to toxic oxygen species; however, their physiological roles are not totally understood. There are several different types of microbial superoxide dismutase including an iron-containing SOD and manganese-containing enzyme that is synthesized only under aerobic conditions. Removal of iron from the growth media causes an increase in the level of manganese superoxide dismutase, while addition of iron salts to the medium abolishes the inductive effect. These studies have led to the suggestion that an iron-responsive repressor protein controls the expression of the iron and manganese SOD gene [57,58]. This putative iron-responsive repressor could exist in either the ferric or the ferrous form, depending on the redox state of the cell, and it was proposed that the active repressor contained Fe(II) whereas the inactive repressor contained either Fe(III) or no iron. No evidence for iron-responsive transcription of the superoxide dismutase genes has been provided. This description fits the behavior of the Fur protein and it is possible that Fur protein also regulates the level of SOD in *E. coli*.

3.1.1.5. Molybdenum: Nitrogenase. Enzymatic reduction of dinitrogen to ammonia is catalyzed by the nitrogenase protein complex. Nitrogenase systems isolated from a variety of microorganisms consist of several components, including a dinitrogenase, a molybdenum iron protein, a dinitrogenase reductase, and finally two molecules of FeMo cofactor. When molybdenum, an essential component of this enzyme complex, is depleted in the media, the *Azotobacter vinelandii* can synthesize an alternative nitrogen fixation system. One alternative nitrogen fixation system has been shown to utilize a vanadium iron cofactor in the place of FeMoCo. Thus a metabolic path decision must be made based on the availability of different metals in the environment. There are two different types of dinitrogenase reductase gene complexes in *A. vinelandii* which are under regulatory control by molybdenum [59]. The same authors have shown that there are two independent *nif* gene clusters, one encoding the conventional molybdenum-containing nitrogenase (*nifHDK* gene cluster) and the other encoding the "alternative" *nifH* gene(s), and that transcription of these sequences is regulated differentially by molybdenum [60]. This regulation of nitrogen fixation systems in *A. vinelandii* by molybdenum at the level of transcription is different from the situation observed in *K. pneumoniae*, where all the nitrogenase genes are apparently expressed in both the presence and absence of molybdate. In vivo transcription of the "conventional" *nifHDK* gene cluster in *A. vinelandii* required at least 50 mM molybdenum in the medium, whereas transcription of the alternative *nifH* gene cluster was not observed unless the molybdenum concentrations were below 50 nM [60]. Thus as molybdenum concentrations drop, the system switches off the transcription of the molybdenum-dependent nitrogenase and switches on the synthesis of the alternative vanadium-dependent nitrogenase. These results strongly implicate the existence of at least one type of molybdenum-responsive metalloregulatory protein. Recent investigations of *nifHDK* deletion mutants grown in the presence of V_2O_5 suggest that growth in the presence of vanadium and growth under molybdenum deficiency conditions are independent phenomena [60]. However, vanadium- and tungsten-dependent regulation of transcription has yet to be demonstrated [61].

3.1.1.6. Molybdenum: Nitrate Reductase. Concentrations of molybdenum in the media have recently been shown to affect the expression of other genes that encode molybdenum-requiring enzymes. A molybdenum cofactor may act synergistically with nitrate in the regulation of the *narCHI* genes which encode the enzyme nitrate reductase [62]. In the absence of oxygen, nitrate is used in preference to fumarate as a principal electron acceptor during the dissimilation of carbon and energy sources. Studies of in vivo transcription from the *narC* operon in *chlD* mutants which are impaired in molybdate transport (or processing) revealed that full induction of the *narC* operon encoding nitrate reductase did not occur unless the growth medium was directly supplemented with 1.0 mM of molybdate.

Wild-type cells containing a functional *chlD* gene did not require molybdenum supplementation, apparently because the molybdenum requirement was met by trace quantities of the metal present as a contaminant in the growth medium. Thus the expression of the nitrate reductase gene requires dual signals from the environment: first, the signal that there is sufficient substrate (i.e., nitrate), and second, that there is sufficient molybdenum for biosynthesis of the molybdenum cofactor essential to the enzymatic activity of nitrate reductase. Obviously synthesis of the nitrate reductase subunits would waste resources if there was a shortage of an essential component of the cofactor. The gene *narL* apparently encodes a regulatory protein which mediates both the nitrate and molybdenum coeffector actions on the *narC* operon [63]. However, proof of a direct role of the *narL* gene product as a molybdenum receptor, nitrate receptor, or activator protein awaits characterization of the *narL* gene(s) and in vitro examination of the purified components.

3.1.1.7. Nickel: Hydrogenase. Nickel is an essential cofactor for many but not all hydrogenases. It is a requirement for growth of many strains, such as *Alcaligenes eutrophus*. Recently nickel has been shown to regulate the biosynthesis of hydrogenase in *Bradyrhizobium japonicum*. These experiments did not show the direct activation of hydrogenase gene expression. Rifampicin inhibition studies suggested that nickel is required for the transcription of a hydrogenase-related gene [64]. This is consistent with the observations that nickel was required for the growth of thermophilic bacteria [65]. Evidence of metal-responsive gene expression awaits more detailed studies of this and other nickel-dependent systems.

3.1.1.8. Tellurium. The genetic structure, organization, and expression of a tellurium resistance determinant which was originally isolated from a strain of *Alcaligenes* has recently been reported. This plasmid-based resistance system is encoded on a 6-kilobase fragment that expresses inducible tellurite resistance [66]. When the tellurite-resistant plasmid pMER610 is transferred to *E. coli*, the same level of resistance to tellurite and tellurate is found as in the original *Alcaligenes* host. A strain of *E. coli* sensitive to 10 μM tellurite became resistant to 1 mM tellurite when the plasmid pMER610 was present in those cells. As with the arsenate, copper, and mercury systems, expression of the tellurite resistance was inducible by prior exposure of the cells to subtoxic levels of tellurite. Transposon insertion analysis suggest that the tellurium resistance determinant has both regulatory and structural elements. However, specific genes have not been identified and a tellurium-responsive metalloregulatory protein has not been characterized. Plasmid-determined resistance to tellurium salts appears to be widespread in nature and was first observed in *P. aeruginosa* and *E. coli* [67]. These plasmids conferred resistance to 1 mM tellurite and increased the tolerance of *P. aeruginosa* and *E. coli* by 10-fold and 100-fold, respectively. Strains with and without plasmids apparently reduce the tellurium salts to the elemental form, producing black colonies. It is not known if the plasmids encode a reductase [68].

3.2. Eukaryotes

There are many examples of metal-responsive gene expression in eukaryotic systems. Circumstantial evidence suggests the existence of trans-acting, metal-responsive, regulatory factors in several of these systems, and one yeast gene product, HAP1, is known to mediate heme-responsive regulation of cytochrome c genes. While specific details concerning the regulation of these genes is beyond the scope of this chapter, several examples of metal-responsive gene regulation in nucleated cells are briefly described below.

3.2.1. Iron

3.2.1.1. Transferrin Receptor. Eukaryotic cells acquire iron through the internalization of diferric transferrin, which is sequestered by a high-affinity cell surface receptor. The expression of the transferrin receptor is regulated by iron availability in the media. At least two genetic elements exist for iron regulation of the transferrin receptor gene. One element is in the DNA upstream of the transferrin receptor gene's transcription start site. The second element requires the presence of specific sequences in the 3' untranslated portion of the transferrin receptor mRNA. Thus two different types of cis-acting elements and perhaps two types of metalloregulatory protein may affect the transferrin receptor gene expression in the presence of iron [69,70].

3.2.1.2. Ferritin. All of the examples of metal-responsive gene expression discussed so far involve regulation of transcription of DNA into mRNA. Ferritin biosynthesis was the first example of translational control of gene expression. Ferritin, a 24-subunit protein containing both heavy and light chains, is the major intracellular storage site for iron. Ferritin levels dramatically increase as cellular iron levels increase, and the principal mode of iron-responsive regulation is at steps in translation of mRNA into protein. Stored ferritin mRNA is translated at low efficiencies into peptides. However, the rates of this translation can increase 40-50 times as cellular iron levels increase [71]. The specific nature of the translational control can vary and may depend on cell type. Several recent investigations have shown that untranslated sequences in both the 5' and 3' regions of the mRNA can mediate the translational regulation [72,73].

An iron-responsive element (IRE) in the 5' end of ferritin mRNA has been delineated by deletion analysis [74]. These sequences are approximately 30 nucleotides long and apparently form stem-loop structures which may be of functional significance [75]. The IRE sequences are highly conserved on an evolutionary time scale. One proposal for the role of these 5' RNA sequences in mRNA translation parallels the role of the Fur operator as discussed above and involves a trans-acting metalloregulatory protein which in this case would bind to mRNA. The scenario suggests that when intracellular iron concentration is low, ferritin synthesis is repressed due to a negative trans-acting molecule(s) binding to the 5' or 3' IREs. Upon iron stimulation, this factor, or group of factors, is modified and its affinity for the regulatory sequence decreases, allowing ribosomes to scan through the 5' untranslated region to form initiated complexes for translation. In this case, the putative metalloregulatory protein would affect the interaction of the nucleic acid with the protein complexes involved with translation (ribosomes) as opposed to those involved with transcription (RNA polymerase). Ferritin is the only system to date in which a metalloregulatory protein acting at a translational level has been implicated [76].

3.2.1.3. Cytochrome c. The transcription of the yeast genes *CYC1* and *CYC7*, which encode the major and minor isologs of cytochrome c, respectively, are induced by increases in the concentration of heme iron in vivo [77,78]. The HAP1 gene encodes a positive activator protein which binds to regulatory regions upstream of both CYC1 and CYC7 genes in a heme-responsive manner in vitro [79,80], supporting earlier suggestions that heme is an effector molecule that induces the synthesis of cytochrome c. Complementation and deletion studies on the HAP1 gene demonstrate that sequences essential for DNA binding and heme response lie in the N-terminal coding region. Isolation and characterization of the corresponding HAP1 protein will provide insights into the molecular basis of receptor recognition of metal-ligand complexes and the mechanism of transcriptional activation.

In eukaryotes, heme is synthesized inside the mitochondria and the HAP1 protein functions inside the nucleus. Thus the heme signal is being relayed from the mitochondria to the nucleus in a mechanism that apparently involves the HAP1 metalloregulatory protein. The HAP1 system is thus an example of protein which mediates a heavy-metal signal between organelles within a cell.

3.2.1.4. Respiratory Proteins and Heme Iron. Intracellular heme iron concentrations are known to affect a variety of other cellular processes. Addition of heme in vitro to heme-depleted nuclei leads to a significant increase in the transcription of cytochrome P-450, a heme-dependent enzyme [81]. Recent studies of mutations in *Hem2*, the structural gene for δ-aminolevulinate dehydratase gene of *Saccharomyces cerevisiae*, have allowed a study of the effects of heme on transcription of a variety of heme-dependent proteins involved in respiration. Transcription of genes coding for cytochrome oxidase as well as genes encoding cytochrome c reductase may be regulated by levels of heme in the cell. A metalloregulatory protein which recognizes heme iron complexes may be responsible for mediating regulation of different respiratory genes. However, no factor has been identified and the involvement of such metalloregulatory protein is speculative at this point [82].

3.2.2. Copper

There are several examples of copper-regulated gene expression in eukaryotic systems. The best understood example of copper-responsive gene regulation is the metallothionein system. Even in this case, a metalloregulatory protein activating (or controlling) the expression of the metallothionein genes has not been identified. Since gene regulation in the metallothionein system is reviewed in Chap. 4 of this volume, detailed discussion of yeast metallothionein gene expression will be omitted from this review. The interested reader is also directed to another recent review of transcriptional regulation in the metallothionein systems [83].

3.2.2.1. Plastocyanin in Cytochrome c-552. Plastocyanin and cytochrome c-552 are electron transfer proteins which can both donate electrons to the photosystem I reaction center in the eukaryotic algae *Chlamydomonas reinhardtii*. However, in the photosynthetic membranes of green eukaryotic algae, plastocyanin appears to be the preferred catalyst. The availability of copper to the cell determines whether a functional copper-containing plastocyanin can be synthesized. When the concentration of available copper is too low to support plastocyanin accumulation at the stoichiometry required for electron transfer, the cell responds by synthesizing an alternative electron carrier, cytochrome c-552. While copper mediation of plastocyanin activity is at the level of the stable protein, cytochrome c-552 is regulated at the level of transcription [84]. These conclusions are based on studies of plastocyanin-less mutant strains of *C. reinhardtii*. This mutant strain can grow photoautotrophically in Cu(II)-deficient but not in Cu(II)-sufficient media. These studies indicate that Cu(II)-dependent expression of cytochrome c-552 synthesis is affected directly by the concentration of Cu(II) rather than indirectly by the level of plastocyanin in the cell. The authors postulate the existence of a Cu(II)-binding regulatory protein which affects the expression of Cu(II)-repressible polypeptides [84-86].

3.2.3. Zinc

Zinc plays critical roles in almost all functions of normal cells and it is intimately involved in the enzymatic machinery which regulates the expression of genes in almost all types of organisms [87]. Zinc can also play a structural role in regulatory proteins. In these cases (e.g., gene 32 protein), it is unlikely that the protein will respond to changes in Zn^{2+} concentrations, even in the presence of biological chelating agents. The zinc is simply bound too tightly in this class of metalloprotein. This again brings up the distinction between metalloproteins, metal-activated proteins, and metalloregulatory proteins. Changes in zinc concentration in the cell may directly regulate the activity of a variety of zinc-responsive enzymes by affecting the levels of the active metalloprotein or metal-activated proteins in the cytoplasm or the nucleus. These essential roles for zinc will make it difficult to establish the existence of a zinc-responsive metalloregulatory protein, i.e., a protein that is designed to act as a zinc receptor and transmit the signal of changes in zinc concentration into the changes in specific gene expression.

The prototypical zinc finger protein TFIIIA loses its DNA- and RNA-binding properties when mild chelating reagents remove zinc from the protein [88]. One could argue that these in vitro observations suggest that TFIIIA may exert a metal-responsive effect on gene expression. In this case TFIIIA may act as a zinc receptor with the role of transmitting signals to specific genes that are to be activated when the zinc concentrations rise. TFIIIA then would be classified as a metalloregulatory protein which transduces an important signal during differentiation. This role for TFIIIA as a receptor in a signal transduction cascade has not been addressed at the physiological level, and in view of the above caveats concerning zinc proteins, it remains a debatable issue. Properties of zinc finger proteins are discussed in Chap. 6 of this volume.

Recent studies on the effects of zinc deficiency in *Euglena gracilis* reveal that protein metabolism is altered selectively, apparently by concurrent activation and repression of specific segments of the genome. With the above caveats concerning multiple roles of zinc in the regulation of gene expression in mind, these results suggest the existence of zinc-responsive metalloregulatory protein(s) [89,90].

4. MODEL COMPLEXES AND METALLOREGULATION

Chemical synthesis of model compounds will play a central role in the elucidation and the molecular recognition of specific heavy metals in the metalloregulatory proteins. This approach may also be valuable in elucidating the molecular basis of metal-induced changes in biopolymer interactions. Recently, a small-molecule model has been synthesized which incorporates three determinants involved in one type of metalloregulation: (1) a metal-binding cavity; (2) a domain which confers specific DNA-binding properties; and; (3) a reporter function group which nicks DNA at the site where the small molecule is bound. The molecule is designed such that when a metal is bound in the pseudomacrocycle site, the two attached dipeptide units would be able to assume a conformation which facilitates specific binding to a given DNA sequence. It was found that the binding of the synthetic molecule to specific DNA sequences was greatly enhanced by binding of heavier alkaline earth cations to the multidentate polyether cavity. This approach may eventually establish the contributions of cation size, coordination number, and hardness to metallospecificity and provide clues as to how metal ions can induce a specific DNA-binding activity in proteins [91].

5. CONCLUSION

Metalloregulatory proteins are the first examples of heavy-metal receptor proteins involved in signal transduction. They provide an excellent basis for examining determinants responsible for molecular recognition of heavy metals and can provide new insights into the fine balance between toxic and essential roles of heavy metals in detoxification, homeostasis, uptake, and metabolism at the cellular level. While only two metalloregulatory proteins have been isolated to date, there are numerous examples of metal-responsive gene expression, and several genes thought to encode metalloregulatory proteins have been identified. Thus this class of metal-activated proteins will likely expand in the next several years.

ACKNOWLEDGMENT

I would like to acknowledge the contributions of my students and coworkers, Jeff Wright, Myung Shin, Diana Ralston, and Betsy Frantz. Their research and ideas have had an important impact on the development of this review. Support from the Camille and Henry Dreyfus Foundation and the Searle Scholars Program of the Chicago Community Trust is gratefully acknowledged. I also thank Steve Watton for expert assistance with artwork.

REFERENCES

1. R. J. P. Williams, *Biol. Rev., 28,* 381 (1953).
2. B. L. Vallee and W. E. C. Wacker, in "Metalloproteins" in "The Proteins" (H. Neurath, ed.), Vol. 5, Academic Press, New York, 1970.

3. R. J. P. Williams, *FEBS. Lett.*, *140*, 3 (1982).

4. E. Frieden, *J. Chem. Ed.*, *62*, 917 (1985).

5. B. Lewin, "Genes II", John Wiley and Sons, New York, Chap. 14, 1985.

6. T. Maniatis, S. Goodbourn, and J. A. Fisher, *Science*, *236*, 1237 (1987).

7. O. Raibaud and M. Schwartz, *Ann. Rev. Genet.*, *18*, 173 (1984).

8. S. Silver, in "Environmental Inorganic Chemistry" (K. J. Irgolic and A. E. Martel, eds.), VCH, Deerfield Beach, FL, 1985, pp. 513-539.

9. C. T. Walsh, M. D. Distefano, M. J. Moore, L. M. Shewchuk, and G. L. Verdine, *FASEB J.*, *2*, 124 (1988).

10. N. L. Brown, *Trends Biochem. Sci.*, *10*, 400 (1985).

11. A. O. Summers, *Ann. Rev. Microbiol.*, *40*, 607 (1986).

12. T. J. Foster, M. Nakahara, A. A. Weiss, and S. Silver, *J. Bacteriol.*, *140*, 167 (1979).

13. N. N. Ni'Bhriain, S. Silver, and T. J. Foster, *J. Bacteriol.*, *155*, 690 (1983).

14. T. J. Foster and N. L. Brown, *J. Bacteriol.*, *163*, 1153 (1985).

15. P. J. Barrineau and A. O. Summers, *Gene*, *25*, 209 (1983).

16. T. O'Halloran and C. Walsh, *Science*, *235*, 211 (1987).

17. T. K. Misra, N. L. Brown, D. Fritzinger, R. D. Pridmore, W. M. Barnes, L. Haberstroh, and S. Silver, *Proc. Natl. Acad. Sci. USA*, *81*, 5975 (1985).

18. P. A. Lund, S. J. Ford, and N. L. Brown, *J. Gen. Microbiol.*, *132*, 465 (1986).

19. A. Heltzel, D. Gambill, W. J. Jackson, P. A. Totis, and A. O. Summers, *J. Bacteriol.*, *169*, 3379 (1987).

20. T. V. O'Halloran, B. Frantz, M. K. Shin, D. M. Ralston, and J. G. Wright, *Cell*, in press.

21. P. A. Lund and N. L. Brown, *Gene*, *52*, 207 (1987).

22. C. W. Ronson, B. T. Nixson, and F. M. Ausubel, *Cell*, *49*, 579 (1987).

23. J. G. Wright and T. V. O'Halloran, unpublished results.

24. W. Stricks and I. M. Kolthoff, *J. Am. Chem. Soc.*, *73*, 1723 (1951).

25. J. E. Penner-Hahn, R. Fronko, H. T. Tsang, J. G. Wright, and T. V. O'Halloran, unpublished results.

26. T. D. Tullius and B. A. Dombroski, *Proc. Natl. Acad. Sci. USA*, *83*, 5469 (1986).

27. N. Irwin and M. Ptashne, *Proc. Natl. Acad. Sci. USA, 84,* 8315 (1987).

28. E. Bienvenue and A. Boudou, *Chem.-Biol. Interact., 48,* 91 (1984).

29. P. J. Godowski, S. Rusconi, R. Miesfeld, and K. R. Yamamoto, *Nature, 325,* 365 (1987).

30. J. B. Neilands, in "Inorganic Biochemistry" (G. L. Eichhorn, ed.), Elsevier, New York, Vol. 1, 1973, p. 167.

31. A. Bagg and J. B. Neilands, *Microbiol. Rev., 51,* 509 (1987).

32. V. Braun, *Trends Biochem. Sci., 75* (1985).

33. K. Hantke, *Mol. Gen. Genet., 197,* 337 (1984).

34. A. Bagg and J. D. Neilands, *J. Bacteriol., 161,* 450 (1985).

35. S. Schaffer, K. Hantke, and V. Braun, *Mol. Gen. Genet., 200,* 110 (1985).

36. J. F. Ernst, R. L. Bennett, and L. I. Rothfield, *J. Bacteriol., 135,* 928 (1978).

37. A. Bagg and J. B. Neilands, *Biochemistry, 26,* 5471 (1987).

38. A. Bindereif and J. B. Neilands, *J. Bacteriol., 162,* 1039 (1985).

39. D. De Lorenzo, S. Wee, M. Herrero, and J. B. Neilands, *J. Bacteriol., 169,* 2624 (1987).

40. S. B. Calderwood and J. J. Mekalanos, *J. Bacteriol., 169,* 4759 (1987).

41. S. B. Calderwood and J. J. Mekalanos, *J. Bacteriol., 170,* 1015 (1988).

42. V. Braun, K. Hantke, K. Eick-Helmerich, W. Köster, V. Pressler, M. Sauer, S. Schäffer, H. Schöffler, H. Staudenmaier, and L. Zimmermann, in "Iron Transport in Microbes, Plants and Animals" (G. Winkelmann, D. van der Helm, J. B. Neilands, eds.), VCH Publishers, Deerfield Beach, FL, 1987, pp. 35-52.

43. J. B. Neilands, K. Komopka, B. Schwyn, M. Coy, R. T. Francis, B. H. Paw and A. Bagg, in "Iron Transport in Microbes, Plants and Animals" (G. Winkelmann, D. van der Helm, J. B. Neilands, eds.), VCH Publishers, Deerfield Beach, FL, 1987, pp. 3-34.

44. J. M. Wood and H. K. Lang, "Environmental Inorganic Chemistry" (K. J. Irgolic and A. E. Martell, eds.), VCH Publishers, Deerfield Beach, FL, 1985, pp. 513-539.

45. B. P. Rosen, U. Weigel, C. Karkaraia, and P. Gangola, *J. Biol. Chem., 263,* 3067-3070 (1988).

46. S. Silver, K. Budd, K. M. Leahy, W. B. Shaw, D. Hammond, R. P. Novick, G. R. Willsky, N. H. Malamy, and H. Rosenberg, *J. Bacteriol., 146,* 983 (1981).

47. R. D. Novick and C. Roth, *J. Bacteriol., 95,* 1335 (1968).

48. H. Nakahara, T. Ishikawa, Y. Sari, I. Kondo, H. Kozukoe, and S. Silver, *Appl. Environ. Microbiol., 975* (1977).

49. R. A. Van Bogelen, P. M. Kelley, and F. C. Neidhardt, *J. Bacteriol., 167,* 26 (1987).

50. S. Lindquist, *Ann. Rev. Biochem., 55,* 1151 (1986).

51. F. C. Neidhardt, R. A. Van Bogelen, and V. Vaughn, *Ann. Rev. Genet., 295* (1985).

52. M. F. Christman, R. W. Morgan, F. S. Jacobson, and B. N. Ames, *Cell, 41,* 753 (1985).

53. T. J. Tetaz and R. K. J. Luke, *J. Bacteriol., 154,* 1263 (1983).

54. D. Rouch, J. Camakaris, B. T. O. Lee, and R. K. J. Luke, *J. Gen. Microbiol., 131,* 939 (1985).

55. D. Rouch, J. Camakaris, and B. T. O. Lee, *J. Cell. Biochem. Suppl. 120,* 328 abstracted (1988).

56. C. L. Bender and D. A. Cooksey, *J. Bacteriol., 169,* 470 (1987).

57. C. S. Moody and H. M. Hassan, *J. Biol. Chem., 259,* 12821 (1984).

58. H. M. Hassan and C. S. Moody, *J. Biol. Chem., 262,* 17173 (1987).

59. R. E. Premakumar, M. Lemos, and P. E. Bishop, *Biochim. Biophys. Acta., 797,* 64 (1984).

60. R. D. Joerger, R. Premakumar, and P. E. Bishop, *J. Bacteriol., 168,* 673 (1986).

61. M. R. Jacobsen, R. Premakumar, and P. E. Bishop, *J. Bacteriol., 167,* 480 (1986).

62. S. Iuchi and E. C. C. Lin, *J. Bacteriol., 169* (1987).

63. S. Iuchi and E. C. C. Lin, *Proc. Natl. Acad. Sci. USA, 84,* 3901 (1987).

64. L. W. Stults, W. A. Sray, and R. J. Maier, *Arch. Microbiol., 146,* 280 (1986).

65. P. Schönheit, J. Moll, and R. K. Thauer, *Arch. Microbiol., 123,* 105 (1979).

66. M. G. Jobling and D. A. Ritchie, *Mol. Gen. Genet., 208,* 228 (1987).

67. A. O. Summers and G. A. Jacoby, *J. Bacteriol., 129,* 276 (1977).

68. D. Taylor and A. O. Summers, *J. Bacteriol., 137,* 1430 (1979).

69. J. L. Casey, B. Dijeso, K. Rao, R. D. Klausner, and J. B. Harford, *Proc. Natl. Acad. Sci. USA, 85,* 1787 (1988).

70. D. O. Owen and L. C. Kühn, *EMBO J. 6,* 1287 (1987).

71. E. C. Theil, *Ann. Rev. Biochem., 56,* 289 (1987).

72. L. F. Dickey, Y. Wang, G. E. Schol, I. A. Wortman III, and E. C. Beal, *J. Biol. Chem., 263,* 3071 (1988).

73. M. W. Hentze, T. A. Rouault, S. W. Caughman, A. Dancis, J. B. Harford, and R. D. Klausner, *Proc. Natl. Acad. Sci. USA, 84,* 6730 (1987).

74. M. W. Hentze, S. W. Caughman, T. A. Rouault, J. G. Barriocanal, A. Dancis, J. B. Harford, and R. D. Klausner, *Science, 238,* 1570 (1987).

75. T. A. Rouault, M. W. Hentze, A. Dancis, W. Caughman, J. B. Harford, and R. D. Klausner, *Proc. Natl. Acad. Sci., USA, 84,* 6335 (1987).

76. M. Aziz and H. N. Monroe, *Proc. Natl. Acad. Sci. USA, 84,* 8478 (1987).

77. L. Guarente and T. L. Mason, *Cell, 32,* 1279 (1983).

78. L. Guarente, B. Lalonde, P. Gifford, and E. Alani, *Cell, 36,* 503 (1984).

79. K. Pfeifer, B. Arcangioli, and L. Guarente, *Cell, 49,* 9 (1987).

80. K. Pfeifer, T. Prezant, and L. Guarente, *Cell, 49,* 9 (1987).

81. V. J. Dwarki, V. N. K. Francis, G. J. Bhat, and G. Padmanaban, *J. Biol. Chem., 262,* 16958 (1987).

82. A. M. Meyers, M. D. Crivellone, T. J. Koerner, and A. Tzagoloff, *J. Biol. Chem., 262,* 16822 (1987).

83. D. H. Hamer, *Ann. Rev. Biochem., 55,* 913 (1986).

84. S. Merchant and L. Bogorad, *EMBO J., 6* (1987).

85. S. Merchant and L. Bogorad, *J. Biol. Chem., 262,* 9062 (1987).

86. S. Merchant and L. Bogorad, *Mol. Cell. Biol., 6,* 462 (1986).

87. B. L. Vallee and K. H. Falchuk, *Phil. Trans. R. Soc. London B, 294,* 185 (1981).

88. J. D. Miller, A. D. McLachlan, and A. Klug, *EMBO J., 4,* 1609 (1985).

89. L. G. Crossley, K. H. Falchuk, and B. L. Vallee, *Biochemistry, 21,* 5359 (1982).

90. M. Czupryn, K. H. Falchuk, and B. L. Vallee, *Biochemistry, 26,* 8263 (1987).

91. J. H. Griffin and P. B. Dervan, *J. Am. Chem. Soc., 109,* 6840 (1987).

92. N. J. Taylor and A. J. Carthy, *J. Am. Chem. Soc., 99,* 6143 (1977).

93. I. G. Dance, S. Choudhury, P. J. Guerney, and A. D. Rae, *Inorg. Chim. Acta, 70,* 227 (1983).

94. G. C. Christou, K. Folting, and J. C. Huffman, *Polyhedron, 3,* 1247 (1984).

4

Yeast Metallothionein: Gene Function and Regulation by Metal Ions

David J. Ecker, Tauseef R. Butt, and Stanley T. Crooke
Department of Molecular Pharmacology
Smith Kline and French Laboratories
P.O. Box 1539
King of Prussia, PA 19406

1. INTRODUCTION: THE YEAST METALLOTHIONEIN SYSTEM AS A MODEL TO STUDY GENE FUNCTION AND METAL ION REGULATION OF GENE EXPRESSION

Metal ions regulate gene expression. Investigators who study this process must be well versed in aspects of both coordination chemistry and molecular biology. Furthermore, they must have a good model system to study metal-regulated genes. In this chapter we cover the recent data on the coordination chemistry and molecular biology of an ideal model system, the copper metallothionein gene from the yeast *Saccharomyces cerevisiae*.

Metallothioneins are a class of low molecular weight, cysteine-rich proteins which bind heavy metals [1,2]. Metallothioneins have been proposed to be involved in various functions, including metal storage and detoxification, development, differentiation, control of cellular metabolism, protection from free radicals, and damage by UV light (reviewed in Ref. 3). Although there may be multiple functions for metallothioneins, it is clear that these proteins can protect cells from the toxicity of certain metal ions [4,5,8]. In many cells the synthesis of metallothioneins is induced at the transcriptional level by the same metals which subsequently bind to the expressed protein. Metal binding by metallothionein results in detoxification (although there are a few caveats to this [8]; see below).

Metallothionein provides a model system for the study of metal-regulated gene expression. The metallothionein system from yeast is especially attractive for several reasons. In the past few years the developments in molecular biology have made genetic manipulation a simple task [6, and references cited therein]. It is possible to delete specific genes from the yeast genome. The effects of the loss of any specific gene can be determined by comparing otherwise identical cells with and without the gene of interest, providing that loss of the gene is not a lethal event. It is also possible to reintroduce an altered form of the gene into the cells that were previously deleted and determine the effects of the modification. Gene cloning in yeast is relatively simple compared with cloning genes from mammalian cells. There are a large number of yeast mutants available, defined and mapped by genetic techniques. The combination of classical genetic techniques and the recent developments of molecular biology have made yeast one of the premiere organisms to study biochemical processes at the molecular level.

An additional advantage of studying metallothionein in yeast is that there is only a single metallothionein gene, while in higher organisms metallothioneins belong to a multigene family [1,2]. Copper resistance in yeast is mediated by induction of this gene. The evidence that this gene mediates copper resistance in yeast is overwhelming. First, copper resistance in yeast has been genetically mapped to the *CUP1* locus, where the yeast metallothionein gene resides [7]. Second, copper-resistant strains of yeast have been shown to have multiple copies of the *CUP1* locus, while copper-sensitive strains have only a single copy. Finally, removal of the copper metallothionein gene from the yeast genome results in dramatically increased copper sensitivity [4,5]. Replacement of the metallothionein gene on a multicopy plasmid, which provides the cells with about 30 copies of the gene, increases the maximum amount of copper that the cells can tolerate from about 10 μM to 2 mM [4,5].

The yeast copper metallothionein gene is, as its name implies, induced only by copper [7]. In contrast, mammalian metallothionein genes are turned on by a number of metals including Cd, Zn, Cu, Hg, Ag, Co, Ni, Bi, as well as glucocorticoids and other hormones [8,2]. A one-metal regulated gene, such as in yeast, provides a more simplified system in which to study gene regulation and the molecular basis of metal-determined specificity. The yeast system provides an evolutionary foundation, on which more complex forms of regulation were likely added.

Finally, the yeast metallothionein gene, when induced by copper, is one of the most efficient transcriptional systems known in yeast [9,10]. This property has been extensively utilized for applications in biotechnology, such as the expression of heterologous proteins in yeast [9-11]. In one case, when the yeast metallothionein promoter sequences were introduced before a heterologous gene, expression of the gene could be induced to greater than 18% of the total cell protein [9]. Although this represents a high-water mark for expression under the metallothionein promoter in yeast, it is sometimes possible to visualize protein products expressed under this promoter on a simple Coomassie-stained protein gel [11].

Several reviews have been published dealing with general aspects of metallothionein [1,12-14], and recent reviews are available that extensively cover mammalian metallothioneins [2] and aspects of yeast metallothionein relevant to biotechnology [10]. A minireview is also available relating to metallothionein function [3]. For information regarding general aspects of yeast genetics, the reader is referred to Refs. 6, 15, and 16. The purpose of this chapter is to provide some perspective and cover the recent data relating to yeast metallothionein, its structure, function, and regulated expression. It is written for bioinorganic chemists, those trained primarily in chemistry but who approach molecular biology seeking new and interesting problems to which they can apply their skills.

2. THE BIOCHEMISTRY AND COORDINATION CHEMISTRY OF COPPER TOXICITY

2.1. Toxic Reactions of Copper

Copper is a metal that is required for life. It is an essential cofactor for numerous enzymes, primarily redox enzymes involved in oxidation reactions and electron transfer [17,18]. However, the same properties that make copper so useful in essential biochemical reactions will also cause toxic effects if the copper concentration in the cell becomes too high or if the metal is not metabolized properly. How does the cell manage to restrain the harmful reactivities of copper while maintaining its availability for essential enzymes? The answer may be in the coordinate saturation of the metal by metallothionein.

Copper has the potential for multiple mechanisms of cytoxicity. Although general mechanisms of copper reactivity with biological molecules have been studied in vitro and in vivo, the cellular target responsible for cell death from copper poisoning is unknown. One mechanism by which copper can cause toxic effects is through coordination reactions. By forming complexes with biological ligands such as proteins, DNA, membranes, or small cellular molecules, the normal functions of these molecules is inhibited. In the case of metal ions whose chemistries are not dominated by redox reactions, biological chelation reactions exclusively determine the toxicity. For example, the chelation reactions responsible for cadmium toxicity have been extensively studied (reviewed in Ref. 19). Several major factors are involved in determining what biological ligands are involved in the chelation mechanism of metal toxicity. The magnitude of the free-energy changes for complex formation is determined by the stereochemical orientation of the ligand, the ligand hardness/softness, the chelate effect, and rate effects. In this regard, both yeast and mammalian metallothioneins are well designed to bind cadmium [23-25]. It is clear that in mammalian cells (but not yeast; see Sec. 2.3) cadmium is normally detoxified by being coordinately saturated in metallothionein. Similar coordinate detoxification by metallothionein of zinc, copper, mercury, and bismuth have been observed in mammalian cells [8].

It is important to note that coordination reactions for nonredox active metals can have a therapeutic effect. Two important metallo-drugs are cis-diamminedichloroplatinum(II), used in cancer treatment [20], and auranofin, a gold compound used in the treatment of rheumatoid arthritis [21]. The mechanisms of action of both of those drugs involve coordination reactions with biological ligands. The metabolism of the gold compound involves coordination to metallothioneins [22].

In the case of copper, much is known about its solution chemistry and coordination reactions in enzymes and copper proteins [17, 18], but little is known about toxic reactions with biomolecules in the cell. These toxic coordination reactions may be far overshadowed by another property of copper, its redox chemistry. It has been shown that Cu(II) can oxidize cysteine residues in vitro [26]. If similar reactions occurred in the cell, this could inactivate essential enzymes.

A highly damaging set of reactions is catalyzed by redox active metals such as iron and copper, involving byproducts of aerobic metabolism. It has been demonstrated in vitro [27] that copper can catalyze a modified Haber-Weiss reaction, in which the oxidized form of copper bound to a cellular protein reacts with superoxide anion, a byproduct of aerobic metabolism. The superoxide reduces copper in the protein-copper(II) complex.

$$\text{protein-Cu(II)} + O_2^- \longrightarrow \text{protein-Cu(I)} + O_2 \qquad (1)$$

Then H_2O_2, also a product of aerobic metabolism, reacts with the reduced copper-protein complex to yield $OH^{\bullet}$ radicals.

$$\text{protein-Cu(I)} + H_2O_2 \longrightarrow \text{protein-Cu(II)} + OH^{\bullet} + OH^- \qquad (2)$$

The highly reactive $OH^{\bullet}$ radicals will attack any local organic molecule. The location of the protein-copper species would determine what is attacked by the $OH^{\bullet}$, which would destroy any vital target in the vicinity. Note that the protein-copper complex acts as a catalyst and may turn over a large number of oxygen metabolites. In in vitro models, this reaction could be demonstrated to destroy enzymes [27]. However, the reaction could be completely halted by the addition of EDTA to the system. These in vitro models provide a theoretical framework and the basis for studies on in vivo toxicity.

This type of redox chemistry has been involved in the therapeutic activities of a class of compounds known as the bleomycins, which are anticancer drugs [28,29]. The drugs bind copper or iron ions and generate reactive hydroxyl radicals near the tumor cell DNA, thereby killing the cells. It is possible that metallothionein can modulate the reactivities of bleomycin by controlling the availability of intracellular copper.

2.2. Protection by Metallothionein

What is the role of yeast copper metallothionein in preventing the toxic reactivities of copper? Two mechanisms can be envisaged. First, the multidentate ligands of metallothionein may coordinately saturate the metal, preventing the metal from binding to other biological ligands or reacting with the byproducts of aerobic metabolism. Second, copper bound to metallothionein may be stabilized in the 1+ oxidation state and therefore may be less catalytic to hydrogen peroxide in the Haber-Weiss reaction. It has been suggested that there is a specific enzymatic mechanism to oxidize the copper in yeast copper metallothionein in order to make Cu(II) available in the cell [39]. There may be additional advantages of sequestering copper in metallothionein that are yet undiscovered, such as providing signals for cellular compartmentalization. With copper in metallothionein, the cell has a nontoxic form of the metal that is still available for metalloenzymes. It has been shown, however, that when yeast cells deleted for the copper metallothionein gene are grown on low-copper medium, there is no apparent effect on cell growth, copper accumulation, or activity in the copper-zinc form of superoxide dismutase [4]. This suggests that copper metallothionein is not necessary for intracellular copper metabolism or the donation of copper to metalloenzymes. Its known role is thus far restricted to protection from the toxicity of excess copper.

The yeast copper metallothionein is a 53 amino protein with an N-terminal glutamine. Biochemical studies have demonstrated that the protein lacks the first eight amino acids predicted from the nucleotide sequence of the gene. This leader sequence, which is unusual for a metallothionein in that it is hydrophobic and contains two aromatic amino acids, is removed by the action of protease [24]. No function has yet been attributed to the leader sequence, and the unprocessed protein apparently has identical biochemical properties as the wild-type processed protein [40]. The mature protein contains 12 cysteine residues and has eight binding sites for Cu(I). All eight copper ions are required to confer maximum stability from proteases. Yeast metallothionein also binds Cd(II), Zn(II), and Ag(I) in vitro [24]. The binding stoichiometry is the same for Cu(I) and Ag(I), while only four molecules of either Cd(II) and Zn(II) are bound.

No crystal structure is yet available for yeast copper metallothionein. Moreover, the mammalian metallothionein crystal structure has been solved for only the zinc/cadmium form of the protein [23]. The coordination geometry of yeast copper metallothionein has been studied by solution coordination and spectroscopic studies. Analysis by EXAFS indicate that each of the eight coppers is trigonally coordinated to sulfur. Some Cu-Cu interactions were also noted. Each of the cysteine thiolates is involved in bridging the eight Cu(I) ions [24, and personal communications].

The relative importance of each of the cysteinyl residues has been probed by Winge and coworkers (personal communications) by biochemical analysis of mutant proteins prepared by site-directed mutagenesis [41]. In the mutant proteins, pairs of cysteines were converted to serines. Four different proteins were prepared, each containing 10 cysteines instead of the usual 12. All of the mutant proteins were at least partially effective in protecting yeast from copper toxicity when expressed in vivo. One of the mutant proteins, cys49→ser, cys50→ser, was very similar to wild-type, with respect to both copper detoxification in vivo and in vitro metal-binding properties. These results are consistent with a study in which the conserved carboxy terminal sequence of yeast copper metallothionein (lys-lys-ser-cys-cys-ser), which is also found in mammalian metallothionein, was deleted from the yeast gene [5]. The truncated gene was nearly as effective as the wild-type gene in protecting yeast in vivo from copper toxicity. It is unclear as to why the loss of the carboxy terminal pair of cysteines does not more severely affect protein function.

2.3. Resistance to Other Metals

It was a paradox as to why the metallothionein from yeast is so specific in protecting cells from copper but not from any other metal. It has been shown that yeast copper metallothionein, when stripped of copper, binds to a number of metals in vitro with high affinity and little specificity [24]. In mammalian cells, metallothioneins are induced and provide protection from a variety of metal ions (see above). However, when the yeast copper metallothionein gene was replaced in vivo with the mammalian metallothionein gene under regulatory control of the yeast metallothionein promoter, the yeast cells were protected from copper [42] but not from any other metal [5]. Because mammalian metallothionein protects cells from a variety of metals, it suggested that the problem is not in the protein but in the regulation of expression of the gene in yeast.

When the yeast copper metallothionein gene is removed from its natural regulatory sequences and expressed under regulation of a constitutive promoter, the yeast cells are resistant to 1,000-fold more cadmium than cells where the gene is under normal regulatory control [5]. This proved that the yeast protein is structurally capable of both binding and protecting cells from the toxic effects of copper and cadmium, but because cadmium does not turn on the gene in yeast, the protein affords no cadmium protection.

Surprisingly, the constitutively expressed metallothionein does not provide cells with any significant protection from the toxicity of a number of other metals tested, including silver and zinc, which have been shown to bind to the yeast protein in vitro [24]. One possible explanation for this observation is that metals that bind the yeast protein in vitro, but are not detoxified in vivo, cause damage at a location such as the yeast cell wall, where metallothionein is not available.

3. *CUP1*: THE COPPER RESISTANCE LOCUS IN YEAST—STRUCTURE AND AMPLIFICATION

The generation of copper resistance in yeast was first described in 1951 [30], when laboratory strains of yeast were grown on plates containing high amounts of copper. After prolonged exposure to copper, some dense brown colonies grew up that were copper-resistant. These investigators inferred that the development of resistance is a two-step process, involving an adaptation to copper followed by rapid proliferation of selected colonies. This interpretation is consistent with our current understanding of the amplification of the yeast copper metallothionein gene as a mechanism of resistance [31-33]. It was shown that this trait was mediated by a single gene when genetic crosses were made between copper-resistant and sensitive strains [34]. The genetic locus where the copper resistance gene mapped was designated *CUP1*. Copper-resistant cells are thus designated *CUP1*^{r} and copper-sensitive cells *CUP1*^{s}.

The *CUP1* locus is located on chromosome VIII, 42 centimorgans distal to the centromere [31]. Copper-sensitive cells, which are killed by concentrations of copper in the medium greater than 10 μM, were found to contain a single copy of the *CUP1* gene [31,32]. In contrast, copper-resistant cells contain 10 or more copies of the gene, amplified on the chromosome at the *CUP1* locus, and are resistant to copper concentrations greater than 2 mM.

The gene for copper resistance was cloned by transforming a copper-sensitive strain with a yeast library and isolating colonies that showed increased copper resistance. The complete nucleotide sequence of the *CUP1* locus has been determined (Fig. 1) [35,36]. The basic repeat unit contains two genes. One encodes for a 246 amino acid protein (named gene X) of unknown function. The other gene encodes a 61 amino acid, cysteine-rich protein. From an analysis of the sequence of this protein, it appears to belong to the class of low molecular weight, cysteine-rich proteins designated metallothioneins in higher eukaryotes. The protein from *Saccharomyces cerevisiae* has been given various names, such as copper

```
BAMHI
!         -450        -440        -430        -420        -410        -400
GGATCCCATTACCGACATTTGGGCGCTATACGTGCATATGTTCATGTATGTATCTGTATT

          -390        -380        -370        -360        -350        -340
TAAAACACTTTTGTATTATTTTTCCTCATATATGTGTATAGGTTTATACGGATGATTTAA

                                                        Spe I
          -330        -320        -310        -300        -290        -280
TTATTACTTCACCACCCTTTATTTCAGGCTGATATCTTAGCCTTGTTACTAGTTAGAAAA

                                  HinfI   I             XbaI
          -270        -260        -250        -240        -230        -220
AGACATTTTTGCTGTCAGTCACTGTCAAGAGATTCTTTTGCTGGCATTTCTTCTAGAAGC

 ------MRE1----->                 ------MRE2---->        <-----

          -210        -200        -190        -180        -170        -160
AAAAAGAGCGATGCGTCTTTTCCGCTGAACCGTTCCAGCAAAAAAGACTACCAACGCAAT

-MRE3----   ------MRE4------->  <----MRE5------

            Hinf I
          -150        -140        -130        -120        -110        -100
ATGGATTGTCAGAATCATATAAAAGAGAAGCAAATAACTCCTTGTCTTGTATCAATTGCA

                                <----MRE5------

          -90         -80         -70         -60         -50         -40
TTATAATATCTTCTTGTTAGTGCAATATCATATAGAAGTCATCGAAATAGATATTAAGAA

          -30         -20         -10          1          10          20
AAACAAACTGTACAATCAATCAATCAATCATCACATAAAATGTTCAGCGAATTAATTAAC

                                             MetPheSerGluLeuIleAsn

          30          40          50          60          70          80
TTCCAAAATGAAGGTCATGAGTGCCAATGCCAATGTGGTAGCTGCAAAAATAATGAACAA

PheGlnAsnGluGlyHisGluCysGlnCysGlnCysGlySerCysLysAsnAsnGluGln

          90          100         110         120         130         140
TGCCAAAAATCATGTAGCTGCCCAACGGGGTGTAACAGCGACGACAAATGCCCCTGCGGT

CysGlnLysSerCysSerCysProThrGlyCysAsnSerAspAspLysCysProCysGly

          150         160         170         180         190         200
AACAAGTCTGAAGAAACCAAGAAGTCATGCTGCTCTGGGAAATGAAACGAATAGTCTTTA

AsnLysSerGluGluThrLysLysSerCysCysSerGlyLysEnd
```

FIG. 1. The DNA sequence of the yeast copper-metallothionein gene and upstream sequences. The bases are numbered with the translational start site as +1. The underlined amino acids are deleted from the mature protein by enzymatic cleavage. The cysteines are highlighted. The amino acids overlined are conserved in metallothioneins from yeast to mammals. The polarity of the 13 base pair repeated sequences (MREs) is arbitrarily defined.

thionein [37], copper chelatin [38], and copper metallothionein. Because it is inducible only by copper, we prefer the term copper metallothionein. Although the *CUP1* locus contains two genes, which are coamplified in copper-resistant cells, only the copper metallothionein gene is turned on by the addition of copper.

Amplification of genes is a well-known mechanism to increase the production of a particular gene product. Fogel et al. [31] found that *CUP1*^{r} strains have from 2 to 14 copies of the *CUP1* locus. Continuous growth of laboratory strains on copper-containing media enhance the copper tolerance of yeast. Some strains selected for copper resistance were found to be disomic for chromosome VIII, indicating that copper resistance can be increased by increasing the *CUP1* copy number by doubling the entire chromosome [32].

Fogel et al. [32] extensively studied the mechanism of gene amplification using the *CUP1* locus as a model. The results indicate that amplification occurs by a nonreciprocal recombination of event called gene conversion. In Fogel's model, a single strand from one of the *CUP1* repeats is misaligned during synaptic pairing with the other chromatid repeat such that the DNA for one repeat is looped out. Subsequent repair replication of the loop leads to an increase of the *CUP1* element, depending on the number of repeats present in the loop. This model is consistent with the observation that strains with a single copy of the *CUP1* locus will not amplify due to the inability of a single gene repeat to form a loop with the other chromatid (Fogel, personal communication).

4. REGULATION OF TRANSCRIPTION OF THE *CUP1* LOCUS

4.1. Yeast Metal Regulatory Elements (MREs)

Regulation of gene expression in yeast and higher eukaryotes depends on RNA polymerase and a group of protein transcription factors which bind DNA near the regulated gene (reviewed in Ref. 43). The transcription factors, also called trans-acting factors, bind to specific DNA sequences known as cis elements upstream of the gene in the promoter region. Specific interactions of the cis and trans elements in the promoter determine the level of gene expression. Eukaryotic promoters are usually complex, having more than one cis element and multiple trans-acting factors [43]. The interactions between cis elements and trans-acting factors may have either positive or negative effects on gene transcription. Cis-element DNA sequences in yeast are known by a variety of names, including upstream activating sequences (UAS), or upstream regulatory sequences (URS).

In the mammalian metallothionein promoter, Stuart et al. [44, 45] identified cis elements near the transcriptional start site of the mouse metallothionein-I gene and named these sequences metal regulatory elements (MREs). The mammalian MREs are short (15 base pair) DNA sequences that are repeated five or more times in the promoter region. They are found as either direct or inverted repeats.

Evidence that these DNA sequences are actually responsible for binding trans-acting factors and thus increasing transcription was provided by two types of experiments. First, DNA regions containing these elements were placed before a "reporter gene" and transfected back into cells [44,45]. This conferred metal-inducible regulation on the reporter gene. Specific mutational analysis of the cis-DNA identified regions responsible for regulation. Second, synthetic 12 base pair MRE sequences placed before a heterologous gene also conferred metal-inducible regulation [45,46]. It was found that the response to metal was relatively independent of the position or orientation of the synthetic MRE, but required at least two copies.

Computer analysis of the DNA sequences upstream of the yeast *CUP1* gene show that there is no sequence homologous to the mammalian MREs. However, there is a set of six repeated sequences that are similar to the mammalian MREs with respect to number, position, orientation, and degree of internal homology (Figs. 1 and 2). These putative 13 base pair MREs are found in both orientations upstream of the yeast copper-metallothionein gene.

There are several reports of analyses of regulation of expression of the yeast copper-metallothionein promoter DNA sequences using reporter genes [4,47-49]. Hamer et al. [4] and Gorman et al. [49]

MRE	SEQUENCE
1	GACATTTTTGCTG
2	GATTCTTTTGCTG
3	GCTCTTTTTGCTT
4*	GCGTCTTTTGCTG
5	GTCTTTTTTGCTG
6	GAGTTATTTGCTT
CONSENSUS:	GANTTTTTTGCTG

*A "CC" PAIR HAD TO BE DELETED FROM MRE4 IN ORDER TO ALIGN

FIG. 2. The repeated 13 base pair putative yeast MREs.

reported experiments in which the sequences found upstream of the copper-metallothionein gene (Fig. 2) were fused to the *E. coli* galactokinase (galK) gene. The *CUP1*/galK fusions were introduced into yeast strains with amplified or deleted *CUP1* loci. In the *CUP1* amplified strains the basal level of galK expression from the *CUP1* promoter was low. Upon addition of copper, galK activity increased a maximum of 10-fold [4] or 15-fold [49]. In contrast, in the strains deleted for the chromosomal *CUP1* locus, therefore lacking copper-metallothionein, expression of galK from the *CUP1* promoter was high in the absence of added copper and could be increased even higher by the addition of copper [4,49]. Moreover, introduction of a second plasmid containing the yeast copper-metallothionein gene under the regulatory control of a constitutive promoter restored the regulation of the *CUP1* promoter in the *CUP1* deleted strain to the levels found in the multicopy *CUP1* strain [49]. This shows that metallothionein in trans can repress the *CUP1* promoter. Thus, yeast copper-metallothionein appears to be a self-regulating gene; the gene is turned on by copper, which is subsequently bound by the expressed protein. This removes the copper and turns off expression of the gene.

A more detailed examination of the *CUP1* promoter has been performed by deletion analysis using the *CUP1*/galK fusions [48] or by making similar fusions using the yeast phosphoglycerate kinase gene as the reporter [47]. In the *CUP1*/galK study, deletions in the region corresponding to MREs 2 and 3, and MREs 4 and 5 were identified as critical to copper induction. A synthetic copy of the DNA corresponding to MREs 4 and 5, which was called UASp, was inserted before a reporter gene in one or two copies and in both orientations. Copper regulation was observed, but two tandem copies were required in order to achieve significant induction. The orientation of the repeats significantly influenced the level of induction [48]. It was also reported that certain mutations downstream of MRE 5 resulted in increased basal transcription, suggesting that there may also be negative control sites present in the promoter region.

In contrast, in the studies with the *CUP1*/phosphoglycerate kinase reporter gene, no constitutive expression was observed in the *CUP1* deleted strain [47]. When the promoter sequences were truncated from the 5' end to -300 from the translational start site (Fig. 1), a high constitutive level of expression was observed, suggesting that a repressor-binding site may exist upstream of the MREs. It is unclear why no constitutive expression of the *CUP1* promoter was observed in the *CUP1* deleted strain as reported in the other studies [4,49].

4.2. Trans-Acting Factors

Although some of the cis elements important for copper regulation of the *CUP1* promoter are known, there is little information available about the trans-acting factors that bind to these elements. Timblin and Bergman have taken a biochemical approach to isolation of these factors (personal communications). Yeast nuclear proteins that bind to a promoter fragment containing MREs 2-5 (Fig. 2) have been purified 11,000-fold by heparin agarose and *CUP1*-specific DNA-cellulose chromatography. Electrophoretic analysis showed that there were two discrete protein-DNA complexes. These proteins were then used in DNA protection (footprint) assays and were found to protect MRE2, MRE3, and part of MRE4. It is clear that these proteins are not copper-metallothionein itself, since the same proteins are also present in yeast strains in which the *CUP1* gene was deleted. Further characterization of these proteins is in progress and should provide important insight into the mechanism of copper activation of the *CUP1* promoter.

A genetic approach is also being used to identify and clone the gene(s) encoding for the *CUP1* trans-acting factors. By isolation of yeast mutants that are hypersensitive to copper but do not map to the *CUP1* locus, it may be possible to identify transacting factor mutants. The genes that encode these trans-acting factors could then be directly cloned by complementation. A potential difficulty may arise with this approach if there are several trans-acting factors that each contribute partially to gene activation. Mutations that give only partial responses are difficult to identify or clone by complementation. In addition, there are also likely to be other genes involved in the response to copper that are not associated with *CUP1* or its regulation. Mutations in membrane receptor proteins for copper, or enzymes of oxygen metabolism such as superoxide dismutase, might be isolated by this procedure. Nevertheless, Fogel and coworkers isolated several mutations which confer copper sensitivity but do not map to the *CUP1* locus (personal communications). One of these mutants, designated *CUP2*, is copper-sensitive despite having multiple intact copies of the *CUP1* locus. Clones that complement the mutation have been isolated and are currently being sequenced. By structural analysis of the gene product it will be determined if these genes really encode trans-acting factors.

4.3. Models for Gene Regulation

The questions remain as to what molecular mechanisms are involved in regulation of the gene. How do the trans-acting factors bind to the cis elements and how does this regulate transcription? Where does copper interact with this system? Several models have been proposed and discussed in the literature to explain regulation of the gene [4,10,14]. In one attractive model, it was proposed that the copper-metallothionein gene product, in its metal-free form, binds directly to the cis elements and represses expression of the gene [14]. Upon addition of copper, the protein binds the copper and is released from the DNA, which causes derepression of the gene.

Although this model is simple and elegant, it has not been borne out by experimental data. First, it was determined that the mammalian metallothionein gene, expressed in yeast under the yeast copper-metallothionein promoter sequences, was properly regulated [42]. The mammalian and yeast proteins have very divergent amino acid sequences, except for a conserved region (lys-lys-ser-cys-cys-ser). Because the MRE sequences are also completely divergent between mammals and yeast, yet the mammalian protein still regulates the yeast gene, it is unlikely that regulation occurs by direct binding of the protein to the MRE. Second, attempts to show direct binding of metallothionein, with or without copper, to the promoter regions have been unsuccessful [42, and Bergman personal communications]. Finally, the direct autoregulatory model would predict that other metals that bind to the protein would also induce the gene. It has been shown that the yeast protein binds cadmium in vitro [24] and that the constitutively expressed protein protects cells from the toxic effects of cadmium in vivo [5]. However, cadmium does not induce the gene.

Another model has been proposed for autoregulation based on the binding of copper to a trans-acting factor, which then binds DNA and stimulates transcription [4,10]. The expressed *CUP1* gene product simply competes with the trans-acting factors for the available intracellular copper [Fig. 3(a)]. Additional features of the model can explain the observation that cadmium actually decreases transcription of the *CUP1* gene [7, and unpublished results from our laboratory]. Because of different preferred coordination geometries between Cd(II) and Cu(I) (the oxidation state of copper most likely inside the cell), cadmium bound to a trans-acting factor designed for copper would be expected to have a different tertiary structure. The Cd(II)-protein complex might bind to DNA and not be able to interact correctly with other transcription factors, resulting in antagonist activity [Fig. 3(b)]. In either case, a model based on copper binding to a trans-

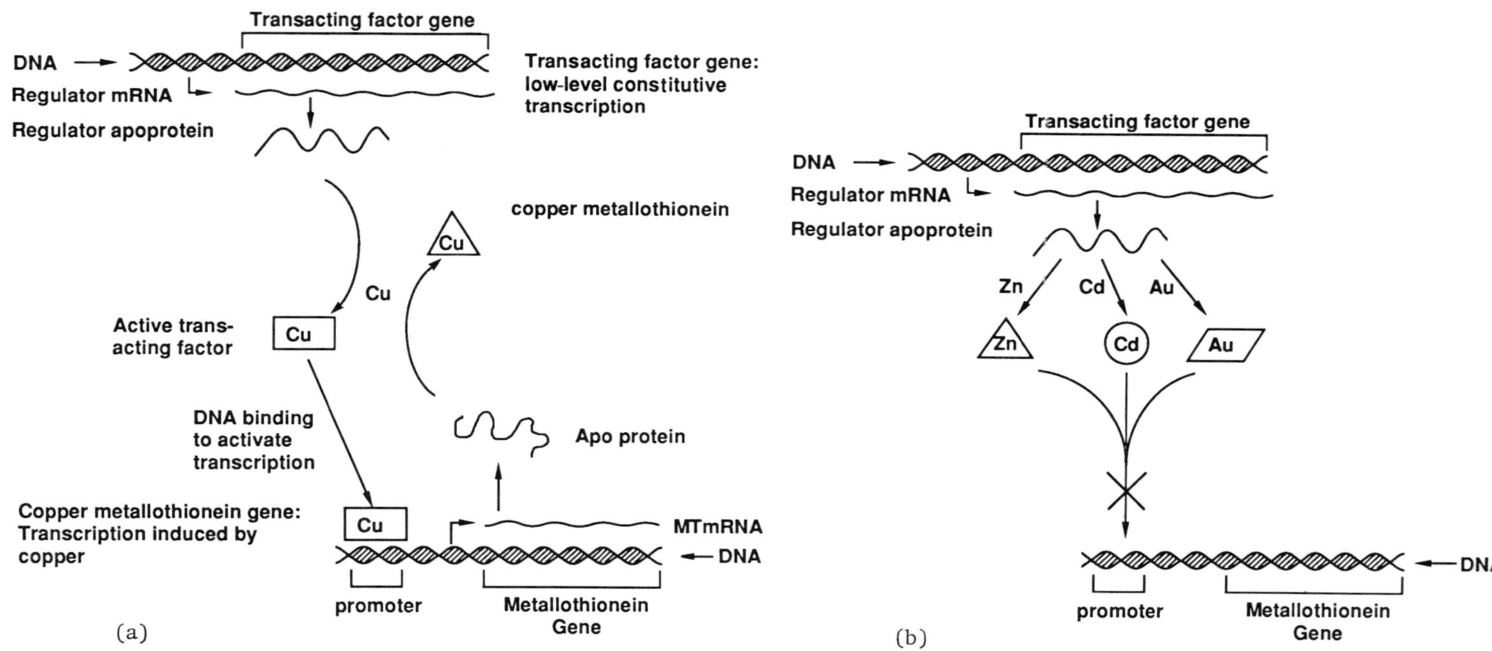

FIG. 3. Model for regulation of the *CUP1* promoter by a copper-binding trans-acting factor. (a) The trans-acting factor, which is the product of a different gene, binds copper and activates transcription of the *CUP1* gene. The expressed protein then competes with the trans-acting for copper and turns off the gene. Details of the model are provided in the text. (b) Competition for the trans-acting factor by other metals. Only copper has the proper charge, size, and preferred coordination geometry to form a complex with the transcription factor with the correct structure to activate the gene. Other metals may bind and compete with copper but do not induce the gene.

acting factor which then binds to DNA and stimulates transcription is attractive and consistent with all of the present data.

It is interesting to speculate on the nature of the structure of this copper-binding transcription factor. The recently discovered zinc finger proteins that bind DNA and specifically control the expression of certain genes (reviewed in this volume) represent a totally new type of structural motif for metal-binding transcription factors. Whether the copper-regulated genes are controlled by similar, "copper finger proteins" or by proteins of a completely different structural motif remains to be determined.

It is also unknown how many copper-regulated genes there are in yeast and higher eukaryotes. It was recently reported that there is a copper-repressible cytochrome c in *Chlamydomonas reinhardtii* [50]. The gene is very tightly repressed by the presence of small amounts of copper in the medium, but induced by copper deprivation. This is exactly the opposite of the *CUP1* gene, yet similar type regulatory proteins may be involved in the control of both genes. In the case of the copper-repressible cytochrome c, the copper-bound trans-acting factor may simply be a negative regulator instead of a positive one.

It is also possible that there are a number of genes in yeast and higher eukaryotes that are not related to copper metabolism but are controlled by copper-binding transcription factors. Given the efficiency of regulation of the *CUP1* gene and nature's propensity to reuse good structural motifs, we consider this a likely possibility.

5. SUMMARY

Copper is an essential metal ion that is toxic in excess concentrations or if improperly metabolized in the cell. The reactions that mediate the toxicity of copper in vivo and the intracellular targets for cytotoxicity are unknown. However, in vitro models suggest that the catalysis of redox reactions by copper may be the dominant mechanism of cytotoxicity. Metallothioneins bind to copper in vivo and inhibit its toxic reactions, possibly by coordinately saturating the metal and stabilizing the Cu(I) oxidation state. Yeast copper metallothionein is functionally similar to mammalian metallothioneins in that it protects yeast from the toxic effects of excess copper. Transcription of the yeast copper metallothionein gene is induced by copper but not by other metals. Regulation of the yeast copper metallothionein gene presents an ideal model system to study the mechanism of metal ion regulation of gene expression. DNA sequences in the promoter region of the yeast copper metallothionein gene have been identified that are responsible for binding the trans-acting factors that regulate expression of the gene. The mechanism of copper regulation of gene expression is currently unknown, but several research groups are close to isolating and cloning the genes that encode the trans-acting factors. Our understanding of the mechanism of the metallothionein gene regulation will be greatly enhanced by the availability of large quantities of copper-specific, trans-acting protein factor.

ACKNOWLEDGMENTS

The authors thank Bill Bergman, Dennis Winge, Tina Etcheverry, and Seymour Fogel for providing unpublished data; Dennis Winge, Jessica Gorman, and Brett Monia for critically reading the manuscript; and Lynne Bonde for editorial and typing assistance.

REFERENCES

1. J. H. R. Kägi and N. Nordberg (eds.), "Metallothionein". Birkhäuser, Basel, 1979.
2. D. H. Hamer, *Ann. Rev. Biochem., 55,* 913-951 (1986).
3. M. Karin, *Cell, 41,* 9-10 (1985).
4. D. H. Hamer, D. J. Thiele, and J. E. Lemontt, *Science, 228,* 685-690 (1986).

5. D. J. Ecker, T. R. Butt, E. J. Sternberg, M. P. Neeper, C. Debouck, J. A. Gorman, and S. T. Crooke, *J. Biol. Chem., 261,* 16895-16900 (1986).

6. K. Struhl, *Nature, 305,* 394-397 (1983).

7. M. Karin, R. Najarain, A. Haslinger, P. Valenzuela, J. Welch, and S. Fogel, *Proc. Natl. Acad. Sci. USA, 81,* 337-341 (1984).

8. D. M. Durham and R. D. Palmiter, *Mol. Cell. Biol., 4,* 484-491 (1984).

9. T. Etcheverry, W. Forrester, and R. Hitzeman, *Biotechnology, 4,* 726-730 (1986).

10. T. R. Butt and D. J. Ecker, *Microbiol. Rev.,* pp. 351-364 (Sept. 1987).

11. D. J. Ecker, M. I. Khan, J. Marsh, T. R. Butt, and S. T. Crooke, *J. Biol. Chem., 262,* 3524-3527 (1987).

12. E. C. Foulkes (ed.), "Biological Roles of Metallothionein", Elsevier, New York, 1982.

13. P. E. Hunziker and J. H. R. Kägi, in "Metallothionein, Part 2, Metallothionein" (P. M. Harrison, ed.), Verlag Chemie, Weinheim, 1985, pp. 141-181.

14. M. Karin and R. I. Richards, *Environ. Health Persp., 54,* 111-115 (1984).

15. J. N. Strathern, E. W. Jones, J. R. Broach (eds.), "The Molecular Biology of the Yeast Saccharmoyces Cerevisiae", Cold Spring Harbor Laboratory, Cold Spring Harbor, New York, 1982.

16. T. D. Petes, *Ann. Rev. Biochem., 49,* 845-876 (1980).

17. H. Sigel (ed.), "Metal Ions in Biological Systems", Vol. 12, Marcel Dekker, New York, 1981.

18. H. Sigel (ed.), "Metal Ions in Biological Systems", Vol. 13, Marcel Dekker, New York, 1981.

19. O. Anderson, *Environ. Health Persp., 54,* 249-266 (1984).

20. A. Pinto and S. J. Lippard, *Proc. Natl. Acad. Sci. USA, 82,* 4616-4619 (1985).

21. S. T. Crooke, R. M. Snyder, T. R. Butt, D. J. Ecker, H. S. Allaudeen, B. Monia, and C. K. Mirabelli, *Biochem. Pharmacol., 35,* 3423-3431 (1986).

22. B. P. Monia, T. R. Butt, D. J. Ecker, C. K. Mirabelli, and S. T. Crooke, *J. Biol. Chem., 261,* 10957-10959 (1986).

23. W. F. Furey, A. H. Robbins, L. L. Clancy, D. R. Winge, B. C. Wang, and C. D. Stout, *Science, 231,* 704-710 (1986).

24. D. R. Winge, R. B. Nielson, W. R. Gray, and D. H. Hamer, *J. Biol. Chem., 260,* 14464-14470 (1986).

25. R. B. Nielson, C. L. Atkin, and D. R. Winge, *J. Biol. Chem., 260,* 5342-5350 (1985).

26. D. C. H. McBrien, *Ciba Found. Symp., 79,* 301-318 (1980).

27. A. Samuni, M. Chevion, and G. Czapski, *J. Biol. Chem., 256,* 12632-12635 (1981).

28. E. A. Sauseville, J. Peisach, and S. B. Horowitz, *Biochemistry, 17,* 2740-2746 (1978).

29. S. M. Hecht (ed.), "Bleomycin: Chemical, Biochemical and Biological Aspects", Springer Verlag, New York, 1979.

30. T. Minagawa, N. Yanagishima, Y. Arakatsu, S. Nagasaki, and J. Ashida, *Botan. Mag., 64,* 65-73 (1951).

31. S. Fogel and J. W. Welch, *Proc. Natl. Acad. Sci. USA, 79,* 5342-5346 (1982).

32. S. Fogel, J. W. Welch, and M. Karin, *Curr. Genet., 7,* 1-9 (1983).

33. S. Fogel, J. W. Welch, and E. J. Louis, *Cold Spring Harbor Symp. Quant. Biol., 49,* 55-65 (1984).

34. A. Brenes-Pomales, G. Lindegren, and C. C. Lindegren, *Nature, 136,* 841-842 (1955).

35. M. Karin, R. Najarain, A. Haslinger, P. Valenzuela, J. Welch, and S. Fogel, *Proc. Natl. Acad. Sci. USA, 81,* 337-341 (1984).

36. T. R. Butt, E. J. Sternberg, J. A. Gorman, P. Clark, D. Hamer, M. Rosenberg, and S. T. Crooke, *Proc. Natl. Acad. Sci. USA, 81,* 3332-3336 (1984).

37. R. Prinz and U. Weser, *J. Physiol. Chem., 356,* 767-776 (1975).

38. R. Premakumar, D. R. Winge, R. D. Wiley, and K. V. Rajagopalan, *Arch. Biochem. Biophys., 170,* 278-288 (1975).

39. H. Hartmann, A. Gartner, and U. Weser, *Hoppe-Seylers Z. Physiol. Chem., 365,* 1355-1359 (1984).

40. C. F. Wright, K. McKenney, D. H. Hamer, J. Byrd, and D. R. Winge, *J. Biol. Chem., 262,* 12912-12919 (1987).

41. C. F. Wright, D. H. Hamer, and K. McKenney, *Nucl. Acid Res., 14,* 8489-8499 (1986).

42. D. J. Thiele, M. J. Walling, and D. H. Hamer, *Science, 231,* 854-856 (1986).

43. K. Struhl, *Cell, 49,* 295-297 (1987).

44. G. W. Stuart, P. F. Searle, and R. D. Palmiter, *Nature, 317,* 828-831 (1985).

45. G. W. Stuart, P. F. Searle, H. Y. Chen, R. L. Brinster, and R. D. Palmiter, *Proc. Natl. Acad. Sci. USA, 81,* 7318-7322 (1984).

46. P. F. Searle, G. W. Stuart, and R. D. Palmiter, *Mol. Cell. Biol.*, *5*, 1480-1489 (1985).

47. T. Etcheverry, W. Forrester, and R. Hitzeman, *Biotechnology*, *4*, 726-730 (1986).

48. D. J. Thiele and D. H. Hamer, *Mol. Cell. Biol.*, *6*, 1158-1163 (1985).

49. J. A. Gorman, P. E. Clark, M. C. Lee, C. Debouck, and M. Rosenberg, *Gene*, *48*, 13-22 (1986).

50. S. Merchant and L. Bogard, *J. Biol. Chem.*, *262*, 9062-9067 (1987).

5

Zinc-Binding Proteins Involved in Nucleic Acid Replication

Joseph E. Coleman and David P. Giedroc
Department of Molecular Biophysics and Biochemistry
Yale University School of Medicine
Sterling Hall, Box 3333
333 Cedar Street, Box 3333
New Haven, CT 06510-8024

1. INTRODUCTION

Zinc is the second most common of the first transition and IIB metal ions in biological systems, iron being the most abundant. The role of iron at the molecular level was recognized early primarily because of the intense color of its most common cofactor form, i.e., Fe-porphyrins. The widespread participation of Zn(II) in the catalytic mechanism of a variety of important enzymes was not appreciated until the 1950s and 1960s when carboxypeptidases A and B, thermolysin, alcohol dehydrogenase, yeast aldolase, and enolase joined the classic example, carbonic anhydrase [1], as zinc metalloenzymes [2,3]. Detailed studies of the mechanisms of action of these enzymes showed Zn(II) either to coordinate a group on the incoming substrate or to activate a coordinated water molecule in certain hydrolysis and hydration reactions [3].

In 1965, *E. coli* aspartate transcarbamoylase, the classical allosteric enzyme, was added to the list of zinc metalloenzymes, since it contained 6 Zn(II) ions per C_6R_6 holoenzyme, where C and R designate catalytic and regulatory subunits, respectively [4]. In contrast to the earlier described zinc enzymes, the Zn(II) does not participate directly in catalysis, since the isolated C_3 trimers do not contain zinc and are completely competent in catalysis [5]. Rather each regulatory subunit contains one Zn(II) ion coordinated to $-S^-$ groups of four Cys residues within the R subunit arranged in a Cys-X_4-Cys-X_{23}-Cys-X_2-Cys sequence near the carboxy terminus [6].

The R-C association is absolutely required for the allosteric regulation of the catalytic activity of aspartate transcarbamoylase by nucleotides which bind to a site on the R subunit [8]. Despite being contained entirely within the regulatory subunit, displacement of the Zn(II) ion with organic mercurials completely dissociates the catalytic and regulatory subunits [7]. The three-dimensional structure of the aspartate transcarbamoylase holoenzyme shows that in each R subunit the four Cys ligands to the Zn(II) are located at the base of a loop of the polypeptide chain whose distal portions carry amino acid residues forming a major part of the R-C interface [6,8]. This structural observation leads to the postulate that removal of the metal ion destabilizes these loops such that the resulting "conformational flux" destabilizes the R-C interface. In other words, a significant portion of the favorable energy of interaction between R and C subunits at the interface appears related to the "rigidity" of the polypeptide structure conferred by the zinc complex. As detailed below, it seems not unlikely that such a "structural" role of Zn(II) in aspartate transcarbamoylase more closely models the characteristics of the Zn(II)-Cys complexes being uncovered in a variety of DNA-binding proteins than do the Zn(II) complexes found at the active centers of more classical zinc metalloenzymes. Such a function contrasts sharply with a model whereby Zn(II) utilizes an open or solvent-occupied coordination position to form a bridged or mixed complex for purposes of catalytic activation [3].

Evidence that Zn(II) may play an important direct or indirect role in processes related to cell division, nucleic acid replication, and transcription has been available for some time, since zinc deficiency states in eukaryotic cells have been shown to be associated with marked disturbances in the cell cycle as well as in DNA and RNA contents of cells [9]. Disturbances in nucleic acid metabolism accompanying zinc deficiency have been studied in most detail in *Euglena gracilis* by Vallee and his colleagues [9,10]. These studies have shown zinc-deficient cells to arrest in S phase with major alterations in the amounts and base composition of the RNAs being synthesized [9,10]. A significant perturbation in the nature of the basic polypeptides bound to the chromatin has also recently been reported ([11] and references contained therein).

In the late 1970s it was widely accepted that a major molecular mechanism underlying such disturbances had been found when it appeared that most DNA and RNA polymerases were zinc metalloenzymes (12,13] for a review). At present, however, the precise contribution of Zn(II) to the structure-function of these polymerases is not clear in that the prototypic DNA polymerase I from *E. coli* [14] as well as the phage T7 DNA polymerase [15] have subsequently been shown not to be zinc metalloenzymes. While the bacterial multisubunit RNA polymerases, like that from *E. coli*, contain two tightly incorporated Zn(II) ions [13,16-23], the small, single-subunit RNA polymerases from bacteriophages do not contain zinc [24]. Clearly, then, the basic enzymatic polymerization of ribo- and deoxyribopolynucleotides does not require zinc as a catalytic cofactor. On the other hand, the largest subunits of the multisubunit RNA polymerases II and III from eukaryotes exhibit remarkable homology to the β' subunit of the *E. coli* enzyme [25-27]. As expected, the second largest subunit of the eukaryotic enzymes exhibit significant sequence homology with the β subunit of the prokaryotic enzyme [28]. The β' subunit appears to tightly coordinate one of the two Zn(II) ions of the polymerase through four Cys-S^- ligands derived from the N terminus [29]. As several eukaryotic RNA polymerases have been found to contain zinc [13,30-32] and comprise with the bacterial enzyme a common molecular architecture consisting of two large and evolutionarily conserved subunits [33], zinc might be envisaged to play at least some of the same roles as in the prototypic RNA polymerase from *E. coli*. Possible functions of Zn(II) in these enzymes will be discussed in Sec. 2.

Of perhaps greater interest relative to the role of zinc at the level of DNA replication and transcription has been the finding in the last several years of zinc firmly incorporated into accessory proteins involved in bacterial and bacteriophage DNA replication, e.g., gene 32 protein from bacteriophage T4 [34,35], repair processes, e.g., *E. coli* UvrA [36], and in eukaryotic transcription, e.g., transcription factor IIIA (TFIIIA) from *Xenopus* oocytes [37,38], the latter required for the efficient transcription of the 5S ribosomal RNA gene. As the DNA sequences for a number of eukaryotic transcriptional regulatory proteins from a variety of species become available, it seems that in many cases, the derived amino acid sequences contain potential zinc-binding sites characterized by Cys_2His_2 or Cys_4 ligand patterns colinear with DNA-binding domains and termed "zinc fingers", a phraseology originally applied to the zinc-binding domains (nine of them) found in TFIIIA [38-43, and 44 for a review]. Various aspects of eukaryotic transcription factors based on the TFIIIA model will be discussed in Chap. 6.

The purpose of this chapter is to discuss the current progress of physicochemical studies which begin to elucidate the roles that such closed zinc complexes have in the structure-function of the several types of zinc proteins now known to be involved in bacterial, bacteriophage, and retroviral replication and transcription processes.

2. PROKARYOTIC RNA POLYMERASES

2.1. Introduction

The RNA polymerase from *E. coli*, the holoenzyme consisting of five subunits, $\alpha_2\beta\beta'\sigma$, was first reported to be a zinc metalloenzyme in 1971 [12]. Since that time extensive investigation of both the *E. coli* and *B. subtilis* enzymes of similar structure has amply confirmed that both proteins contain two tightly incorporated Zn(II) ions per molecule [16-23,29]. In contrast to many Zn(II) enzymes, but not unlike ATCase [7], these intrinsic Zn(II) ions cannot be exchanged with exogenous ^{65}Zn in the medium, and until recently it was not possible to remove either Zn(II) ion from RNA polymerase except by dissociation and denaturation of the multisubunit enzyme in 6 M urea [19-23].

2.2. Removal, Substitution, and Exchange of One or Both Intrinsic Zn(II) Ions of Prokaryotic RNA Polymerases

The functions of the two intrinsic Zn(II) ions in the *E. coli* and *B. subtilis* RNA polymerases are not well defined. Since the core polymerases, lacking the σ subunit, contain the full complement of zinc, the Zn(II) does not appear to participate directly in the promoter recognition function conferred by the σ subunit. Limited information on Zn(II) binding has been provided by denaturation of the RPase to the point where the subunits could be separated by chromatographic techniques [16-21]. Such procedures were then followed by renaturation and assessment of Zn(II) binding by the individual subunits. Summarizing all the data collected by such studies, it seems reasonably certain that one of the Zn(II) ions is incorporated into the β' subunit, while the other is less tightly bound and may be present at or near the interface of the β,β' subunits [16-18].

2.2.1. Discrimination Between Zn(II) Sites Based on Chemical Reactivity and Metal Ion Exchange

Zinc proteins in which the Zn(II) ions are coordinated to Cys-S^- ligands can often be titrated with organic mercurials with the release of the protein-bound Zn(II) as the Hg-R competes for the Zn(II) coordination site. For example, the six Zn(II) ions of aspartate transcarbamoylase are sequentially released by p-hydroxymercuriphenylsulfonate (PMPS) and the released Zn(II) ions can be detected directly by the formation of a colored compound when complexed with the indicator dye, 4-(2-pyridylazo)resorcinol (PAR), which absorbs at 500 nm [7]. The latter procedure applied to the RNA polymerase from *E. coli*, either the holo- or core enzyme, results in the release of only one of the two Zn(II) ions in the protein as shown in Fig. 1 [29].

In the absence of denaturation, the other Zn(II) ion of *E. coli* RPase is not accessible to mercurials [29]. The *E. coli* enzyme

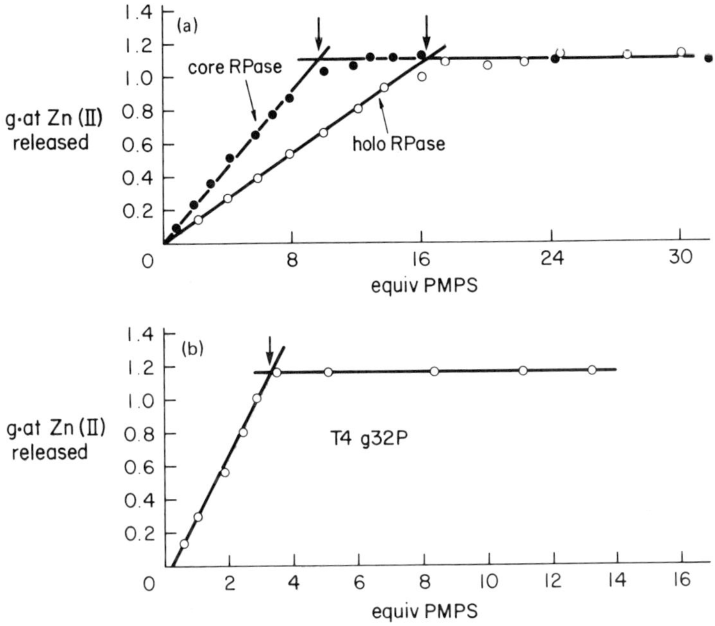

FIG. 1. (a) Release of Zn(II) from *E. coli* RNA polymerase (RPase) holoenzyme (○) and core enzyme (●) upon titration with PMPS. Zn(II) was detected by formation of the complex with 4-(p-pyridylazo)resorcinol (PAR), Zn(II)-PAR_2, which absorbs at 500 nm and was converted to g at Zn(II) using $\varepsilon_{500nm} = 6.6 \times 10^4\ M^{-1}\ cm^{-1}$ [7]. (b) Release of Zn(II) from T4 g32P upon modification of cysteine SH^- groups [34] with PMPS.

contains surface sulfhydryl groups which can be modified with an organomercurial [45] which accounts for the consumption of 8 to 16 PMPS molecules before the Zn(II) is completely released. Since the number of PMPS molecules required to displace the Zn(II) is larger when the holoenzyme is used, it is possible that the binding of the β subunit does stabilize the accessible Zn(II) complex. The PMPS-accessible Zn(II) site has been termed the B site, while the "buried" site has been termed the A site [29].

2.2.2. *Visible Absorption Spectroscopy of Co_2 and Co_1 RNA Polymerases*

While the Zn(II) ions are not exchangeable in vitro, it is possible to grow *E. coli* on a Co(II)- or Cd(II)-enriched medium and isolate active RNA polymerases containing a full complement of these metal ions in place of the native Zn(II) ion [22,23]. The optical absorption of the Co(II) species provides a valuable probe of the nature of the coordination sites within the enzyme. The absorption spectrum of a biosynthesized Co(II) enzyme before and after titration of the enzyme with PMPS to form the one Co(II) species are shown in Fig. 2. The low-energy absorption maxima at 760 nm (ε = 800 M^{-1} cm^{-1}), 710 nm (ε = 900 M^{-1} cm^{-1}), 602 nm (ε = 1,500 M^{-1} cm^{-1}) clearly identify at least one of the Co(II) sites in the enzyme as consisting of tetrahedral coordination geometry around the Co(II) [29,46]. These visible absorption bands accompanied by the broad absorption centered near 600 nm are very similar to those seen for Co(II)-substituted in the tetrathiolate sites of aspartate transcarbamoylase [47], the structural Zn(II) site of alcohol dehydrogenase [48], the zinc-substituted Cu(II) site at rubredoxin [49], and the cluster thiol Zn(II)/Cd(II) sites of metallothionein [50] as well as the model compound tetrathio(phenolato)cobaltate(II) [51]. Like the latter well-characterized systems, Co(II) in RNA polymerase also shows the expected intense $S^- \longrightarrow$ Co(II) charge transfer bands below 400 nm with extinction coefficients well above 10^3. In the presence of the latter, the "d-d" transitions of the Co(II) ion are expected to "borrow" some intensity from these bands and in fact, like the other tetrathiolate Co(II) complexes, the low-energy transitions are more intense than for other Co(II)-substituted Zn(II) metalloenzymes containing mixed-N,O donor sets (cf. [52]).

As in the case of the native Zn(II) RPase, one of the two Co(II) sites is accessible to PMPS and can be displaced by this reagent. The displaced Co(II) ion can then be completely removed from the enzyme sample by dialysis against EDTA (see Sec. 2.2.3). Removal of the accessible Co(II) ion causes only a slight change

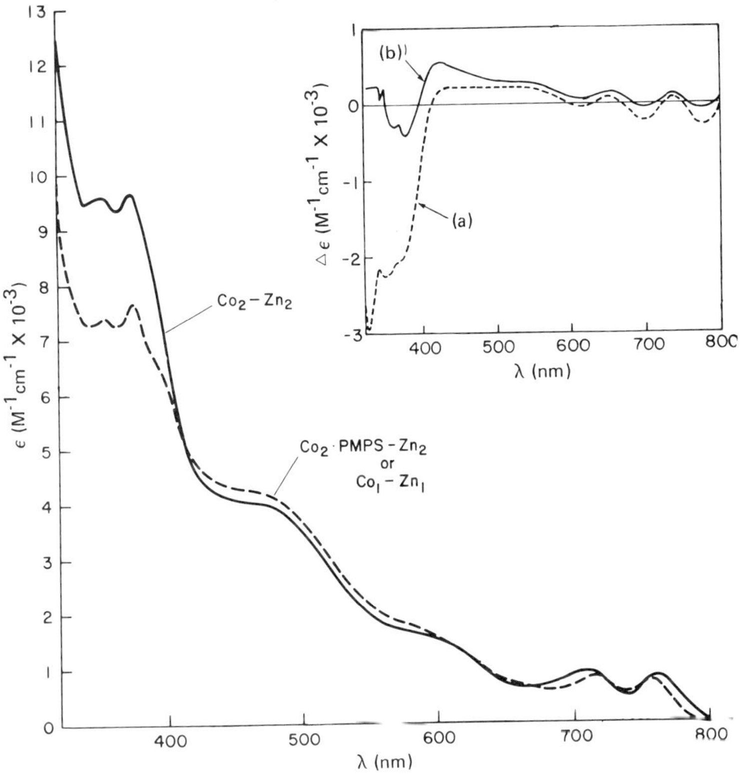

FIG. 2. Visible absorption spectra of Co(II)-substituted *E. coli* RPases. (——) Co_2 RPase holoenzyme biosynthesized by growth of *E. coli* on Co(II)-enriched media. (---) After removal of one of the Co(II) ions with PMPS, EDTA, and thiols to form Co_A RPase. (Inset) Difference visible absorption spectra. (Curve a) Co_2 RPase-PMPS complex (Co_A RPase) less Co_2 RPase. Conditions: 10 mM Tris-HCl, 0.2 M NaCl, 5% v/v glycerol pH 8. (Curve b) Co_2 RPase-PMPS complex plus DTT (to reverse modification and give Co_2 RPase) less untreated Co_2 RPase. (Data plotted from Ref. 29.)

in the position and intensities of the low-energy absorption bands, but a substantial loss in the intensity of the charge transfer transitions in the near-UV (Fig. 2). This charge transfer intensity is ≥90% restored upon removal of the PMPS from RPase which provides evidence for reformation at the "B-site" coordination sphere upon reversal of the RPase-PMPS complex regenerating a Co_2 RPase (Fig. 2, inset). Thus while the accessible Co(II) ion may well be liganded to at least one $-S^-$ ligand, the single "buried" Co(II) ion is the one responsible for the tetrahedral-tetrathiolate-like absorption spectrum. This Co(II) site, which corresponds to the A-site Zn(II), is quite likely to be one composed of four Cys residues that occur in the extreme N terminus of the amino acid sequence of the β' subunit, Cys-containing sequences which are also found in the largest subunits of a number of eukaryotic polymerases (see Sec. 2.3).

2.2.3. Functional Characteristics of Fully Active One-Metal and Mixed-Metal Hybrid RNA Polymerases

Not only is it possible to displace the accessible Zn(II) [Co(II) or Cd(II)] ion from RPase with PMPS, but it is also possible to remove the displaced Zn(II) [Co(II) or Cd(II)] from the enzyme solution by dialysis against EDTA and then to remove all the reacted organomercurial groups from the enzyme with an excess of thiol [29]. Within the currently available precision with which the activity of *E. coli* RNA polymerase can be measured for either the holo- (σ-dependent) or core enzymes, the derivatives containing only the "buried" metal ion appear to have full polymerase activity. The data are summarized in Table 1. Unfortunately, activities by standard assay techniques are not the same from one native RPase preparation to another, so comparisons can only be made within one preparation, not between different preparations. Specific elongation rates, however, determined on highly specific σ-dependent promoters like those of T7 bacteriophage DNA, are more comparable from one preparation to another [53] (see Table 1). From the data summarized in Table 1 it is possible to conclude that the accessible or B-site Zn(II) in *E. coli* RNA polymerase does not function directly in the catalytic reaction or in binding the template. Likewise the recognition of σ-specific promoters appears to proceed normally in the absence of the PMPS-accessible Zn(II) in that activity on the A_1, A_2, and A_3 promoters of phage T7 is normal in the absence of this

TABLE 1

Metal Contents and Transcriptional Activities of Various Metalloderivatives of *E. coli* RNA Polymerase

Enzyme	Zn(II) (g at/mol)	Co(II) (g at/mol)	Cd(II) (g at/mol)	Specific act on calf thymus DNA[a] (nmol UMP incorp/ 10 min/mg protein)	Specific act on T7 DNA[b] (nmol UMP incorp/ min/mg protein)	Elongation rate on T7 DNA (nuc/sec^{1})
Core RPase	1.85	—	—		137	
Core RPase	1.80	—	—	793	—	
Core Zn_ARPase	0.86	—	—	—	115	
Core Zn_ARPase	1.00	—	—	729	—	
Holo RPase	2.16	—	—	—	190	
Holo Zn_ARPase	0.99	—	—	—	191	
Holo RPase	1.70	—	—	622	223	17.9
Holo Zn_ARPase	0.90	—	—	609	166	16.8
Holo Co_2RPase	—	2.00	—	318	73	20.8
Holo Co_ARPase	—	1.00	—	265	50	21.7
Holo Cd_2RPase[c]	0.30	—	1.70	168	60	12.6
Holo Cd_ARPase[c]	0.45	—	0.60	183	64	13.3

[a] nmol UMP incorporated/10 min/mg protein. Values are ±10-15%.

[b] nmol UMP incorporated/min/mg/protein as determined with the quantitative assay [17]. Conditions: 2 μg of enzyme (44 nM) and 78 μg T7 DNA (3 nM) in 1 ml.

[c] In contrast to the Co(II) enzyme, biosynthesis of the Cd(II) enzyme appears to be more easily prone to Zn(II) incorporation, largely into the nontitratable A site as 1 g at Cd(II) is removed in formation of a mixed Me_A species. Thus, the elongation rates for the Cd(II) enzymes may be upper limits and Cd(II) incorporation might result in a moderate decrease in the elongation rate.

Zn(II) ion (Table 1). The assays using T7 DNA as template have been coupled with identification of the lengths of the ^{32}P-labeled transcripts by gel electrophoresis. The radioautography of the transcripts produced at the sequential time points uniformly showed progressive increase in average mRNA length over the first 6 min which then level off at the expected average length of 7,133 bases expected from initiation at the three leftward A promoters and termination at the *E. coli* terminator sequence at 7,720 base pairs on the T7 genome [53]. At least 90% of the mRNA consisted of this transcript after 6-8 min of reaction. The native Zn_2 as well as the Zn_A, Co_2, Co_A, Cd_2, and Cd_A enzymes all showed the same pattern of mRNA development with correct termination. The only differences were the small variations in time development of the mRNA as reflected in the slightly different calculated elongation rates reported in Table 1.

It is not possible, however, to state that the accessible Zn(II) has no function, since in vivo the enzyme recognizes a large spectrum of promoters, many of them involving other accessory proteins like the catabolite gene activator protein (CAP), repressors, and probably other regulatory proteins. None of these "modified" RNA polymerase-promoter interactions has been tested with the Zn_A RPase.

Since the basic polymerization reaction appears unaffected by the removal of the B-site Zn(II), it has been possible to probe the function of the "buried" A-site metal ion directly. Wu and Wu and coworkers first suggested, based on a variety of denaturation-reconstitution experiments, that one of the two intrinsic metal ions is catalytically critical and provided evidence to suggest that the nucleotide base of an initiating ATP molecule bound directly to the inner coordination shell of the Co(II) ion in a reconstituted Co_1Zn_1 hybrid [20,21,54] (see Sec. 2.2.4). In contrast, spectral data presented above provide evidence that the A-site Co(II), the only site that can still be postulated to directly affect polymerization activity, is located in a tetrahedral-tetrathiolate complex [29]. If the coordination is to four $Cys\text{-}S^-$, then such an electron-rich complex would tend to restrict expansion of the coordination sphere and therefore not be likely to be accessible to solvent or additional monodentate ligands from solution.

2.2.4. $^{35}Cl^-$ *NMR Quadrupolar Broadening Experiments as a Probe of Solvent-Accessible Zn(II) Coordination Sites in* E. coli *RNA Polymerase*

One method which has been used to assess the presence of an open or solvent-accessible inner coordination site in a protein-bound zinc ion is to replace the intrinsic Zn(II) with a paramagnetic species. Mn(II) is most desirable in terms of its large paramagnetic contribution to relaxation of solvent protons, but Co(II) is a better mimic of the functional properties of Zn(II) sites, and the somewhat smaller contribution of Co(II) to the paramagnetic relaxation of bulk solvent can be detected if a coordinated solvent molecule is in fast exchange. The exchanging water molecule enhances the longitudinal relaxation rate ($1/T_1$) of bulk water protons relative to that in the presence of the native diamagnetic Zn(II) protein (for a review, see Ref. 55).

Another experimental approach to access the same information, i.e., an open coordination site, is based on the ability of metal ion-bound solvent molecules to be displaced by the anion Cl^-. Due to significant quadrupolar broadening of the $^{35}Cl^-$ NMR line (spin = 3/2) upon formation of the metal-chloride bond, simple observation of excess $^{35}Cl^-$ NMR linewidth of the bulk Cl^- as a function of the zinc content of a metalloprotein can provide evidence of a solvent (anion)-accessible coordination site [56]. This technique has the advantage over the paramagnetic technique outlined above in that the native Zn(II) protoin can be used at significantly lower protein concentrations without relying on the metal ion substitution studies.

In the case of *E. coli* RPase, Chatterji and Wu [20,21,54] showed that the $1/T_1$ of bulk water protons in the presence of the native Zn_2 RPase (0.26 sec^{-1}) was significantly less than that obtained with the denatured, renatured, and reconstituted Co_1Zn_1 RPase (0.39 sec^{-1}) at a 24-MHz applied magnetic field. Evidence was presented which suggested that up to two water molecules were bound directly to the Co(II) nucleus, one of which apparently could be replaced with substrate ATP or dinucleotide ApA, which gave rise to relaxation of the base protons as well as the ^{31}P nucleus of the bound initiator nucleotide [21]. This suggested that the metal ion was present at the initiation site of the enzyme and acted as a catalytic cofactor [54].

Based on these data, biosynthetically prepared Co_2 RPase should exhibit a similar or greater degree of paramagnetic relaxation of water protons, depending on whether the nonexchanged Zn(II) ion in the above reconstituted metal hybrid species is accessible to solvent. Using similar experimental conditions (10 μM RPase, 20-MHz field), we were unable to detect significant deviation in the water proton $1/T_1$ values obtained for the Co_2, Co_A, Zn_2, and Zn_A RPases prepared as described (Sec. 2.2.2) (0.31-0.33 sec^{-1}) in direct contrast to the results obtained for the reconstituted enzyme [21].

To avoid the use of the metal-substituted enzyme, we assessed the $^{35}Cl^-$ NMR linewidth in the presence of the native core Zn_2 RPase during and after the chemical modification procedure which gives rise to a Zn_1 or Zn_A RPase as shown in Fig. 3. Upon addition of Zn_2 RPase to buffered 0.2 M NaCl, moderate line broadening occurs which is within the range expected for increased viscosity of the solution and/or weak Cl^- binding to solvent accessible amino acid side chains [57]. With modification of titratible Cys-SH groups with PMPS, however, dramatic broadening of the NMR resonance occurs, consistent with removal of $-S^-$ ligand atoms from the B-site Zn(II) which creates a coordination site(s) available for anion (Cl^-) binding. ^{65}Zn exchange experiments suggest that the PMPS-modified RPase still contains B-site Zn(II) bound to the enzyme at this point through nonsulfhydryl ligands [29]. Upon addition of the Zn(II) chelator EDTA, which strips the enzyme of B-site Zn(II) [29], the excess linewidth is reversed to a level comparable to that of the native Zn_2 enzyme. This is not changed significantly upon reversal of the RPase-PMPS complex with excess thiol and subsequent dialysis and redetermination of the linewidth associated with the reconstituted Zn_A RPase.

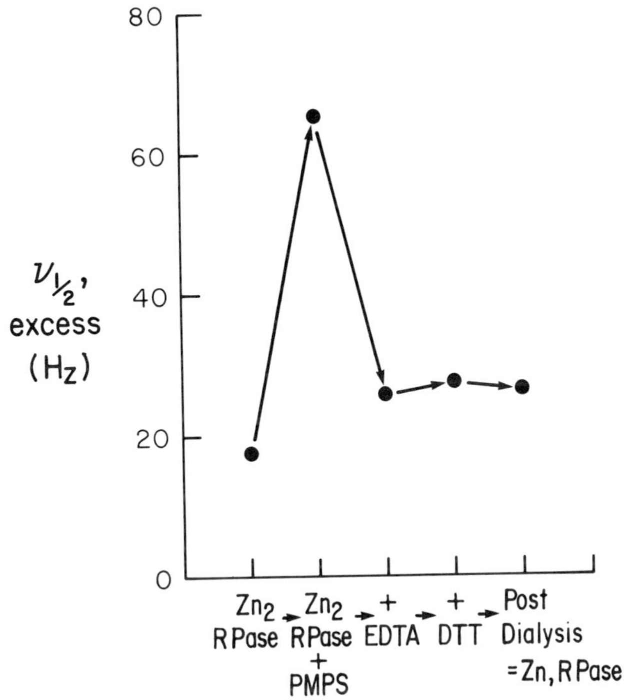

FIG. 3. Excess ^{35}Cl NMR linewidth of bulk Cl^- ($\nu_{1/2}$ observed - $\nu_{1/2}$ unperturbed Cl^-) in the presence of 25 μm Zn_2 core RPase as a function of the chemical modification steps in formation of the Zn_1 core enzyme. $\nu_{1/2}$, unperturbed refers to ^{35}Cl NMR linewidth of the buffer TNG (10 mM Tris-HCl, pH 8, 5% v/v glycerol, and 0.2 M NaCl). The following refers to indicated manipulations of Zn_2 RPase. Untreated → plus 30 eq of PMPS to form the RPase-PMPS complex → plus 3 mM EDTA to chelate Zn_B from Zn_2 RPase → plus 5 mM DTT to reverse sulfhydryl modification of RPase → following dialysis to return Zn_A RPase to TNG buffer [0.87 g at Zn(II) found]. Experiments were carried out at 19.59 MHz with a Bruker CXP-200 wide-bore spectrometer equipped with a low-frequency, broad-band, tunable probe externally locked with a capillary containing D_2O. 5,000-6,000 scans were collected at 298°K using 0.9-sec pulse delay and a spectral width of 8,000 Hz.

Since the Zn_A and Zn_2 enzymes have comparable $\nu_{1/2}$, excess values and knowing that the Zn at site A is tightly incorporated and probably tetrathiolate, direct coordination of the Cl^- (and presumably solvent water) at either metal site is unlikely under native conditions. B-site Zn(II) becomes capable of efficiently binding Cl^- only when S^- coordinating donor atom(s) are chemically modified with the organomercurial and stripped from the inner coordination shell. EDTA then chelates the Zn(II) from the enzyme and in so doing expulses the anion from the metal ion.

Thus, if the A-site metal ion plays a direct role in catalysis, its role may not be to coordinate a group on the incoming substrate following solvent molecule extrusion as proposed [54], but to provide rigid conformational constraints to an essential domain of the β' subunit or its interface with other subunit(s). In support of this conclusion is the finding that the addition of Mg-ATP causes little significant perturbation of the intense d-d absorption bands arising from Co(II) in the A site [29].

2.3. Subunit Structure Relative to Zn(II) Ions, Cloning of Individual Subunits, and Relationship to Subunits of the Eukaryotic RNA Polymerases

The denaturation data on the *E. coli* RPase suggest that one of the two Zn(II) ions, probably the most tightly bound one, is contained within the β' subunit [16-18,22]. To address this question directly, we have undertaken cloning and overexpression of the isolated β' subunit. Figure 4(a) shows that when the *rpoC* gene is placed downstream from the strong hybrid *E. coli* promoter *tac*, massive overproduction of the β' subunit occurs upon promoter induction [58]. As was found for the overexpression of the β subunit [58,59], nearly all of the synthesized subunit is sequestered in the insoluble fraction when standard subcellular fractionation schemes are employed thus precluding direct zinc analysis on the subunit preparation. However, renaturation of the cloned β subunit in the presence of sufficient β' and α subunit gives rise to a transcriptionally active RPase [59]. In the case of β', "refolding" of the β' polypeptide in the presence of Zn(II), followed by purification on heparin-agarose [Fig. 4(b)], produces a reasonably soluble subunit preparation which has stably incorporated 0.9 g at Zn(II)/mol [58]. The N terminus of

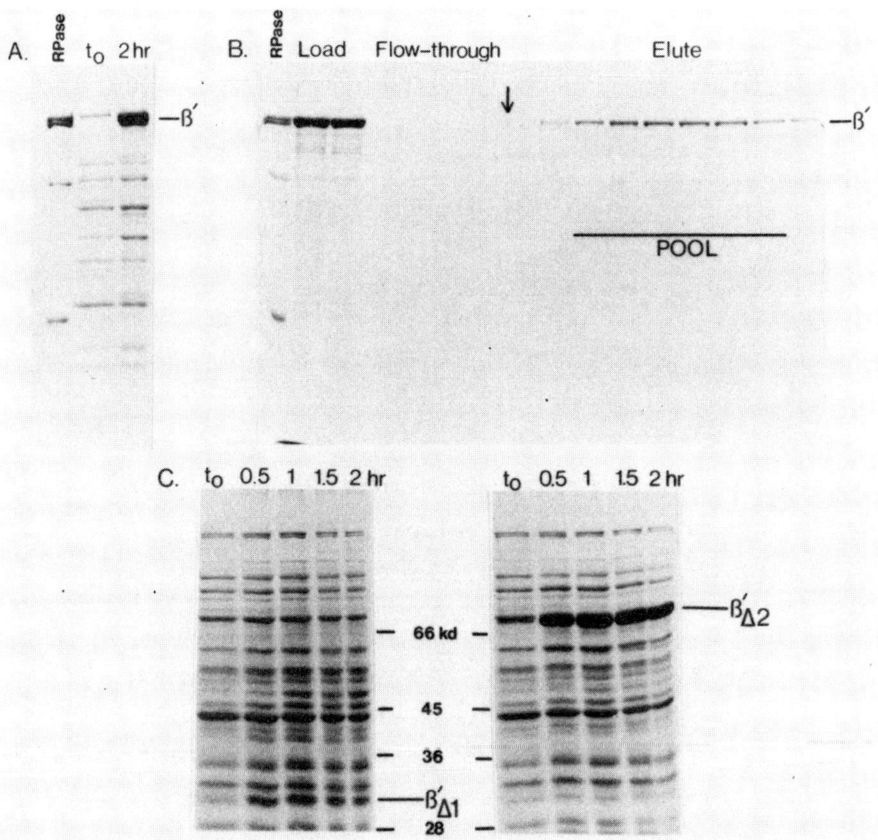

FIG. 4. Overproduction and purification of *E. coli* RPase β' subunit and two deletion C-terminal peptides. (A) Induction of *E. coli* JM101 cells harboring ptac β' with IPTG (construct to be described elsewhere). t_0, cell extract before induction; t = 2 hr. cell extract taken 2 hr following induction of *tac* promoter with IPTG; RPase std., *E. coli* RPase holoenzyme with β,β' subunit doublet, σ subunit, and α subunit shown with decreasing molecular weight. (B) Chromatography of urea-solubilized, renatured β' fraction (load) on heparin agarose. The flow-through fractions and fractions eluted with a 0.6 M NaCl step gradient are indicated as are the eluted fractions pooled for characterization (Pool). RPase std., as in (A). (C) Time course of induction of *E. coli* JM101 cells harboring the β' C-terminal deletion plasmids ptac $\beta'_{\Delta 1}$ (left panel) and ptac $\beta'_{\Delta 2}$ (right panel) where a translation stop oligonucleotide was cloned into the unique *Bgl*II and *Cla*I sites in the *rpo*C gene in ptac β', respectively. These plasmids direct the synthesis of $\beta'_{\Delta 1}$, predicted to contain the N-terminal 248 amino acids (left panel), and $\beta'_{\Delta 2}$, predicted to contain the N-terminal 642 amino acids (right panel). The migration of molecular weight standards are as indicated in kD.

TABLE 2

Cysteine-Rich Sequences Found Toward the Amino Termini of the Largest Subunits of Various RNA Polymerases

Large subunit		Ref.
E. coli (β')	^{58}Cys-X_{11}-Cys-X-Cys-X_{12}-Cys-X_2-Cys^{88}	60
Vaccinia	^{48}Cys-X_2-Cys-X_6-Cys-X_{30}-Cys-X_2-Cys^{92}	27
Yeast pol(II)	^{67}Cys-X_2-Cys-X_6-Cys-X_{29}-Cys-X_2-Cys^{110}	25
Yeast pol(III)	^{67}Cys-X_2-Cys-X_6-Cys-X_{29}-Cys-X_2-Cys^{110}	25
Dros pol(II)	^{66}Cys-X_2-Cys-X_6-Cys-X_{29}-Cys-X_2-Cys^{109}	26

the β' subunit contains a Cys-rich amino acid sequence between residues 58 and 88, Cys-X_{11}-Cys-X-Cys-X_{12}-Cys-X-X-Cys (Table 2) [60]. Cys-X-X-Cys (cf. [61]) and, to a lesser extent, Cys-X-Cys [62] represent known linear amino acid arrangements in which the Cys side chain sulfur atoms participate in forming the vertices of a number of tetrahedral Zn(II)-binding sites and thus might encompass the A-site Zn(II) coordination sphere of RPase. Such a cluster occurs nowhere else in the sequence of the β' subunit [60]. In preliminary attempts to probe the possible coincidence of the A-site metal ion-binding site and this Cys-rich region, we have constructed two plasmids which direct the synthesis of C-terminal deletions of the β' subunit which correspond to N-terminal 248 and 642 amino acid species of β' [Fig. 4(c)]. Experiments are under way to assess the zinc-binding and functional properties of these truncated β' polypeptides.

It has recently been shown with the cloning and sequencing of the genes which encode the largest subunits of the vaccinia virus RNA polymerase, the yeast RNA polymerases II and III, and the *Drosophila* RNA polymerase II that there exist up to five distinct regions of amino acid sequence homology, four of which display considerable relatedness to the β' subunit of the *E. coli* enzyme and together cover ~70% of the *rpoC* gene [25-27]. Among the eukaryotic and poxvirus sequences, the N-terminal homologous region includes a cluster of Cys residues much like the same region in the β' subunit although not significantly related at the amino acid level. As this region of β' is proposed to function in the binding of the structural A-site Zn(II) in the prokaryotic enzyme, such regions might function in zinc binding in the multisubunit eukaryotic polymerases perhaps via the two conserved Cys-X-X-Cys sequences (Table 2).

Since one cannot reversibly remove the A-site metal ion from the prototype *E. coli* RNA polymerase without first unfolding the polypeptides and dissociating the subunits, it remains difficult to probe the precise function of this tightly incorporated metal ion. Some possible functional roles of a "structural" metal ion in a protein binding to DNA might be inferred from the accumulated data on T4 gene 32 protein (see Sec. 3) from which it has been possible to reversibly remove a tetrahedral, thiolate-coordinated Zn(II) ion. Under such conditions it can be shown that the metal ion modulates important features of the *inter*molecular protein structure associated with productive DNA binding. A similar possibility exists for the RNA polymerases. As photoaffinity labeling [63] and neutron small-angle scattering experiments [64] have provided evidence that the polymerase active site residues are proximate to or contributed by the β,β', and σ subunits, Zn(II) might modulate critical subunit-subunit (i.e., β-β') interactions which establish the integrity of the catalytic machinery. A detailed understanding awaits further work.

3. GENE 32 PROTEIN FROM BACTERIOPHAGE T4, AN ACCESSORY PROTEIN OF DNA REPLICATION THAT CONTAINS A STRUCTURAL ZINC ION

3.1. Introduction

The product of gene 32 of bacteriophage T4 (g32P) is one of a class of proteins that bind to single-stranded ssDNA (or ssRNA) in a stoichiometric fashion and are referred to as helix-destabilizing or ssDNA-binding proteins (ssbs) (see [54-56] for reviews). Gene 32 protein is known to play roles in DNA replication, recombination, and repair as well as autoregulating its own synthesis by binding to a

5' AT-rich region of its own mRNA and inhibiting translation [65-73]. Gene 32 protein contains a single Zn(II) ion which contributes to the structural stability of the molecule and in so doing significantly affects the ability of the protein to bind to ssDNA [34,35].

The function of g32P in DNA replication is associated with the ability of the protein to quickly cover the single-stranded regions of DNA that arise just behind the T4 DNA replication forks and thus protect these regions from endogenous nucleases as well as stabilizing a particular ssDNA conformation that is most appropriate to serve as substrate for other catalytic proteins such as the T4 DNA polymerase. Gene 32 protein is the classic ssDNA-binding protein, and its structure and function have been probed by a large number of techniques, the results of which are summarized briefly below.

The basic structural features of this protein of 301 amino acids are depicted in Fig. 5 by diagramming on the amino acid sequence the domain structure of the protein as well as a number of other amino acids centrally involved in the structure or function of the protein. Gene 32 protein (M_r 33,488) binds in a highly cooperative manner to ssDNA [68,74,75]. Estimates of the intrinsic binding constant K_a from poly[d(A-T)] melting experiments range from 10^8 to 10^{11} depending on conditions [35]. K_a is the product of two terms: K_{int}, which describes the binding of the protein to a single site on a nucleotide lattice, and ω, which represents the additional affinity provided by the cooperative binding of contiguous g32P molecules to ssDNA containing multiple sites [74,75]. K_{int} can be calculated to be approximately 10^5-10^7 depending on the lattice and method of measurement, while ω is approximately 10^3 and is essentially independent of NaCl concentration [69,74]. The cooperativity factor ω derives from protein-protein interactions between contiguously bound g32P molecules.

Limited proteolysis has been used to demonstrate that g32P is a three-domain protein ([76-79]; see [80] and [81] for reviews). Trypsin readily cleaves off a domain (B) of 21 amino acids from the N terminus of the native protein and a C-terminal domain (A) of 48 amino acids [76,77]. The resulting trypsin-resistant core, residues 22-253, termed g32P-(A+B) contains all the determinants required to

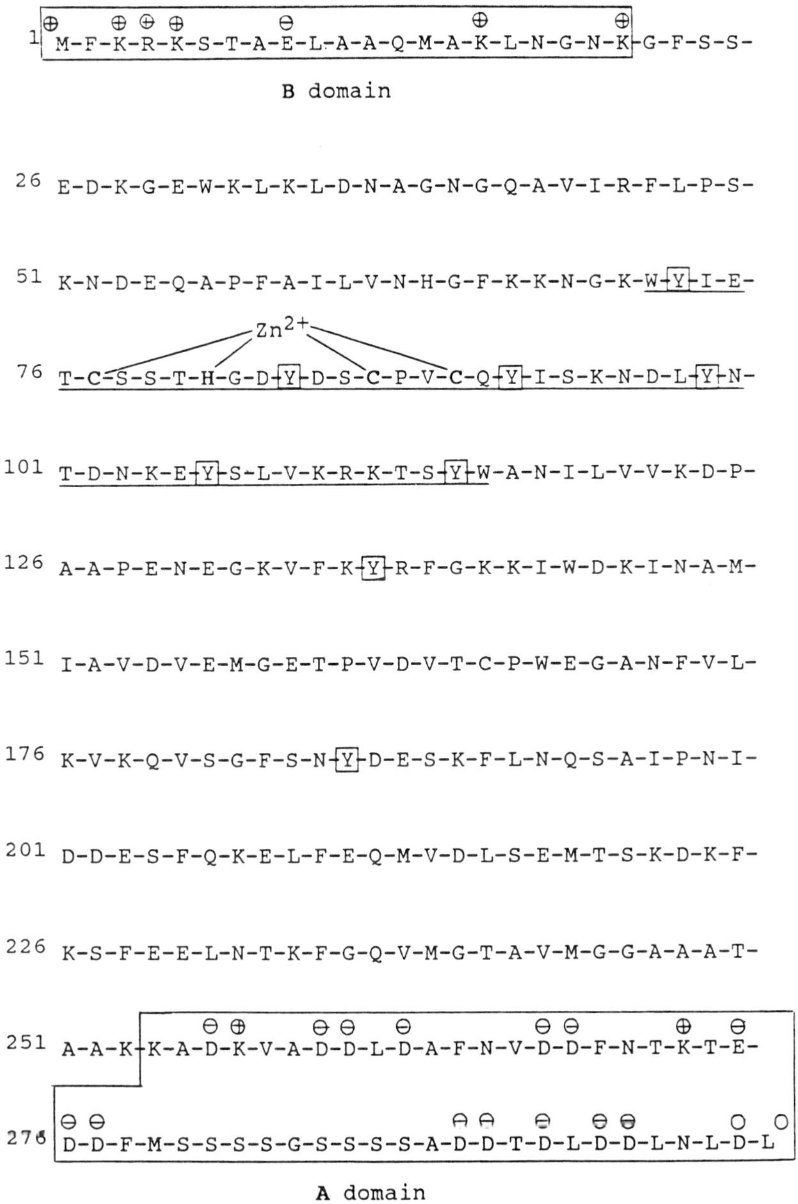

FIG. 5. The primary structure of the gene 32 protein (g32P) from bacteriophage T4 [98]. The N-terminal B domain of 21 amino acids and the C-terminal A domain of 48 amino acids are indicated by the boxes with the charges on each indicated above the sequence. The positions of the eight Tyr residues in the molecule are boxed and the likely Zn(II)-binding residues are indicated by the lines to the Zn(II). g32P-(A+B) (g32P core, g32P*) refers to species formed following proteolytic removal of the A and B domains (cf. [76]).

interact with ssDNA and oligonucleotides, but the core binds noncooperatively to ssDNA [78,79]. The highly positively charged amino terminal B domain (Fig. 5) enables the protein to bind cooperatively to DNA and increases the intrinsic association constant by approximately three orders of magnitude [79]. Limited proteolysis studies on g32P in the presence and absence of ssDNA has shown that the first nine amino acids, Met-Phe-Lys-Arg-Lys-Ser-Thr-Ala-Glu-, are essential for cooperative binding and this region of the protein is resistant to proteolysis if the protein is first bound cooperatively to poly(dT), but are readily proteolyzed off if the protein is bound to $d(pT)_8$, that is, noncooperative binding [77]. As such, loss of the N-terminal nine amino acid residues reduces the affinity for poly(dT) by 200 fold [77]. The larger, negatively charged C-terminal A domain (Fig. 5) appears to interact with other proteins of the T4 replisome [82], as well as providing a kinetic barrier to the melting of native dsDNAs, a role that may be important in the in vivo function of the molecule [68,83].

Despite the loss of cooperative protein-protein interactions, g32P-(A+B) (also known as g32P core or g32P*) and native g32P have similar affinities for $d(pT)_8$, allowing the conclusion that most, if not all, of the nucleotide-binding site is made up of amino acid residues derived from the core domain. Since the intact protein undergoes oligomerization in solution, even in the absence of nucleotides, it has been difficult to study native g32P by the recently available NMR techniques for probing protein structure. However, g32P-(A+B) does not oligomerize and it has been possible to study g32P-(A+B)-oligonucleotide interactions by a variety of one-dimensional NMR difference spectral techniques [80,81,84,85]. These include deuteration of specific sets of amino acid residues to increase the resolution by eliminating overlapping resonances, since even in the core form the 232 amino acid residues constitutes a polypeptide large enough for high-resolution NMR studies. However, with a protein overproduced by a bacterial vector, it is possible to selectively deuterate given types of residues by overproducing the protein in a host auxotrophic for a given amino acid with supply of the desired labeled or unlabeled amino acid(s). Such procedures

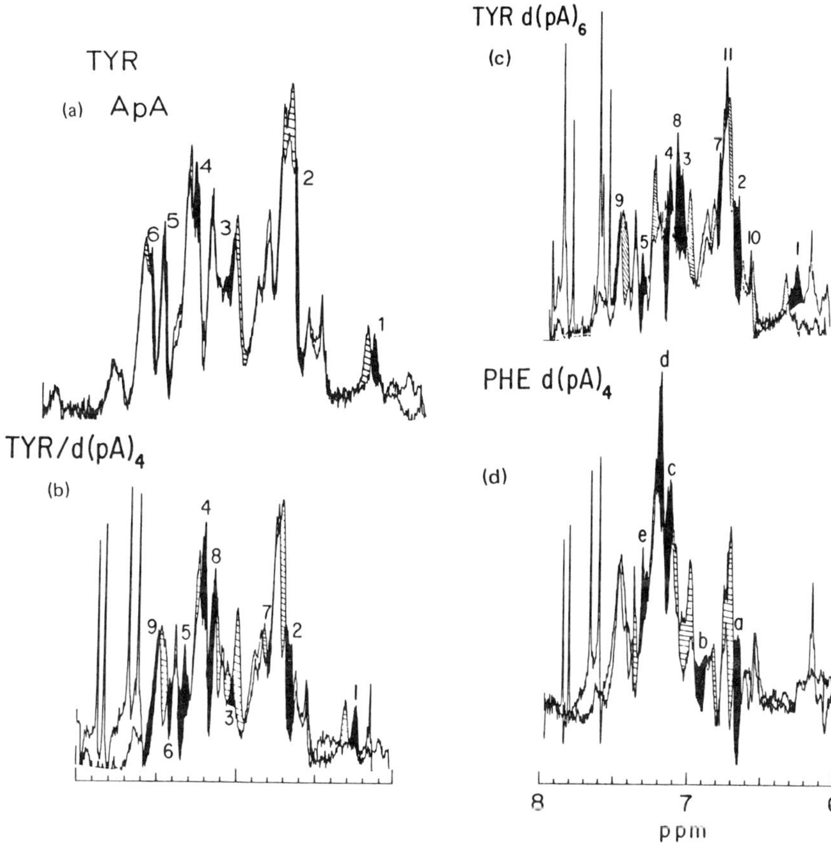

FIG. 6. (a) Aromatic ^{1}H NMR (500-MHz) difference spectra of the g32P*-ApA complex. Spectra are of the Tyr resonances alone and are derived from a difference spectrum constructed from the spectrum of a protein containing protonated Tyr minus that of a protein containing perdeuterated Tyr residues (envelope of hatched peaks) superimposed on the Tyr resonances of the g32P*-ApA complex generated in the same manner, i.e., as a difference spectrum formed from the complexes containing protonated vs. perdeuterated Tyr. (b) Same spectra for the g32P*-d(pA_4) complex. (c) Same spectra for the g32P*-d(pA_6) complex. (d) Phe difference spectra for the g32P*-d(pA_4) complex generated in the same manner as the Tyr spectra but with protein containing perdeuterated Phe. (Data taken from Ref. 84.)

greatly improve the NMR spectra [84]. An example in the case of g32P is shown in Fig. 6.

It is possible to conclude from the ^{1}H NMR studies of g32P and its oligonucleotide complexes that the side chains of several Tyr residues, possibly five, interact with the bases of a bound hexanucleotide, e.g., $d(pA)_6$ (Fig. 6) [84]. The protons of two Phe residues also appear to closely approach some of the bases of the bound nucleotide as well [84]. The nucleotide $d(pA)_8$, thought to extend over the entire DNA binding surface, causes similar NMR shifts (mostly upfield, although a few downfield shifts occur). The magnitudes of these shifts are greater for purines than for pyrimidines and support the contention that the shifts are ring current-induced. The magnitudes of the shifts of the Tyr and Phe protons on oligonucleotide binding are not large enough, however, to suggest a complete base Tyr or base Phe stacking model for the DNA-binding interaction. Rather it appears that the aromatic residues participate in forming relatively hydrophobic base-binding pockets. The degree to which base rings are actually stacked on aromatic amino acid rings awaits more detailed investigation. The participation of the Tyr residues in DNA binding was indicated prior to the NMR work by chemical modification studies which showed that nitration of five of the eight Tyr residues in the protein abolished DNA binding [86]. Binding of DNA prior to the introduction of the nitrating agent, tetranitromethane, prevented the reaction [86].

Recent site-directed mutagenesis of the individual Tyr residues has confirmed that several of them are involved in DNA binding [87]. DNA binding by g32P can be assessed easily by monitoring the ability of bound protein to invade double-stranded regions of the partially double-stranded alternating copolymer poly[d(A-T)] as measured by a hyperchromic increase in the absorbance of the polynucleotide at 260 nm (helix → coil transition). A series of melting profiles for poly[d(A-T)] induced by native g32P and three Tyr → Ser (or Ala) mutants are shown in Fig. 7 [87]. Native g32P has a dramatic ability to destabilize double-stranded poly[d(A-T)], reducing the melting temperature from 65° to 17°C under the conditions of Fig. 7, since the equilibrium of the system, even at relatively low temperature, is in favor of the ssDNA-g32P complex. While two of the mutant proteins, Tyr-115 → Ser-115 and Tyr-137 → Ala-137, have not lost all of their ability to induce the melting of poly[d(A-T)], they

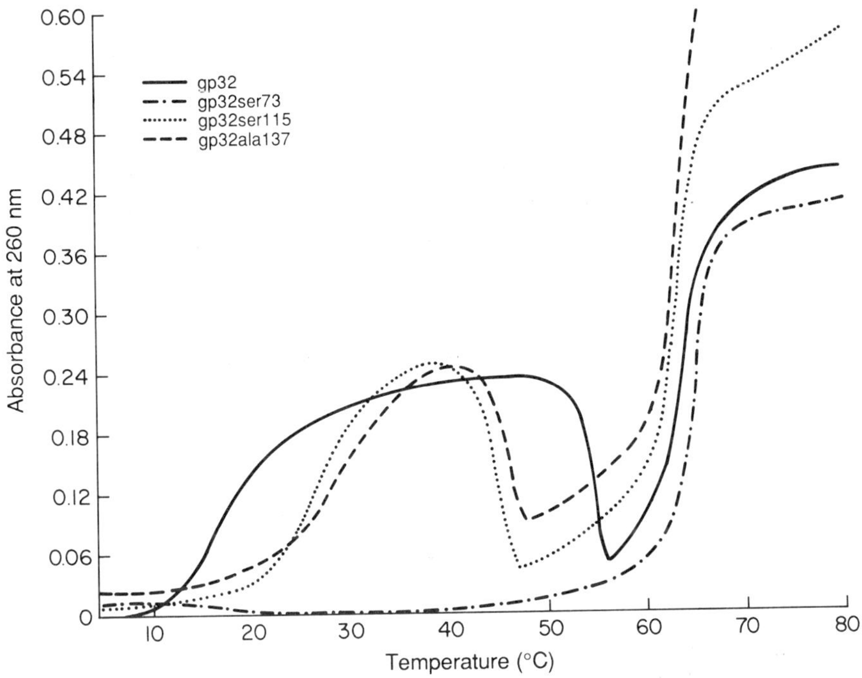

FIG. 7. Effects on the thermal denaturation of the partially double-stranded alternating copolymer, poly[d(A-T)], as followed by optical absorption at 260 nm induced by g32P and site-directed mutants of g32P. Conditions: 150 mM NaCl, 10 mM Tris-HCl, 0.01 mM EDTA, pH 8, and a pro concentration of 1.2 x 10^{-5} M. (——) Native g32P, (-·-) Tyr73 → Ser73 mutant, (···) Tyr115 → Ser115 mutant, (---) Tyr137 → Ala137 mutant. (Data taken from Ref. 87.)

reduce the melting temperature by only 30°C rather than 40°C as observed for the native protein (Fig. 7). Thus Tyr residues 115 and 137 appear to directly contribute to the binding affinity of g32P for ssDNA. Their loss leads to a ~10-fold decrease in the K_a describing the DNA binding, without evidence of other major changes in protein structure. This is supported by ^{1}H NMR spectra of the Tyr-115 → Ser-115 mutant, which show the loss of the signals from the 2,6 and 3,5 protons of Tyr-115 without other general changes in the ^{1}H NMR spectrum [84]. The aromatic proton signals from Tyr-115 are identified from the difference spectrum between native and mutant protein as among those that undergo upfield shifts upon nucleotide binding.

In marked contrast to the above two mutants, the Tyr-73 → Ser-73 mutant has lost all ability to destabilize ds poly[d(A-T)] and the protein does not appear to bind to ssDNA by all criteria applied [87]. The protein as isolated is also devoid of Zn(II). Thus while Tyr-73 may be involved in forming the DNA-binding surface, it appears to have a much more general role in maintaining the structure of the protein. This structural role may be quite closely related to the structure of the Zn(II)-binding domain and we will return to the structure of the protein around Tyr-73 in Sec. 3.3. With this introduction to the general structural features of g32P, the specific roles of the intrinsic Zn(II) ion in determining the structure and function of this protein will be discussed.

3.2. Identification of Gene 32 Protein as a Zinc Metalloprotein

3.2.1. Zinc Content of Intact Gene 32 Protein and Its DNA-Binding Core

The analytical data demonstrating that g32P contains 1 g at/mol of tightly incorporated Zn(II) are summarized in Table 3 [34]. The intrinsic Zn(II) ion is not removed by 10 mM EDTA and is only slowly chelated by extensive dialysis against 10 mM 1,10-phenanthroline [34]. Zinc contents of g32P-(A+B) are identical to those of the intact protein (Table 3); hence the coordinating ligands to the Zn(II) ion are contained within the structure of the DNA-binding core of the protein.

3.2.2. Reversible Removal of the Zn(II) Ion from Gene 32 Protein

While chelating agents are relatively ineffective in removing the intrinsic Zn(II) from g32P, reaction of the protein with the organomercurial reagent under the conditions described earlier for *E. coli* RNA polymerase (see Sec. 2.2) results in the titration of three of

TABLE 3

Metal Contents of Various g32P and g32P-(A+B) Preparations

Treatment	Metal content (g at/mol protein)			Designation
	Zn(II)	Cd(II)	Co(II)	
Untreated	1.01 ± 0.05	—	—	Zn(II)
10 mM EDTA	1.03 ± 0.06	—	—	—
10 mM dipicolinic acid	1.06	—	—	—
PMPS/EDTA/thiol	0.05 ± 0.03	—	—	Apo
10 mM o-phenanthroline, 48 hr	0.06 ± 0.02	—	—	Apo
Apo + Zn(II)	1.07 ± 0.10	—	—	Zn(II)-R
Zn(II) g32P + 1 mM $CdCl_2$	0.11 ± 0.06	0.83 ± 0.10	—	Cd(II)
Zn(II) g32P + 0.1 M $CoCl_2$	0.08 ± 0.03	—	0.94 ± 0.05	Co(II)
Proteolysis ⟶ core g32P				
Untreated	0.99 ± 0.02	—	—	Zn(II)core
10 mM EDTA	0.98	—	—	—

Source: Data summarized from Refs. 34, 35, and unpublished observations.

the four SH groups in the molecule and the simultaneous release of one Zn(II) ion as detected by color development with PAR [Fig. 1(b)] [34,46]. As with RPase, the Zn(II) can then be totally removed from the protein solution by dialysis against 10 mM EDTA and the bound PMPS released from the protein by dialysis against excess thiol [34, 35]. A stable apo g32P can then be maintained either in the presence or absence of thiol reagents. It is possible to reincorporate 1 g at/mol of Zn(II), Co(II), or Cd(II) into apo g32P by adding excess metal ion in the presence of thiol followed by extensive dialysis during which the metal content falls to 1 g-at/mol protein [34]. The intrinsic Zn(II) ion can also be exchanged with Cd(II) or Co(II) by dialysis of the native protein against a large excess of either of these metal ions (Table 3).

3.3. Contribution of the Zn(II) Ion to the Structure of Gene 32 Protein

3.3.1. Limited Proteolysis as a Probe of Gene 32 Protein Structure

The great stability of the DNA-binding core of g32P (amino acid residues 22-253) could have been an earlier clue to the possibility that the protein contained a metal ion than the recognition of the Cys-77-X_3-His-81-X_5-Cys-87-X_2-Cys-90 as a possible metal-binding cluster that finally led to the zinc analysis [34,46]. It is, in fact, the high structural stability conferred on the core domain that protects this domain from proteolysis and allowed the exploration of function with the limited proteolysis studies. As shown in Fig. 8(a), treatment of the native g32P with trypsin at 25°C leads to the appearance of stable g32P-(A+B) after 10 min with essentially no further proteolysis during the next 40 min [34,35]. In marked contrast, trypsinolysis of the apoprotein at 25°C leads to extensive proteolysis of the entire protein with only a transient appearance of a small amount of the band corresponding to g32P-(A+B) [Fig. 8(a)].

If the apoprotein is reconstituted with either Co(II) or Cd(II), both ions are equally as effective as Zn(II) in protecting the DNA-

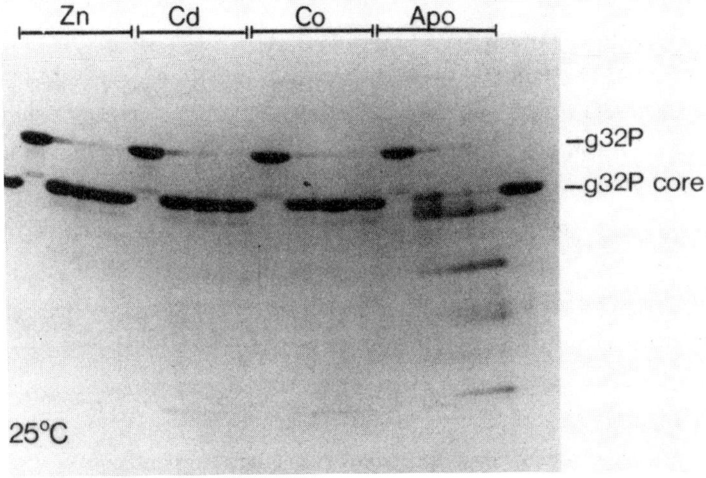

(a)

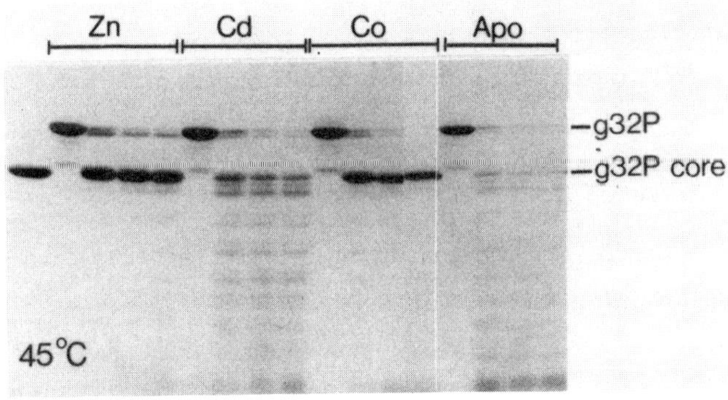

(b)

FIG. 8. Limited proteolysis of metallo g32Ps. (a) g32Ps were treated with trypsin at a 25:1 (w/w) g32P:trypsin ratio at 25°C. Aliquots were removed at the times indicated below and prepared for SDS-PAGE. Lanes 1 and 18 are g32P-(A+B) standards; lanes 2-5, Zn(II) g32P digested for 0, 10, 30, and 50 min; lanes 6-9 Cd(II) g32P digested over the same time course; lanes 10-13, Co(II) g32P digested over the same time course; lanes 14-17, Apo g32P digested over the same time course. (b) Limited proteolysis of metallo g32Ps as in (a) except at 45°C and the g32P/trypsin ratio was 100:1. Lanes correspond exactly to those in (a). (Data taken from Ref. 34.)

binding core from proteolysis at ambient temperature [Fig. 8(a)]. If the protein is exposed to trypsin at 45°C, the core containing Cd(II) becomes moderately more susceptible to proteolysis while the Zn(II) and Co(II) derivatives remain resistant, suggesting that the larger Cd(II) ion has not stabilized the protein to the same degree as the two smaller metal ions [34]. On the other hand, the apoprotein at 45°C is rapidly hydrolyzed to small peptides [Fig. 8(b)]. From these experiments, one can postulate that the coordinated Zn(II) ion must significantly reduce the conformational motion inherent in the nonchelated polypeptide, thus making individual peptide bonds less likely to assume a transiently "unfolded" conformation that exposes them to the active site of an attacking protease. These conclusions are supported directly by differential scanning calorimetry data (Sec. 3.3.3) which show that the metal ion stabilizes the protein against thermal unfolding. As will be shown below in detail, the binding of both the Co(II) and Cd(II) ions confer on g32P DNA-binding affinities and binding modes that are essentially indistinguishable from those of the native Zn(II) protein.

3.3.2. Coordination Chemistry of the Metal-Binding Site in Gene 32 Protein

The release of the Zn(II) ion concurrently with the titration of three of the four Cys-SH groups in g32P led to the postulate that the three closely spaced Cys residues—77, 87, and 90—were ligands to the Zn(II) ion (Fig. 5) [34]. Within this cluster of potential Cys ligands is found His-81, which could be a suitable ligand donor in constitution of a S_3N tetrahedral ligand field around the metal ion. Indeed, when Co(II) is substituted for Zn(II) in g32P, the protein has an optical absorption spectrum diagnostic of Co(II) in a tetrahedral ligand field with the addition of strong ($\varepsilon > 10^3$) charge transfer bands in the near-UV below 400 nm typical of $-S^- \rightarrow$ Co(II) charge transfer (Fig. 9) [88]. The relative positions of the d-d absorption bands at 680 nm (ε = 480 M^{-1} cm^{-1}), 645 nm (ε = 660 M^{-1} cm^{-1}), and 608 nm (ε = 430 M^{-1} cm^{-1}) are overall quite similar to those observed for Co(II) substituted at the active center of

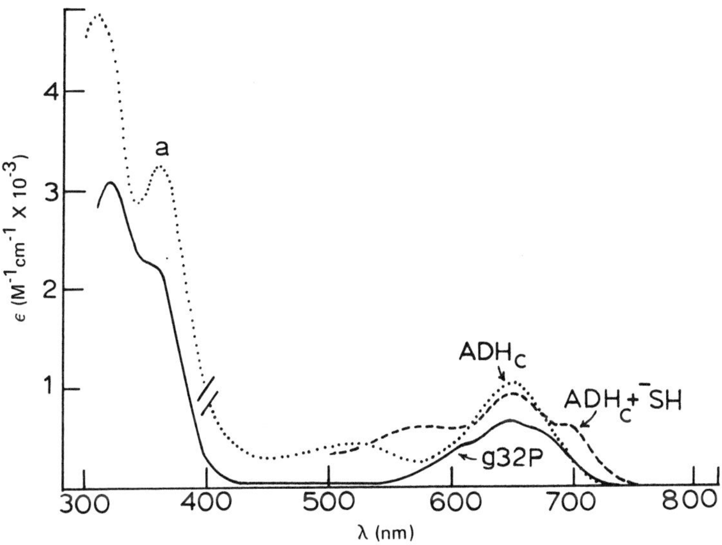

FIG. 9. Visible absorption spectra of Co(II)-substituted g32P (——); liver alcohol dehydrogenase, catalytic sites (····); liver alcohol dehydrogenase, catalytic sites, in the presence of the inhibitor $^{-}$SH (---). Conditions: 100 μM Co(II) g32P corrected for absorption of the Zn(II) g32P, 10 mM Tris-HCl, 5% (v/v) glycerol, 0.2 M NaCl, pH 8. (Data taken from Ref. 34.)

alcohol dehydrogenase with the addition of an $^{-}$SH ligand from solution replacing the usually coordinated H_2O, creating S_3N-distorted tetrahedral coordination around the Co(II) (Fig. 9) [48]. Splitting of the $^4A_2 \rightarrow {}^4T_1(P)$ visible transition of alcohol dehydrogenase into three components with energy separations greater than that expected from spin-orbit coupling alone (>1000 cm^{-1})[89] is consistent with significant distortion in metal-to-ligand bond length and/or angle in the tetrahedral geometry as has been observed crystallographically [90,91]. In the case of g32P, however, rather less splitting of the absorption envelope suggests comparably less deviation from perfect tetrahedral symmetry [88].

^{113}Cd NMR of ^{113}Cd(II) substituted at Zn(II)-, Cu(II)-, and Ca(II)-binding sites in metalloproteins has been a valuable probe of structure surrounding such macromolecular sites [92,93]. The range. of chemical shift observed for the $^{113}Cd(II)$ ion bound in various protein molecules of defined three-dimensional structure is over 800 ppm, from -125 for certain symmetrical Ca(II)-binding sites with oxygen ligands to 750 ppm for sites made up of four $-S^-$ ligands [chemical shifts relative to $^{113}Cd(II)$ perchlorate] [92]. The magnitude of $^{113}Cd(II)$ chemical shifts has proved to be generally rather diagnostic of the chemical nature of the ligands. $^{113}Cd(II)$ coordination complexes in metalloproteins containing sulfur ligands have signals extending from 370 ppm for coordination spheres containing one sulfur ligand [94] to 751 ppm for a tetrathiolate complex formed by the four $Cys-S^-$ ligands at the structural Zn(II) site of alcohol dehydrogenase [95]. There have not been good examples of $^{113}Cd(II)$ in macromolecular sites containing three sulfur ligands, but two $Cys-S^-$ ligands to $^{113}Cd(II)$, as in the $^{113}Cd(II)$-substituted active site of alcohol dehydrogenase, create a ^{113}Cd signal at 519 ppm [95]. The ^{113}Cd NMR signal of $^{113}Cd(II)$-substituted g32P is shown in Fig. 10 and has a chemical shift of 637 ppm, which places it well within the range expected for the presence of three $Cys-S^-$ as ligands [96]. Thus, a S_3N ligand donor set as proposed is consistent with the spectroscopic properties of both the Co(II)- and $^{113}Cd(II)$-substituted g32Ps.

The optical absorption spectrum of Co(II) g32P and the NMR signal of $^{113}Cd(II)$ g32P can both be used to probe for changes in the environment around the metal ion when the protein binds to nucleotides. Binding of either $p(dT_8)$ or poly(dT) to the Co(II) protein causes no significant changes in the energies or the intensities of the d-d electronic transitions [97]. Likewise binding of $d(pA)_6$ to the $^{113}Cd(II)$ g32P causes only a 2.5 ppm downfield shift of the resonance (Fig. 10, spectrum B). At a subsaturating concentration of $d(pA)_6$ (0.5 M equivalents), the observed resonance is positioned approximately between the free and bound lines (spectrum C), indicative of fast exchange on the NMR time scale between bound and free forms of ^{113}Cd g32P (K_{off} 220 sec^{-1}) [96]. This suggests that while the Cd(II) ion senses the binding of the nucleotide, it

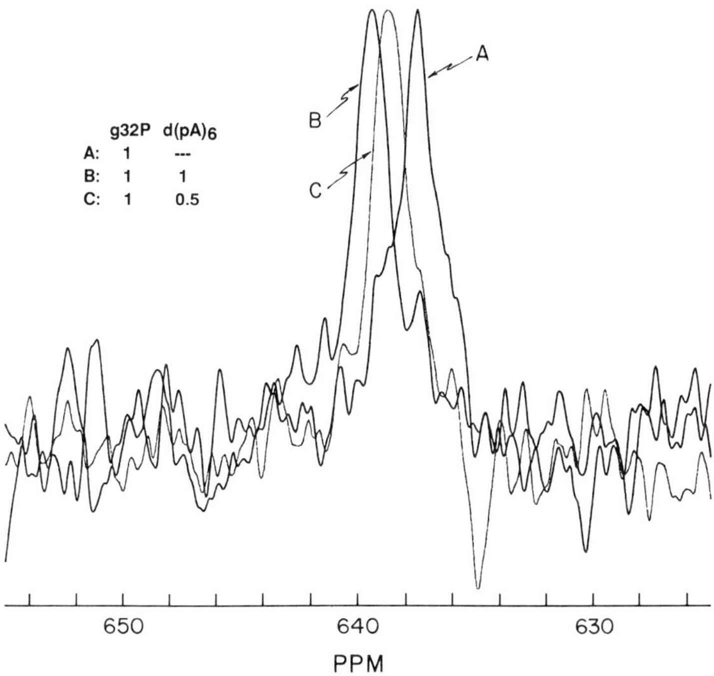

FIG. 10. ^{113}Cd NMR (110.9 MHz) of gene 32 protein. The effect of $d(pA)_6$ binding on the chemical shift of ^{113}Cd in g32P-(A+B). The resonances labeled A, B, and C represent free core protein, a 1:1 nucleotide-protein complex, and a 0.5:1 nucleotide-protein complex, respectively. NMR conditions: 0.7 mM g32P-(A+B) in 50 mM sodium phosphate, 30 mM NaCl, pH 8, 298°K, pulse angle = 45°, sweep width = 10,000 Hz, recycle time delay = 1.5 sec. (From D. P. Giedroc, J. E. Coleman, I. M. Armitage, and B. A. Johnson, in preparation.)

does not appear likely that the DNA ligand induces any large changes in the coordination geometry which exclude the possibility of expansion from four coordination to five coordination or ligand substitution/exchange upon oligodeoxyribonucleotide binding.

3.3.3. Does Zn(II) Induce Any Change in the Average Conformation of Gene 32 Protein?

3.3.3.1. Circular Dichroism of Gene 32 Protein. The circular dichroism (CD) of native g32P in the far-UV shows a spectrum of a folded polypeptide that does not contain a large amount of α helix [86]. Although a moderate negative CD band is present at 222 nm [86], α helix is probably not the predominant secondary structure. Indeed secondary structure prediction algorithms applied to g32P suggests 36% α helix, 18% β sheet, and 46% random coil [98]. Removal of the intrinsic Zn(II) by the PMPS method does not produce large alterations in the CD spectrum, since the CD spectrum of the apoprotein is similar to that published for the native g32P [99] [Fig. 11(a), insert]. Small but significant changes do occur, however, in the secondary structure as can be shown by the difference CD spectrum formed by subtracting the apoprotein spectrum from that obtained for the same apoprotein sample reconstituted with Zn(II). Such a CD difference spectrum shows alterations in both the far-UV CD contributed by the peptide bond chromophores and the near-UV where dyssymmetry of the aromatic chromophores would be expected to contribute [Fig. 11(a)]. The far-UV difference has a maximum just above 222 nm, with the reconstituted Zn(II) protein having a molar ellipticity approximately $1 \cdot 10^3$ deg cm^2 $dmol^{-1}$ greater than the apoprotein. If one assumes 100% α-helix formation, for example, the magnitude of such a change is consistent with 7 ± 1 helical residues suggesting that Zn(II) induces the formation of moderate amounts of additional secondary structure. The same CD changes result upon reconstitution of the apoprotein with Cd(II) [Fig. 11(a)], ruling out the possibility that the induced negative ellipticity at 222 nm results from $S^- \rightarrow$ Zn(II) charge transfer bands, which themselves can exhibit considerable optical activity in a dyssymmetrical protein environment.

The initial CD studies of g32P showed that the Trp chromophores of the folded protein experience sufficiently dyssymmetrical environments as to induce relatively large CD bands in the region of 250-320 nm [86]. The detailed changes in these CD bands as a function of the presence or absence of Zn(II) are shown in Fig. 11(b). There must be significant changes in the conformation or microenvironment of the Trp residue(s) when Zn(II) is removed, since relatively large changes occur in these CD bands [Fig. 11(b)] [99].

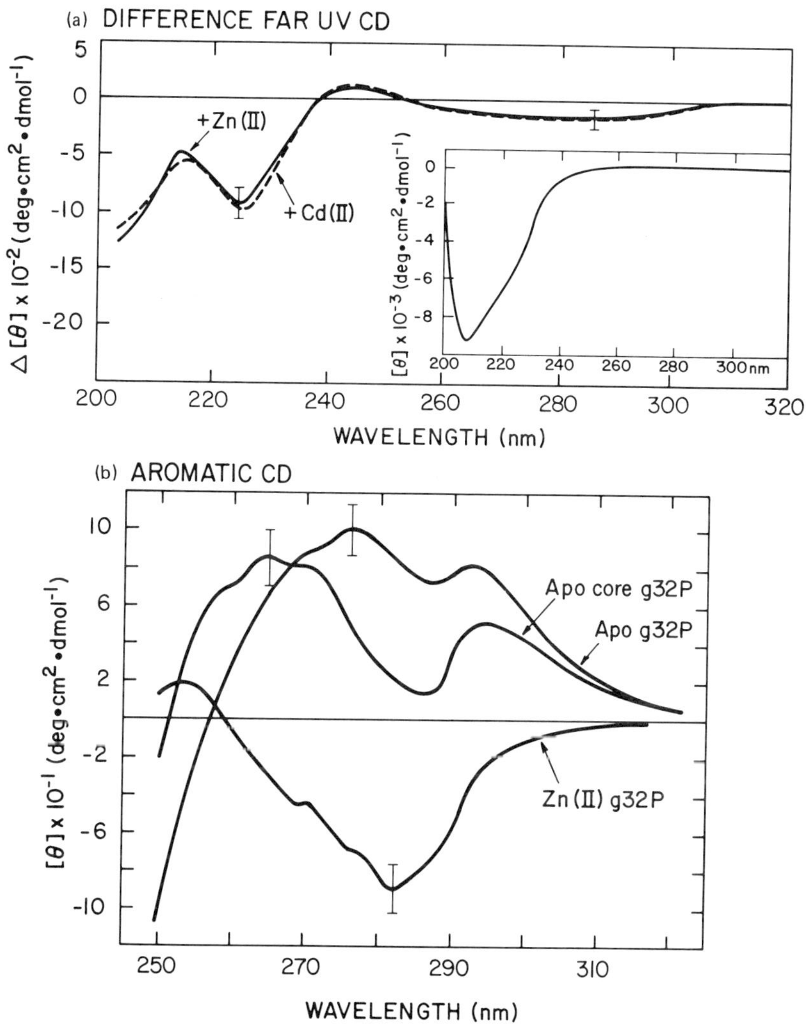

FIG. 11. Effect of Zn(II) or Cd(II) reconstitution on the circular dichroism (CD) of apo gene 32 protein. (a, inset) Ultraviolet CD of 10 μM metal-free apo g32P prepared by the PMPS method. The major part of the figure is the difference CD ($\Delta\theta_M$ in deg cm^2 dmol^{-1}) of apo g32P + Zn(II) [or Cd(II)] less that of apo g32P. Conditions: 50 mM sodium phosphate, 30 mM NaCl, pH 8. (b) Aromatic CD spectra of 300 μM native Zn(II) g32P, apo g32P, and the apo g32P-(A+B). Buffer conditions are as in (a). (D. P. Giedroc and J. E. Coleman, in preparation.)

Of the five Trp residues in the molecule, two (Trp-72 and Trp-116) flank the Zn(II)-binding amino acid sequence and are located adjacent to Tyr-73 (shown to have an important role in maintaining the protein structure by the Tyr-73 → Ser-73 mutant) and Tyr-115 (shown to participate directly in DNA binding by the Tyr-115 → Ser-115 mutant) (see Sec. 3.1 above). Thus it does not seem unduly speculative to suggest that Zn(II) binding influences the local protein conformation and, more particularly, that it may do so within the nucleotide-binding domain. A third Trp, Trp-31, which is rather close in the primary structure to the N-terminal B domain, may be sensing removal of Zn(II) from the g32P-(A+B) in that apo g32P-(A+B) and apo g32P exhibit distinctly different average Trp CD chromophores [Fig. 11(b)]. As will be domonstrated (Sec. 3.4), the loss of Zn(II) from the N-terminal region of the nucleotide-binding core influences the thermodynamics of g32P binding to ssDNA in much the same way as loss of the N-terminal B domain of the polypeptide, i.e., both result in a reduction of cooperative binding interactions. The above conformational probes suggest that the Zn(II) domain and the B domain may have some mutual influence on the conformation of the N-terminal half of the protein where a large part of the nucleic acid-binding domain appears to be located.

3.3.3.2. Differential Scanning Microcalorimetry of Gene 32 Protein: The Effect of Zinc. Differential scanning microcalorimetry has been performed on a number of zinc metalloproteins and their corresponding metal-free apoproteins [100-103]. The results have given an excellent picture of the quantitative thermodynamic effects of metal binding to these proteins. One very obvious qualitative result has been the finding that the Zn(II)-binding factor contributes largely to the thermal stability of these proteins, raising the midpoint of the thermally induced unfolding T_m by 20-25°C [100-103]. The thermal transition also becomes much narrower, i.e., the unfolding is more cooperative. The large negative free energy of stabilization of these zinc metalloproteins has been associated with a large enthalpy term ΔH, contributed by formation of the Zn(II) complex and reflected in a much larger enthalpy of unfolding for the native zinc proteins compared to the apoproteins [100-103].

Zn(II) binding to apo g32P can qualitatively be said to bring about the same result: the apo g32P undergoes a thermal transition with a T_m of 49°C, while the thermal transition of the Zn(II) g32P has a T_m of 56°C [Fig. 12(a)] [104,105]. The latter is a more cooperative transition than that observed for the apoprotein. There is, however, a significant departure in the data compared to most Zn(II) metalloproteins. The enthalpy of unfolding of the Zn(II) protein is 121 kcal/mol compared to 72 kcal/mol for the zinc-free apoprotein, a difference of 49 kcal/mol. The latter ΔH value is similar to those observed for other Zn(II) metalloproteins, but it is associated with only a 6°C rise in T_m [105]. Thus in contrast to other Zn(II) metalloproteins, the relatively large enthalpy term associated with zinc binding is not accompanied by a proportionately large thermal stabilization of the folded protein. It therefore seems reasonable to conclude that Zn(II) binding is associated with some conformational changes in the molecule that bring about a significant decrease in the entropy of the folded protein, significantly reducing the increased thermal stability expected from the formation of a highly stable Zn(II)-thiolate complex. Such conformational changes need not be large and the thermodynamic picture appears to fit well with the detection of small but significant changes in conformation induced by Zn(II) and detected by the circular dichroism.

While not obtainable with the same degree of accuracy as ΔH, the entropies of denaturation calculated for the apo g32P and Zn(II) g32P are 0.26 and 0.42 $kcal \cdot K^{-1} \cdot mol^{-1}$, respectively, consistent with the conclusion that the Zn(II) protein has a considerably smaller entropy in the *folded* form than does the apoprotein [105].

Binding of a ligand such as a polynucleotide in a cooperative fashion to many molecules of a protein would be expected to stabilize that protein against thermal unfolding. Indeed, Zn(II) g32P complexed with poly(dT) results in a T_m of 60°C for the thermal unfolding of the protein and a much more highly cooperative transition than observed

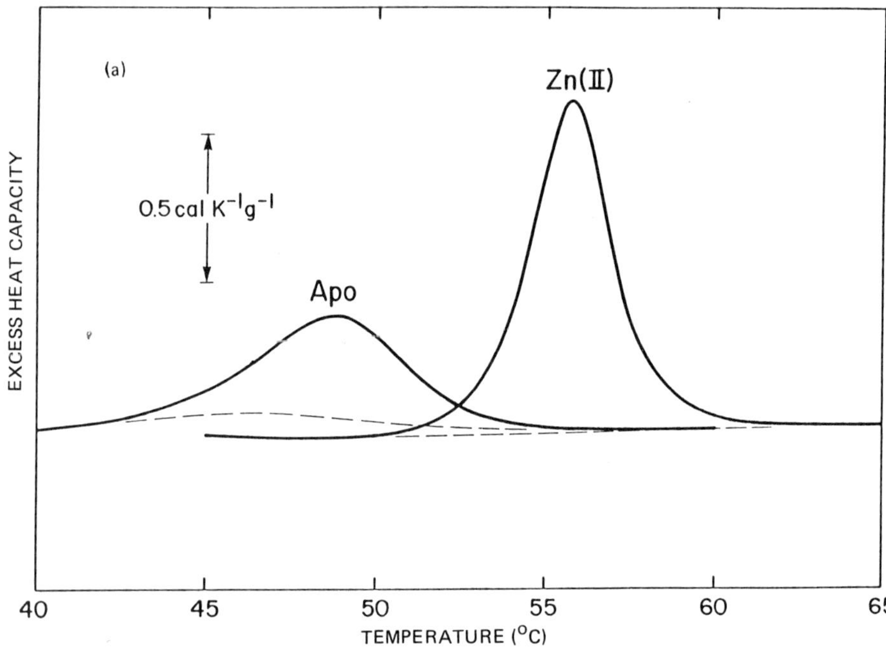

FIG. 12. Effects of Zn(II) and poly(dT) on the thermal unfolding of gene 32 protein as measured by differential scanning calorimetry (DSC). (a) DSC scans of apo and Zn(II) g32P. (b) DSC scans of apo, Zn(II), Co(II), and Cd(II) g32Ps in the presence of saturating (stoichiometric) concentrations of poly(dT). (Data from Ref. 105.)

for the protein alone [Fig. 12(b)]. The unfolding of the Co(II) g32P in the complex with poly(dT) suggests that the Co(II) derivative may actually form a slightly more stable complex with poly(dT) than the native Zn(II) protein. The Cd(II) protein in complex with poly(dT) unfolds at a significantly lower temperature, confirming the moderately lower stability of the unliganded Cd(II) protein first indicated by the proteolysis experiments (Fig. 8).

The apo g32P binds to ssDNA as shown by a number of criteria including electron microscopy [104], quenching of the protein fluorescence by nucleotides [34], and destabilization of ds poly[d(A-T)] under conditions of very low salt [35]. The mode of binding of poly(dT) to the apoprotein, however, is dramatically different from

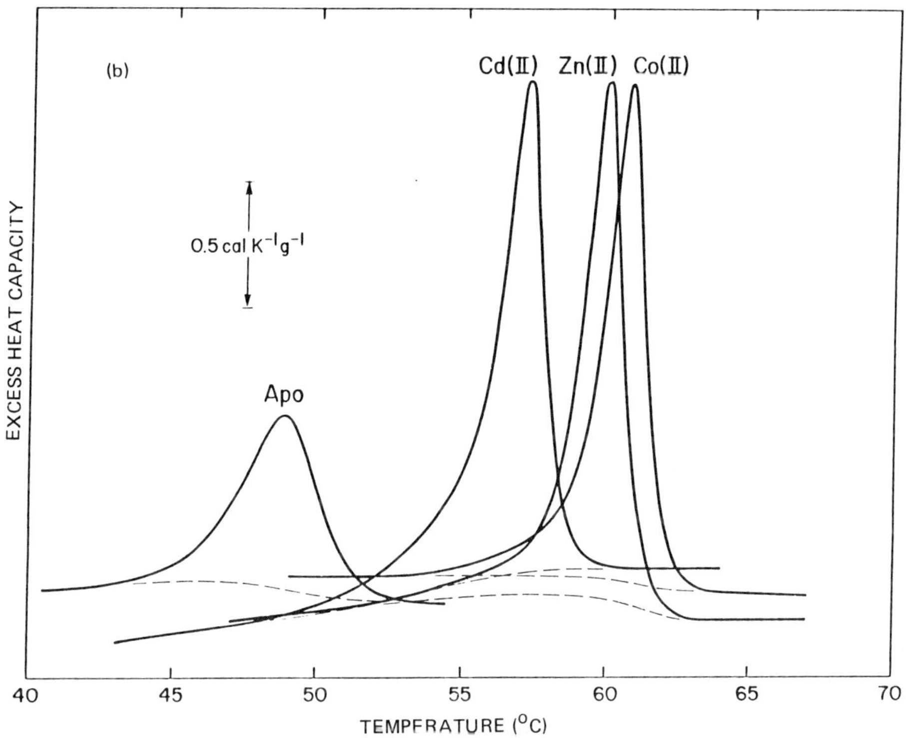

that of the metalloprotein, since poly(dT) does not significantly stabilize the apoprotein to thermal unfolding [Fig. 12(b)]. The T_m remains 49 ± 1°C although the binding of poly(dT) is reflected in a moderate increase in the denaturation/enthalpy reflecting polynucleotide dissociation upon thermal unfolding of apo g32P [Fig. 12(b)]. Clearly, the interaction of polynucleotides with the g32P in the absence of Zn(II) has been radically altered from the interaction characteristic of the metalloprotein. The quantitative documentation of these differences and their possible functional significance are covered in the following Sec. 3.4.

3.4. The Intrinsic Zn(II) of Gene 32 Protein Makes Possible a Mode of Binding to Single-Stranded Polynucleotides Essential for Its Function

The binding constants of gene 32 protein for a variety of oligo- and polynucleotides can be measured by several methods including quenching of the protein fluorescence by nucleotide binding, enhancement of the fluorescence of polyethenoadenylic acid by binding of g32P, and can also be calculated from the melting of ds poly[d(A-T)] induced by the binding of g32P [35]. The association constants K_a determined by these methods using a variety of oligo- and polynucleotides as ligands for g32P and comparing the binding constants of the apo, Zn(II), Co(II), and Cd(II) g32Ps are collated in Table 4. The general conclusion can be stated as follows: Removal of the Zn(II) ion maximally reduces the K_a for ssDNA by approximately three orders of magnitude. This maximum differential binding can best be gauged from the poly[d(A-T)] melting data, since fluorescence quenching titrations with poly(dT) cannot accurately measure the large K_a values for the Zn(II) protein, and represent only a minimum value. On the other hand, the fluorescence-quenching titrations demonstrate that upon binding to a single-site nucleotide lattice, e.g., $d(pT)_8$, there is relatively little difference in K_a between the apo and

TABLE 4

Association Constants (M^{-1}) and ω Values Derived From the Binding of Apo and Metallo g32Ps to a Variety of Oligo- and Polynucleotide Lattices

g32P	$d(pT)_8$	$d(pT)_{16}$	ω	poly(dT)	poly[d(A-T)]
Zn:	$4 \cdot 10^5$	$5 \cdot 10^7$	3,800	$>10^8$	$5 \cdot 10^{10}$
Cd:	$4 \cdot 10^5$	$3 \cdot 10^7$	2,000	$>10^8$	$13 \cdot 10^{10}$
Co:	$5 \cdot 10^5$	$4 \cdot 10^7$	1,500	$>10^8$	$24 \cdot 10^{10}$
Apo:	$8 \cdot 10^4$	$3 \cdot 10^5$	4	$3 \cdot 10^7$	$4 \cdot 10^7$

Source: Data from Ref. 35.

metallo g32Ps (Table 4). When even a two-site lattice is used, e.g., $d(pT)_{16}$, a large difference in K_a is observed between the Zn(II) and apo g32Ps (Table 4). All data taken collectively are consistent with a change in the cooperativity factor ω from 10^3 to ≤ 50, as shown in Table 4. This assumption is compatible with the data derived from the melting of poly[d(A-T)] where the K_as for all the metallo g32Ps are at least 10^3-fold greater than the K_a for the apoprotein (Table 4) [35]. Finally, RNA titrations using polyethenoadenylic acid carried out at low NaCl concentrations and curve-fit to ascertain K_{int} and ω [75] also indicate a lower binding constant of the apo g32P for this lattice, a decrease almost exclusively associated with the cooperativity parameter ω (Table 4).

It is thus a reasonable model that the Zn(II) domain of the protein, like the N-terminal B domain, contributes importantly to the cooperativity of g32P binding to ssDNA and therefore to the modulation of the protein-protein interactions essential to the protein-DNA interaction. The qualitative features of the melting profiles for poly[d(A-T)] induced by the metallo and apo forms of g32Ps are of interest in this regard and are shown in Figure 13. At 150 mM NaCl, which affords reasonable stability to the double helix, $T_m = 65°C$, Zn(II), Co(II), Cd(II), and apo g32P reconstituted with Zn(II) are all extremely effective in melting poly[d(A-T)], decreasing the melting temperature to below 22°C [Fig. 13(a)] [35].

The apo g32P will not melt poly[d(A-T)] under these conditions [Fig. 13(b)]. Melting of poly[d(A-T)] by apo g32P might begin at higher temperature under these conditions, but the apo protein unfolds before this can be tested. By lowering the NaCl concentration to 25 mM, which destabilizes the double helix, T_m - 52°C, the apo g32P can be shown to destabilize poly[d(A-T)] and induce a very broad melting curve, probably with a T_m falling just below 30°C [Fig. 13(c)]. The apoprotein unfolds, however, before this melting curve is complete. Another species binding noncooperatively to DNA, g32P-B, also shows this broad "noncooperative" type of melting curve which can be completed because the Zn(II) domain of g32P-B is intact

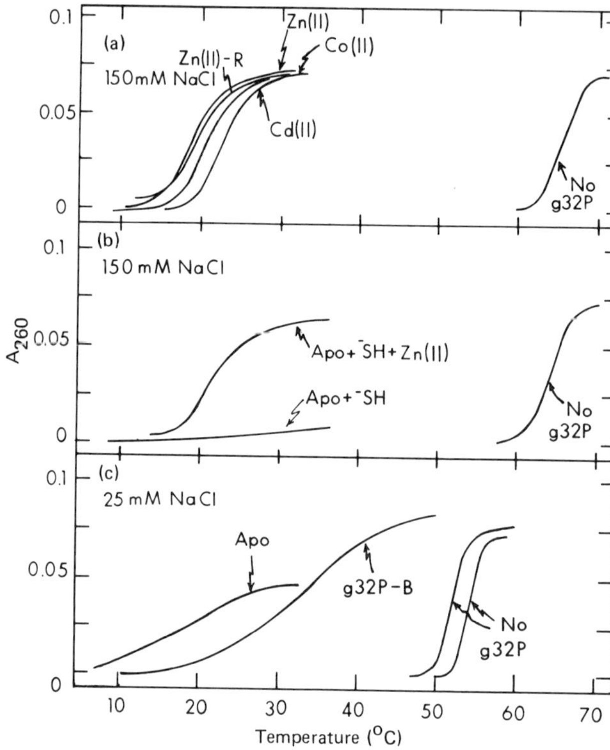

FIG. 13. Thermal denaturation of poly[d(A-T)] by various gene 32 proteins as detected by hyperchromism at 260 nm. (a) Melting of poly[d(A-T)] at 150 mM NaCl in the absence and presence of Zn(II) (native), Cd(II), Co(II), and apo g32P reconstituted with Zn(II), g32P Zn(II)-R. (b) Thermal denaturation of poly[d(A-T)] at 150 mM NaCl by apo g32P + thiol + Zn(II) and by apo g32P + thiol. (c) Thermal denaturation of poly[d(A-T)] at 25 mM NaCl by apo g32P, g32P-B. In all parts of the figure g32P was usually present in a concentration twofold over nucleotide-binding sites (n = 7) except for g32P-B, 4.5-fold excess over the nucleotide sites. (Data taken from Ref. 35.)

and the protein does not unfold during the transition [Fig. 13(c)] [79]. While these broad, noncooperative melting curves cannot be directly translated into decreased cooperativity parameters, to a first approximation it seems likely that g32P-B and the apo g32P have formally similar alterations in their mode of binding to ssDNA.

Obviously, the molecular details of the structural changes of these two species must be different (see below).

As in vitro experiment demonstrating the potential functional consequences of this change in binding mode when Zn(II) is removed from g32P can be set up using the native and apo g32Ps to protect single-stranded fd DNA from digestion by *N. crassa* endonuclease [106]. Such an experiment is shown in Fig. 14 [35]. Conditions were set up

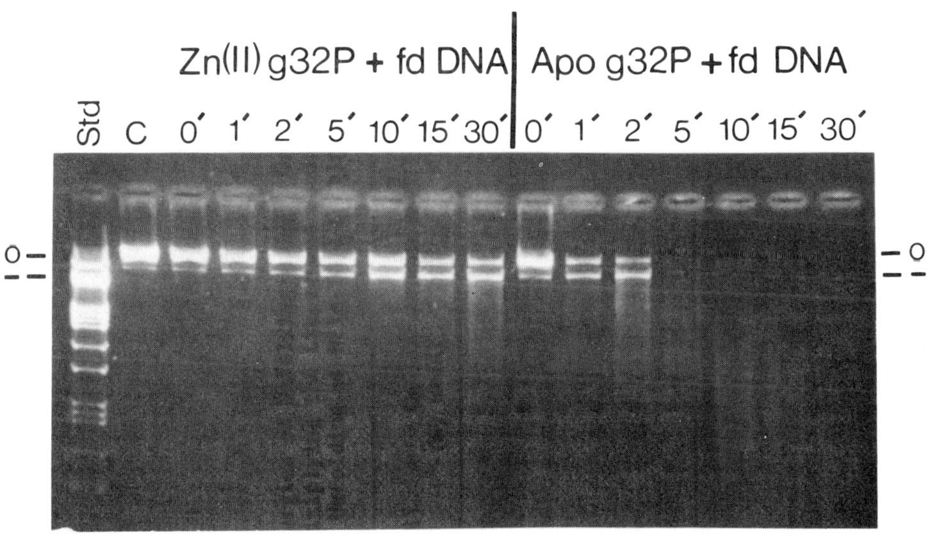

FIG. 14. Endonuclease digestion of g32P-fd DNA complexes as monitored by agarose gel electrophoresis. Digestions by *N. crassa* endonuclease of g32P-fd DNA complexes formed with native Zn(II) g32P or apo g32P were carried out for 1, 2, 5, 10, 15, and 30 min. The migration of intact closed-circular and that of once-nicked fd DNA species are designated by a circle and a line as shown. St. ds DNA molecular weight ladder derived from digestion of T7 DNA with *HpaI*; C, fd DNA standard; 0', controls, i.e., DNAs recovered from 30 min, 37°C, incubation of fd DNA with Zn(II) g32P or apo g32P as indicated with no exogenous nuclease added. (Data taken from Ref. 35.)

so that both native g32P and the apoprotein completely saturated the fd DNA circle. The native protein protects the DNA from significant digestion under these conditions for up to 30 min, while the same degree of digestion of the DNA is observed in the presence of the apoprotein within 2 min and the DNA is completely digested within just 10 min.

3.5. Conclusions

Gene 32 protein appears to require both an intact N-terminal B domain and a calorimetrically rigid Zn(II) domain within the N-terminal region of the core DNA-binding domain in order to initiate cooperative binding to a polynucleotide. Both domains must be intact for cooperative binding to ssDNA. Thus each is required but not sufficient for the protein-protein interactions between adjacent g32P molecules bound on a long nucleotide lattice that are responsible for the cooperative feature of DNA binding. A schematic model of the possible relationship of these domains is shown in Fig. 15. It is possible in a situation where several simultaneous surface interactions are required (i.e., between domain B and the neighboring g32P molecule, between the core domain and domain B, and finally between the core domain and the DNA) that the Zn(II) confers on the core domain, in contact with both the B domain and the nucleotide, a rigidity or damping of its motion that greatly stabilizes these multiple surface interactions, perhaps not unlike the function of Zn(II) in stabilizing the R-C interface of aspartate transcarbamoylase.

4. DO OTHER SINGLE-STRANDED NUCLEIC ACID BINDING PROTEINS USE ZN(II)-BINDING DOMAINS?

4.1. Introduction

Single-stranded DNA- or RNA-binding proteins which include T4 g32P now constitute a large class of proteins with representative members

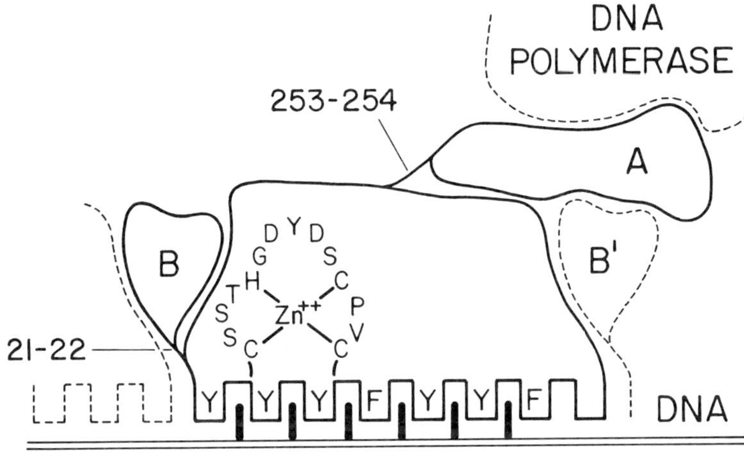

FIG. 15. Schematic model of T4 gene 32 protein incorporating the Zn(II) domain. The N-terminal B domain (residues 1-21) interacts with the adjacent g32P molecule, while the C-terminal A domain (residues 254-301) interacts with the DNA polymerase [69]. The Zn(II) domain, encompassing residues 77-90, while near the nucleotide-binding domain, which must include Tyr-115 and Tyr-137 as well as several other Tyr and Phe residues (see text), plays a major role in stabilizing the interaction with the adjacent g32P monomer perhaps through direct contact with the B domain on either monomer.

extending up the evolutionary tree from bacteriophages to mammalian cells [65,67]. As primary structures of individual proteins of this group are determined, one can begin to ascertain whether any general structural motifs can be identified which would characterize these proteins as a class. Such structural motifs might include the use of aromatic amino acid side chains to form base-binding pockets or the incorporation of zinc-binding domains. While the participation of aromatic amino acids in nucleotide interaction has been clearly identified in one other example, i.e., gene 5 protein from bacteriophage fd [107], and may be involved in the nucleotide-binding surface of a small protein, A1, which binds to heteronuclear RNA [108,109], Zn(II) is not present in the two other most thoroughly studied proteins of this group, fd gene 5 protein and the SSB from *E. coli* [34]. The latter appears to be involved in DNA replication in *E. coli*, much as g32P is involved in T4 DNA replication [66,67]. However, it is apparent that the ssDNA-binding modes of g32P and SSB vary significantly from one another in that SSB binds as a tetramer with limited cooperativity between dimers of tetramers in contrast to the highly cooperative monomer-monomer polymerization of g32P along the ssDNA [110,111].

Another group of low molecular weight, basic proteins derived from the *gag* gene (see Sec. 4.2) of retroviruses function in vitro as single-stranded nucleic acid-binding proteins [112-117], whose in vivo role is best described as binding to the genomic RNA during encapsidation. Upon infection of a new cell or even within the originally infected cell, these proteins may in fact play a regulatory role in the balancing mRNA production vs. the production of RNA for packaging of daughter virions. These *gag*-derived proteins contain one or two repeats of a conserved distribution of three Cys and one His residues in a 14 amino acid sequence which is found in a spacing nearly perfectly inverted to that in g32P [116], suggesting the possibility that these sequences may also represent Zn(II) domains. It is this class of protein which will be discussed below.

4.2. Genetic and Biochemical Studies of the Low Molecular Weight Nucleic Acid-Binding Proteins from Retroviral and Retrovirallike Elements

All retroviruses contain at least three open-reading frames which encode polyproteins which are proteolytically processed to form the structural and enzymatic activities required for replication and propagation of the virus. These include *gag*, which encodes the core structural proteins which constitute the inner shell of the virus and encapsulate the template RNA dimer, *pol*, which encodes the enzymatic activities of reverse transcriptase, protease, ribonuclease, and the integrase, and *env*, which encodes the outer membrane surface glycoproteins which function in extracellular recognition events. Eukaryotic transposable or reverse-transposing elements are characterized by two open reading frames which possess limited but significant sequence homology with the *gag* and *pol* reading frames of retroviruses [118]. These elements also form retroviruslike particles containing element RNA and reverse transcriptase which accumulate in the nuclei of infected cells [119-122]. Examples of such endogenous transposable or mobile elements include the copia [123], copialike [124], and I-factor [125] elements from *Drosophila*, the yeast Ty elements [126], mammalian intercisternal-associated particle (IAP) elements [127], and the Tx elements from *Xenopus* oocytes [128]. Data obtained with the yeast Ty element provide evidence for reverse transcription in these elements as a group as the presence of an RNA intermediate was conclusively demonstrated [129] and thus these can be generally considered retroviruslike or RNA-based mobile genetic elements.

The ultimate or penultimate processed C-terminal product of the *gag* polyprotein from all retroviruses is invariably an extremely basic protein which generally ranges in size from 54 to >90 amino acids. As it is the major virus protein associated with RNA in detergent-disrupted virions [112], it has been termed the basic, low molecular weight nucleic acid-binding protein. When compared as a group across divergent evolutionary lines, these proteins do not exhibit striking sequence homology to one another but they do conserve a Cys-X_2-Cys-X_3-N-His-X_4-Cys 14 amino acid sequence where X is unspecified and N is most often Gly, but occasionally Ser, Thr, or Asp. In most instances but not all, this sequence appears as a tandem repeat separated by a very basic, hydrophilic stretch of 5-12 amino acids (Table 5). A subset of the sequenced aforementioned transposable elements are predicted to contain one or two copies of this Cys-rich domain within their *gag*-related open reading frames and include the *D. melanogaster* copia and I-factor elements, the hamster IAP element and the *Xenopus* oocyte Tx elements (Table 5). Finally, the cauliflower mosaic virus (CaMV), a plant DNA virus which appears to replicate via reverse transcription [130], contains a coat protein gene upstream from the reverse transcriptase locus in which is also found a three-Cys/one-His domain [131].

TABLE 5

Primary and Predicted Secondary Structures of ..Cys-X_2-Cys-X_4-His-X_4-Cys.. 14 Amino Acid Sequence Elements Found in the *gag* Gene of Retroviruses and the *gag*-like ORFs of Various Eukaryotic Transposable or Other Putative RNA-Based Reverse-Transcribing Elements as Compared to the Zinc-Binding Domain of Phage T4 g32P

gag or gag-like polyprotein	Sequence	Reference
avian:		
RSV	. . G L C Y T C G S P G H Y Q A Q C P K K R K S G N S R E R C Q L C N G M G H N A K Q C R K . .	149
	* * * * * * * * * * * * * *	
	t t t t t t t t t t t t β β β β t t t t t t t t t t t t t t t t t t t t t t t t t t t	
mammalian:		
Mo-MLV	. . D Q C A Y C K E K G H W A K D C P K K P R G P[R G P R P Q A S L L]	116
	t t t t t t t t t t t t t t t t t t t t t t t t t t	
FeLV	. . D Q C A Y C K E K G H W V R D C P K R P R - -[K K P - A N S T L L]	150
HTLV-I	. . Q P C F R C G K A G H W S R D C T Q - P R - P P P G P C P L C Q D P T H W K R D C P R . .	151
	β β β β t t t t t t t t t t t t t t t t t t t t t t t t t t t t t t t t t t t t t t t	
HTLV-II	. . Q P C F R C G K F G H W S R D C T Q - P R - P P P G P C P L C Q D P S H W K R D C P Q . .	152
	β β β β t t t t t t t t t t t t t t t t t t t t t t t t β β β β t t t t t t t t t t t t t	
BLV	. . G P C Y R C L K E G H W A R D C P T K A T G P P P G P C P I C K D P S H W K R D C P T . .	153
	* *	
lentiviruses:		
visna	. . Q K C Y N C G K P G H L A R Q C R Q G - - I I C H H C G K R G H M Q K D C R Q . .	154
EIAV	. . Q T C Y N C G K P G H L S S Q C R A P - K - V C F K C K Q P G H F S K Q C R S . .	155
HIV-1	. . V K C F N C G K E G H T A R N C R A P R K K G C W K C G K E G H Q M K D C T S . .	156
	t t t t t t t t t t t t t t t t t t t t t t t t t t t t t t a a a a a a a a	
SIV-I_{MAC}	. . I K C W N C G K E G H S A R Q C R A P R R N G C W K C G K M D H V M A K C P N . .	157
STLV-III_{AGM}	. I K C W N C G K E G H S A R Q C R A P R R Q R C W K C G K M D H V M A K C P D . .	158
HIV-2_{ROD}	. . F K C W N C G K E G H S A R Q C R A P R P N G C W K C G K P G H I M T N C P D . .	159
	* * * * * * * * * * * * * * * *	
DNA virus:		
CaMV ORF IV	. C R C W I C N I E G H Y A N E C P N . .	131
eukaryotic transposable elements:		
copia ORF I	. . V K C H H C G R E G H I K K D C F H . .	123
	+ + + + + + + + + + + + (visna)	
	β β β β t t t t t t t a a a a a a a	
Tx 1,2	. . V R C F L C K N L G H T R Q S C P K . .	128
I Factor	. . L R C K K C L R F G H P T P I C K S . .	125
IAP-18	. . K A C F N C G R M G H L K K D C Q A P E R T R E S K L C Y R C G K G Y H R A S E C G I . .	127
	β β β β β β t t t t t t t a a a a a t t t t t t t t t β β β β t t t t t t t t t t t t t	
T4 g32P:	. . T E C S S T H G D Y D S C P V C Q Y . .	98
	β β t t t t t t t t t t β β β β β β	

*Proceeding down the table, the symbol compares the avian and mammalian *gag* amino acid sequences toward one another, the mammalian sequences as a group, and finally the lentivirus sequences as a group, and denotes identities and conservative substitutions in each case.

+Compares the copia ORF I with the visna *gag* amino acid sequence and denotes identities and conservative substitutions.

a, β, and t denote secondary structure prediction of α helix, β sheet, and β turn, respectively.

The amino acid sequences of these 14 residue units derived from these sources are compiled in Table 5 and are compared to the likely Zn(II) domain of T4 g32P. As already indicated, the sequence of amino acids in g32P, Cys-X_3-His-Gly-X_4-Cys-X_2-Cys, is very nearly a precise inversion of the retroviral and transposable element sequences. With regard to the possible Zn(II)-binding capacities of such a sequence, random three-dimensional structure-generating routines utilizing the distance and bond angle constraints associated with tetrahedral ligation of a Zn(II) ion to the three Cys and one His side chains as applied to the first repeat of the HIV-1 *gag* sequence show that this sequence of amino acids can accommodate such a structure without trivial violation of boundary conditions [132]. The generated structures globally resemble a restrained "loop" structure tethered to the metal by the His side chain [132]. Such a role in metal ion coordination might explain the considerable evolutionary pressure to maintain not only the Cys side chains but the His donor nitrogen atom as well in completing a tetrahedral metal-binding site. Similar extended loop structures about a Zn(II) chelate can be generated for both the forward and inverted polypeptide sequences of the experimentally validated three-Cys/one-His Zn(II) domain in g32P [96].

In a further extension of the similarity between the g32P Cys-rich domain and those of the retroviral and retrotransposable elements, secondary structure prediction algorithms as applied to the primary structures of the *gag* and *gag*-like regions through the Cys-rich domain show a related secondary structure prediction. As shown in Table 5, minimal absolute sequence conservation exists between the putative zinc ligands taken as a group, but all repeats are characterized by a propensity of turn prediction at least through the middle portion of the repeat with some deviation to a generally weak probability of β sheet or α helix toward either the N terminus (i.e., ..Cys-X_2-Cys..) or the C terminus (i.e., ..His-X_4-Cys..), respectively, in some cases (Table 5). Such a secondary structure would seem well suited to accommodate a loop structure around a Zn(II) ion in g32P and, by extension, the retroviral and like sequences. In addition, these domains as a group are by and large polar with glycine, proline, hydrophilic, and charged amino acids overrepresented and bulky hydrophobics underrepresented. One or two consistently placed amino acid side chains are also found within every repeat and may function in ss nucleic acid binding via side-chain/base intercalation, evidence for which has been put forward for Trp-35 in Mo-MLV p10 binding [133]. As such, these domains are likely to be found at or very near the protein surface like many β- and ω-type turns found in proteins of known structure where their roles or putative roles in macromolecular recognition (as described for g32P above) can be asserted [134,135].

Several reports describing in vitro deletion or insertion mutagenesis of these Cys-rich regions within a proviral clone and subsequent transfection into cultured cells show that these mutant proviruses are largely normal for virion particle assembly but are unable to package the RNA and are therefore noninfectious [136,137]. Specifically, deletion of part of all of one of the two Cys repeats in RSV p12 and the single repeat of Mo-MLV p10 resulted in nearly normal particle production (provided the *gag-pol* reading frame was maintained) but were totally defective in RNA packaging. As these defects could be complemented with wild-type *gag* gene, the defect was not at the level of virion RNA structure but within the p12 (or p10) protein itself. In addition, reverse transcriptase levels were normal in all mutants which shows that a functional *pol* gene was present. Finally, the replication-defective simian sarcoma virus contains a seven amino acid insertion between the His and the third Cys residue [138]. On the other hand, a two amino acid insertion between the Cys and His residues of the N-terminal RSV p12 repeat only moderately affected packaging and 35S → 70S dimer RNA formation [137]. A two amino acid insertion outside of the Cys-His repeat in RSV p12 gene gave rise to essentially wild-type virion production [137].

An intact *gag* gene polyprotein appears to be required for the maturation of the retroviruses [139], and the proteolytic processing of the *gag* gene does not appear to occur until particle assembly and budding are well under way [140]. Following processing, the major protein may well be the small basic nucleic acid-binding protein since detergent extraction of the pure virions results in the isolation of a ribonucleoprotein particle enriched in this protein [112]. However, it is unknown whether the major determinants that function in specific aspects of RNA binding during encapsidation actually reside in the *gag* polyprotein, the processed protein itself, or both.

There is conflicting evidence as to the precise nature of the *gag* gene processed protein-RNA interaction as derived from in vivo and in vitro experiments. Meric, Darlix, and Spahr [141,142] showed that when mature Rous sarcoma virions are subjected to UV irradiation and the RNA-protein complexes isolated, the major protein bound is RSV p12. It appears bound, however, at specific sites within the RNA genome which conform to two consensus six-nucleotide sequences containing pyrimidine (typically uridine) and adenine residues. A model was constructed which showed that these binding regions perhaps stabilize the dimer linkage structure and directly affect the 35S $\rightarrow$ 70S RNA dimerization step during the packaging reaction in forming a compact ribonucleoprotein core with large regions of unprotected RNA [141]. The low crosslinking efficiency reported may be of particular importance here. In any case, strong evidence was presented for intermolecular p12-p12 interactions [141].

In contrast, a number of in vitro polynucleotide-binding experiments carried out with RSV p12 (or Mo-MLV p10) purified by denaturation of mature virions, isolation of the viral core, purification and renaturation of highly purified components seem to suggest that the basic nucleic acid-binding proteins bind to homologous and heterologous RNAs as well as synthetic ribo- and deoxyribohomopolymers with approximately equal affinities, i.e., true nonspecific binding [115, 117]. Karpel et al. ([117] and references contained therein) reported stoichiometric and saturable (nonspecific) binding of Mo-MLV p10 and RSV p12 to two synthetic polynucleotides, poly(U) and polyethenoadenylic acid. These studies showed an occluded site size of about six residues and a $K_a = 10^6$-10^7 M^{-1} at low NaCl concentrations.

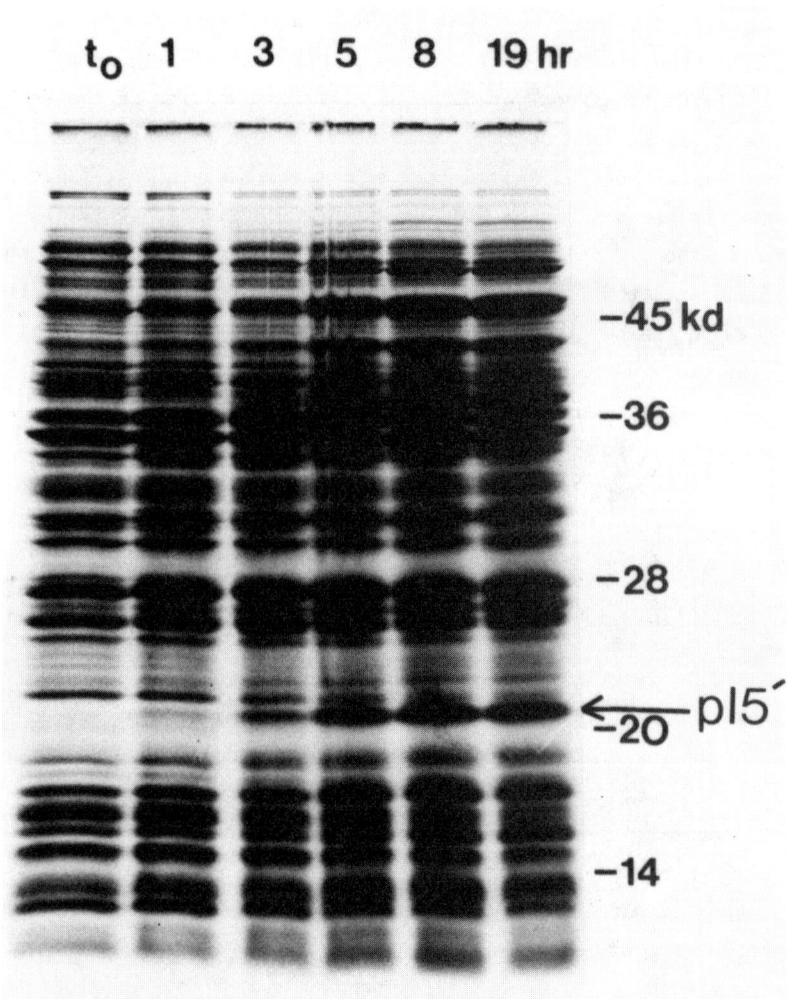

FIG. 16. Subcloning and overexpression of HIV_{gag} p15(p7/p9) in *E. coli* AR120 cells. Time course of nalidixic acid induction of HIV_{gag} p15(p7/p9) fusion protein, p15', under transcriptional control of the phage λ P_L promoter. Upon extraction and purification of the band labeled p15' and subjecting to automated N-terminal sequencing, 38 cycles showed a perfect match for residues expected at the N terminus.

Surprisingly, this binding was noncooperative and was unaffected by oxidation state, carboxyamidomethylation, or performic acid oxidation of the three Cys residues within the 14 amino acid Cys-rich domain of Mo-MLV p10 [117]. Others have reported essentially similar binding characteristics of a totally synthetic Mo-MLV p10 with the Cys side chains completely protected [143]. This protein was devoid of divalent metal ion.

We point out that the metal-free apo g32P binds with a substantially reduced cooperativity approaching unity and exhibits an affinity for polyethenoadenylic acid quite comparable to the above preparations of Mo-MLV p10 [34]. To address whether these retroviral and retrovirallike domains function in Zn(II) binding, which would thereby enhance the overall association of these proteins for polynucleotides through metal ion-induced cooperative interactions between monomers, we have subcloned and overexpressed the *gag* p15 (p7/p9) protein from retrovirus HIV-1 as a fusion protein containing approximately eight amino acids derived from the 3' end of the p24 major capsid protein gene utilizing a bacterial expression system under transcriptional control of the phage λ P_L promoter (details of this construction will be described elsewhere). Promoter induction of *E. coli* cells harboring this plasmid directs the expression of a protein, termed p15', with the expected mobility of ~20 kD and an N-terminal sequence as shown in Fig. 16. Preliminary attempts to purify this protein show it to be associated with an insoluble nucleoprotein complex upon cell disruption. Denaturation using either low pH or urea treatment followed by renaturation in the presence of metal ions generates variable degrees of nucleic acid-binding activity associated with nondialyzable zinc [132]. Purification and characterization experiments are in progress as is construction of alternative expression plasmids in the hope of circumventing the insolubility problem.

5. CONCLUSIONS

Evidence has been presented in Sec. 3 which documents that zinc performs an exclusively structural role in stabilizing g32P against thermal denaturation and proteolytic degradation, and is associated with the induction of some degree of polypeptide folding which enables a conformation essential for high-affinity, cooperative binding to ssDNA. In the simplest sense, the Zn_A metal ion of *E. coli* RPase (Sec. 2) and that of g32P implicate zinc in the stabilization of protein-protein (or subunit-subunit) interactions as the principal role in these two systems rather than protein-nucleic acid interactions per se. A similar scenario might operate in the *E. coli* UvrA protein, a zinc-containing ATPase [36] which binds to chemically damaged double-stranded DNA in a ATP-dependent manner but acts productively in a concerted fashion in the excision of the breakpoint region only in a complex with the other two subunits of the multisubunit UvrABC excision repair system, UvrB and UvrC [144]. This contrasts with a direct role of the two ..Cys-X_2-Cys-X_{19-20}-Cys-X_2-Cys.. putative Zn(II) binding "fingers" in DNA binding which has previously been argued [36].

We suggest that zinc also plays a stabilizing role and may be partly responsible for inducing a specific conformation in the "zinc finger" domains of eukaryotic transcription factors which enables high-affinity binding. There are not, as yet, analytical data on zinc content of the eukaryotic transcription factors other than TFIIIA [145], but in this case, zinc protects TFIIIA domains against proteolysis [38,145]. The finding of a zinc-remedial mutant in the case of yeast GAL4, a transcriptional activator of galactose metabolism [42], coupled with demonstration of zinc-enhanced in vitro binding of bacterially expressed zinc finger protein fragments derived from HeLa cell SV40 promoter transcriptional activator SP1 and yeast mother-daughter cell-specific regulatory protein SW15 to their respective dsDNA elements [146,147] suggest that, in general, a conformation required for specific DNA recognition is at least partially stabilized by zinc. It should be recognized that in all cases amino acid side chains which make actual contact with the DNA remain unknown. The physicochemical studies with g32P show that it is unlikely that the Zn(II) is directly involved in the nucleotide binding. Thus, the possibility that the metal binding domains perform a more accessory function, e.g., local or global stabilization of protein structure which facilitates spacially adjoining regions to make the essential contacts with the DNA and thus is indirectly but intimately coupled to the DNA binding site, must be entertained. In this context, it is useful to point out that a single finger domain derived from the TFIIIA molecule appears unable to bind specifically to its recognition sequence [145] while multifinger domain fragments of TFIIIA and two three-finger domains (SW15 and SP1) all exhibit sequence-specific in vitro binding [146,147]. In the case of GAL4, sequence-specific binding has been obtained with a single proposed Cys_4 finger but inactive mutant proteins were isolated which had point mutations up to 15 amino acids C-terminal to the putative finger [148]. It could be that the conformation of the intervening connecting region (which appears rather well conserved among the Cys_2His_2 domains [147]) plays more than a tethering role in terms of protein-DNA contacts. In fact, in g32P, the zinc binding domain appears just N-terminal, but might be considered coincident at another level of resolution, to the current description of the ssDNA binding groove (Sec. 3.1).

Be it stabilization of protein-protein or protein-nucleic acid interactions, a conformationally "rigid" zinc-binding domain might be considered to play an essentially equivalent thermodynamic function in each case. Specifically, a metal chelate positioned at the macromolecular interface might be predicted to maximize the energy of interaction between the two components via significant reduction in the configurational entropy of the immediate polypeptide chain as has been previously discussed with regard to the R-C interaction in *E. coli* aspartate transcarbamoylase. Such a prediction is corroborated in the thermodynamics associated with the unfolding of gene

32 protein as the large enthalpy change accompanying zinc binding does not all appear in thermal stabilization of the protein which suggests a significant difference in the entropies of the folded forms of the apo and Zn(II) g32Ps [105]. In each of these DNA binding proteins and/or domains discussed above, Zn(II) coordination would appear to be responsible for helping to establish some rather subtle features of protein conformation.

ABBREVIATIONS

CD	circular dichroism
C-terminal	carboxy-terminal
ds	double-stranded
DSC	differential scanning calorimetry
DTT	dithiothreitol
EDTA	ethylenediaminetetracetic acid
g-at	gram atom
g32P	gene 32 protein
g32P-B	g32P residues 22-301
g32P-(A+B)	g32P core, g32P*: g32P residues 22-253
HIV-1	human immunodeficiency virus-1
IPTG	isopropanyl-β-D-thiogalactopyranoside
kD	kilodalton
Mo-MLV	Moloney murine leukemia virus
N-terminal	amino terminal
NMR	nuclear magnetic resonance
PMPS	p-hydroxymercuriphenylsulfonic acid
RPase	RNA polymerase
RSV	avian Rous sarcoma virus
SDS-PAGE	sodium dodecyl sulfate polyacrylamide gel electrophoresis
ss	single-stranded
Tris-HCl	tris(hydroxymethyl)aminomethane hydrochloride
UV	ultraviolet

ACKNOWLEDGMENTS

We thank Drs. Bruce A. Johnson and Kenneth R. Williams for stimulating discussion and permission to quote unpublished data. The work in the authors' laboratory was supported by grants GM21919 and DK09070 from the NIH to J.E.C. D.P.G. acknowledges the receipt of a NIH postdoctoral fellowship 1-F32-GM10972.

REFERENCES

1. D. Keilin and T. Mann, *Nature, 144,* 442 (1939).
2. J. F. Chlebowski and J. E. Coleman, in "Metal Ions in Biol. Systems", Vol. 6 (H. Sigel, ed.), Marcel Dekker, New York, 1976, p. 1.
3. I. Bertini, C. Luchinat, W. Maret, and M. Zeppezauer (eds.), "Zinc Enzymes", Progress in Inorganic Biochemistry and Biophysics, Vol. 1, Birhäuser, Boston, 1986.
4. J. P. Rosenbusch and K. Weber, *J. Biol. Chem., 246,* 1644 (1971).
5. J. C. Gerhart and H. K. Schachman, *Biochemistry, 4,* 1054 (1965).
6. H. L. Monaco, J. L. Crawford, and W. N. Lipscomb, *Proc. Natl. Acad. Sci. USA, 75,* 5276 (1978).
7. J. B. Hunt, S. H. Neece, H. K. Schachman, and A. Ginsberg, *J. Biol. Chem., 259,* 14793 (1984).
8. R. B. Honzatko, J. L. Crawford, H. L. Monaco, J. E. Ladner, B. F. P. Ewards, D. R. Evans, S. G. Warren, D. C. Wiley, R. C. Ladner, and W. N. Lipscomb, *J. Mol. Biol., 160,* 219 (1982).
9. K. H. Falchuk, A. Krishan, and B. L. Vallee, *Biochemistry, 14,* 3439 (1975).
10. K. H. Falchuk, C. Hardy, L. Ulpino, and B. L. Vallee, *Proc. Natl. Acad. Sci. USA, 75,* 4275 (1978).
11. M. Czupryn, K. H. Falchuk, and B. L. Vallee, *Biochemistry, 26,* 8263 (1987).
12. M. C. Scrutton, C. W. Wu, and D. A. Goldwait, *Proc. Natl. Acad. Sci. USA, 69,* 2497 (1971).
13. J. E. Coleman, in "Zinc Enzymes" (T. G. Spiro, ed.), John Wiley and Sons, New York, 1983, p. 220.
14. L. J. Ferrin, A. S. Mildvan, and L. A. Loeb, *Biochem. Biophys. Res. Commun., 112,* 723 (1983).

15. I. Slaby, B. Lind, and A. Holmgren, *Biochem. Biophys. Res. Commun., 122,* 1410 (1984).

16. S. M. Halling, F. J. Sanchez-Anzaldo, R. Fukuda, R. H. Doi, and C. F. Meares, *Biochemistry, 16,* 2880 (1977).

17. M. M. Hanna and C. F. Meares, *Proc. Natl. Acad. Sci. USA, 80,* 4238 (1983).

18. M. M. Hanna and C. F. Meares, *Biochemistry, 22,* 3546 (1983).

19. D. Chatterji and F. Y.-H. Wu, *Biochemistry, 21,* 4651 (1982).

20. D. Chatterji and F. Y.-H. Wu, *Biochemistry, 21,* 4657 (1982).

21. D. Chatterji, C. W. Wu, and F. Y.-H. Wu, *J. Biol. Chem., 259,* 284 (1984).

22. D. C. Speckhard, F. Y.-H. Wu, and C. W. Wu, *Biochemistry, 16,* 5228 (1977).

23. C. W. Wu, F. Y.-H. Wu, and D. C. Speckhard, *Biochemistry, 16,* 5449 (1977).

24. G. C. King, C. T. Martin, T. T. Pham, and J. E. Coleman, *Biochemistry, 25,* 36 (1986).

25. L. A. Allison, M. Mayle, M. Shales, and C. J. Engles, *Cell, 42,* 599 (1985).

26. J. Biggs, L. L. Searles, and A. L. Greenleaf, *Cell, 42,* 611 (1985).

27. S. S. Broyles and B. Moss, *Proc. Natl. Acad. Sci. USA, 83,* 3141 (1986).

28. D. Sweetser, M. Nonet, and R. A. Young, *Proc. Natl. Acad. Sci. USA, 84,* 1192 (1987).

29. D. P. Giedroc and J. E. Coleman, *Biochemistry, 25,* 4969 (1986).

30. P. Petranyl, J. J. Jendrisak, and R. R. Burgess, *Biochem. Biophys. Res. Commun., 74,* 1031 (1977).

31. K. H. Falchuk, B. Mazus, L. Ulpino, and B. L. Vallee, *Biochemistry, 15,* 4468 (1976).

32. H. Lattke and U. Weser, *FEBS Lett., 65,* 288 (1976).

33. B. Bartholomew, M. E. Dahmus, and C. F. Meares, *J. Biol. Chem., 261,* 14226 (1986).

34. D. P. Giedroc, K. M. Keating, K. R. Williams, W. H. Konigsberg, and J. E. Coleman, *Proc. Natl. Acad. Sci. USA, 83,* 8452 (1986).

35. D. P. Giedroc, K. M. Keating, K. R. Williams, and J. E. Coleman, *Biochemistry, 26,* 5251 (1987).

36. R. F. Doolittle, M. S. Johnson, I. Husain, B. Van Houten, D. C. Thomas, and A. Sancar, *Nature, 323,* 451 (1986).

37. J. S. Hanas, D. Hazuda, D. R. Bogenhagen, F. Y.-H. Wu, and C. W. Wu, *J. Biol. Chem., 258,* 14120 (1983).

38. J. Miller, A. D. McLachlan, and A. Klug, *EMBO J., 4,* 1609 (1985).

39. H. Blumberg, A. Eisen, A. Sledziewski, D. Bader, and E. T. Young, *Nature, 328,* 443 (1987).

40. R. Miesfeld, S. Rusconi, P. J. Godowski, B. A. Maler, S. Okret, A.-C. Wilkström, J. A. Gustafsson, and K. Yamamoto, *Cell, 46,* 389 (1986).

41. V. Giguere, N. Yang, P. Segui, and R. M. Evans, *Nature, 331,* 91 (1988).

42. M. Johnston, *Nature, 353,* (1987).

43. G. Diakun, L. Fairall, and A. Klug, *Nature, 324,* 689 (1986).

44. A. Klug and D. Rhodes, *Trends Biochem. Sci., 12,* 464 (1987).

45. J. P. Mending and S. Beychok, *Biochem. Biophys. Res. Commun., 51,* 711 (1973).

46. D. P. Giedroc, K. M. Keating, C. T. Martin, K. R. Williams, and J. E. Coleman, *J. Inorg. Biochem., 28,* 155 (1986).

47. R. S. Johnson and H. K. Schachman, *J. Biol. Chem., 258,* 3528 (1983).

48. W. Maret, I. Anderson, H. Dietrich, H. Schneider-Bernlöhr, R. Einarsson, and M. Zeppezauer, *Eur. J. Biochem., 98,* 501 (1979).

49. S. May and M. H. Kuo, *Biochemistry, 17,* 3333 (1978).

50. M. Vasak, J. H. R. Kägi, B. Holmquist, and B. L. Vallee, *Biochemistry, 20,* 6659 (1981).

51. D. Swenson, N. C. Baenziger, and D. Coucouvanis, *J. Am. Chem. Soc., 100,* 1932 (1978).

52. J. E. Coleman and R. V. Coleman, *J. Biol. Chem., 247,* 4718 (1972).

53. M. J. Chamberlin, W. C. Nierman, J. Wiggs, and N. Neff, *J. Biol. Chem., 254,* 10061 (1979).

54. F. Y.-H.Wu, in "Zinc Enzymes", Vol. 1, Progress in Inorganic Biochem. and Biophys. (I. Bertini, C. Luchinat, W. Maret, M. Zeppezauer, eds.), Birkhäuser, Boston, Basel, 1986, p. 535.

55. A. S. Mildvan, *Acc. Chem. Res., 10,* 246 (1977).

56. T. Bull, J.-E. Norne, P. Reimarsson, and B. Lindman, *J. Am. Chem. Soc., 100,* 4643 (1978).

57. P. Gettins and J. E. Coleman, *J. Biol. Chem., 259,* 11036 (1984).

58. D. P. Giedroc and J. E. Coleman, in preparation.

59. J. D. McKinney, J. Lee, R. E. O'Neill, and A. Goldfarb, *Gene, 58,* 13 (1987).

60. Y. A. Ovchinnikov, G. S. Monastyrskaya, V. V. Gubanov, S. O. Guryev, I. S. Salomatina, T. M. Shuvaeva, V. M. Lipkin, and E. D. Sverdlov, *Nucleic Acids Res., 10,* 4035 (1982).

61. J. M. Berg, *Science, 232,* 485 (1986).

62. W. F. Furey, A. H. Robbins, L. L. Clancy, D. R. Winge, B. D. Wang, and C. D. Stout, *Science 231,* 704 (1986).

63. A.-Y. M. Woody, C. R. Vadev, R. W. Woody, and B. E. Haley, *Biochemistry, 23,* 2843 (1984).

64. P. Stockel, R. May, I. Strell, Z. Cejka, W. Hoppe, H. Neumann, W. Zillig, and H. L. Crespi, *Eur. J. Biochem., 112,* 419 (1980).

65. B. M. Alberts, J. Barry, P. Bedringer, R. L. Burke, U. Hibner, C.-C. Liu, and R. Sheridan, *ICN-UCLA Symp. Mol. Cell Biol., 19,* 449 (1980).

66. J. E. Coleman and J. L. Oakley, *CRC Crit. Rev. Biochem., 7,* 247 (1980).

67. J. W. Chase and K. R. Williams, *Ann. Rev. Biochem., 55,* 103 (1986).

68. B. M. Alberts and L. Frey, *Nature, 227,* 1313 (1970).

69. S. C. Kowalczykowski, D. G. Bear, and P. H. von Hippel, *The Enzymes, 14A,* 373 (1981).

70. M. Russel, L. Gold, H. Morrissett, and P. Z. O'Farrell, *J. Biol. Chem., 251,* 7263 (1976).

71. H. M. Krisch, A. Bolle, and R. H. Epstein, *J. Mol. Biol., 88,* 89 (1974).

72. P. H. von Hippel, S. C. Kowalczykowski, N. Lonberg, J. W. Newport, L. S. Paul, G. Stormo, and L. Gold, *J. Mol. Biol., 162,* 795 (1982).

73. P. Gauss, K. B. Krassa, D. S. McPheeters, M. A. Nelson, and L. Gold, *Proc. Natl. Acad. Sci. USA, 84,* 8515 (1987).

74. R. C. Kelly, D. E. Jensen, and P. H. von Hippel, *J. Biol. Chem., 251,* 7240 (1976).

75. S. C. Kowalczykowski, L. S. Paul, N. Lonberg, J. W. Newport, J. A. McSwiggen, and P. H. von Hippel, *Biochemistry, 25,* 1226 (1986).

76. H. Moise and J. Hosoda, *Nature, 259,* 455 (1976).

77. K. R. Williams and W. H. Konigsberg, *J. Biol. Chem., 253,* 2463 (1978).

78. K. R. Williams, L. O. Sillerud, D. E. Schafer, and W. H. Konigsberg, *J. Biol. Chem., 254,* 6426 (1979).

79. E. K. Spicer, K. R. Williams, and W. H. Konigsberg, *J. Biol. Chem., 254,* 6433 (1979).

80. J. E. Coleman, K. R. Williams, G. C. King, R. V. Prigodich, Y. Shamoo, and W. H. Konigsberg, in "Protein Engineering", Alan R. Liss, New York, 1987, p. 323.

81. J. E. Coleman, K. R. Williams, G. C. King, R. V. Prigodich, Y. Shamoo, and W. H. Konigsberg, *J. Cell. Biochem.*, *32*, 305 (1986).

82. R. L. Burke, B. M. Alberts, and J. Hosoda, *J. Biol. Chem.*, *255*, 11484 (1980).

83. D. E. Jensen, R. C. Kelly, and P. H. von Hippel, *J. Biol. Chem.*, *251*, 7215 (1976).

84. R. V. Prigodich, Y. Shamoo, K. R. Williams, W. H. Konigsberg, and J. E. Coleman, *Biochemistry*, *25*, 3666 (1986).

85. R. V. Prigodich, J. Casas-Finet, K. R. Williams, W. H. Konigsberg, and J. E. Coleman, *Biochemistry*, *23*, 522 (1984).

86. R. A. Anderson and J. E. Coleman, *Biochemistry*, *14*, 5485 (1975).

87. Y. Shamoo, W. J. Roberts, J. E. Coleman, K. R. Williams, and W. H. Konigsberg, in "Protein Structure, Folding and Design 2", Vol. 69 (D. L. Oxender, ed.), Alan R. Liss, New York, 1987, p. 385.

88. A. B. P. Lever, in "Inorganic Electronic Spectroscopy", Elsevier, Amsterdam, 1968, Chap. 9.

89. R. W. Lane, J. A. Ibers, R. B. Frankel, G. C. Papaefthymiou, and R. H. Holm, *J. Am. Chem. Soc.*, *99*, 84 (1977).

90. H. Eklund, B. Nordström, E. Zeppezauer, E. Söderlund, I. Ohlsson, T. Boiwe, B. O. Söderberg, O. Tapia, and C. I. Bränden, *J. Mol. Biol.*, *102*, 27 (1976).

91. G. Schneider, E. Cedergren-Zeppezauer, S. Knight, H. Eklund, and M. Zeppezauer, *Biochemistry*, *24*, 7503 (1985).

92. I. M. Armitage and J. D. Otvos, in "Biological Magnetic Resonance", Vol. 4 (L. J. Berliner and J. Reuben, eds.), Plenum Press, 1982, p. 79.

93. I. M. Armitage, A. J. M. Schoot-Uiterkamp, J. F. Chlebowski, and J. E. Coleman, *J. Magn. Reson.*, *29*, 375 (1978).

94. H. R. Engeseth, D. R. McMillin, and J. D. Otvos, *J. Biol. Chem.*, *259*, 4822 (1984).

95. B. R. Bobsein and R. J. Myers, *J. Am. Chem. Soc.*, *102*, 2454 (1980).

96. D. P. Giedroc, J. E. Coleman, I. M. Armitage, and B. A. Johnson, in preparation.

97. D. P. Giedroc and J. E. Coleman (unpublished observations).

98. K. R. Williams, M. B. LoPresti, and M. Setoguchi, *J. Biol. Chem.*, *256*, 1754 (1981).

99. D. P. Giedroc and J. E. Coleman, in preparation.

100. J. F. Chlebowski and S. Maybrey, *J. Biol. Chem.*, *252*, 7042 (1977).

101. J. F. Chlebowski, S. Maybrey, and M. C. Falk, *J. Biol. Chem.*, *254*, 5745 (1979).

102. J. R. Lepock, L. D. Arnold, B. H. Torrie, B. Andrews, and J. Kruur, *Arch. Biochem. Biophys.*, *241*, 243 (1985).

103. H. R. Engeseth and D. R. McMillin, *Biochemsitry*, *25*, 2448 (1986).

104. K. M. Keating, D. P. Giedroc, L. D. Harris, L. R. Ghosaini, K. R. Williams, J. M. Sturtevant, and J. E. Coleman, in "Protein Structure, Folding, and Design 2", Vol. 69 (D. L. Oxender, ed.), Alan R. Liss, New York, 1987, p. 35.

105. K. M. Keating, L. R. Ghosaini, D. P. Giedroc, K. R. Williams, J. E. Coleman, and J. M. Sturtevant, *Biochemistry*, *27*, 5240 (1988).

106. M. J. Curtis and B. M. Alberts, *J. Mol. Biol.*, *102*, 793 (1976).

107. G. C. King and J. E. Coleman, *Biochemistry*, *26*, 2929 (1987).

108. F. Cobianchi, R. L. Karpel, K. R. Williams, V. Notario, and S. H. Wilson, *J. Biol. Chem.*, *263*, 3307 (1988).

109. B. M. Merrill, K. L. Stone, F. Cobianchi, S. H. Wilson, and K. R. Williams, *J. Biol. Chem.*, *263*, 1063 (1988).

110. W. Bujalowski and T. M. Lohman, *Biochemistry*, *25*, 7799 (1986).

111. W. Bujalowski and T. M. Lohman, *J. Mol. Biol.*, *195*, 897 (1987).

112. N. L. Davis and R. R. Rueckert, *J. Virol.*, *10*, 1010 (1972).

113. S. Oroszlan, C. W. Long, and R. V. Gilden, *Virology*, *72*, 523 (1976).

114. M. Schulein, W. N. Burnette, and J. T. August, *J. Virol.*, *26*, 54 (1978).

115. B. J. Smith and J. M. Bailey, *Nucleic Acids Res.*, *7*, 205 (1979).

116. L. E. Henderson, T. D. Copeland, R. C. Sowder, G. W. Smythers, and S. Oroszlan, *J. Biol. Chem.*, *256*, 8400 (1981).

117. R. L. Karpel, L. E. Henderson, and S. Oroszlan, *J. Biol. Chem.*, *262*, 4961 (1987).

118. G. R. Fink, J. D. Boeke, and D. J. Garfinkel, *Trends Genet.*, *2*, 118 (1986).

119. D. J. Garfinkel, J. D. Boeke, and G. R. Fink, *Cell*, *42*, 507 (1985).

120. J. Mellor, M. H. Malim, K. Gull, M. F. Tuite, S. McCready, T. Dibbayawan, S. Kingsman, and A. J. Kingsman, *Nature*, *318*, 583 (1985).

121. E. L. Kutt, N. A. Wivel, and K. K. Lueders, *Cancer Res.*, *28*, 2137 (1968).

122. T. Shiba and K. Saigo, *Nature*, *302*, 119 (1983).

123. S. M. Mount and G. M. Rubin, *Mol. Cell. Biol., 5,* 1630 (1985).

124. K. Saigo, W. Kugimiya, Y. Matsuo, S. Inouye, K. Yoshioka, and S. Yuki, *Nature, 312,* 659 (1984).

125. D. H. Fawcett, C. K. Lister, E. Kellett, and D. J. Finnegan, *Cell, 47,* 1007 (1986).

126. J. Clare and P. Farabaugh, *Proc. Natl. Acad. Sci. USA, 82,* 2829 (1985).

127. M. Ono, H. Toh, T. Miyata, and T. Awaya, *J. Virol, 55,* 387 (1985).

128. D. Carroll, D. S. Knutzon, and J. E. Garrett, in "Mobile DNA" (M. M. Howe and D. E. Berg, eds.), ASM Press, in press.

129. J. D. Boeke, D. J. Garfinkel, C. A. Styles, and G. R. Fink, *Cell, 40,* 491 (1985).

130. T. Hohn, B. Hohn, and P. Pfeiffer, *Trends biochem. Sci., 10,* 205 (1985).

131. S. N. Covey, *Nucleic Acids Res., 14,* 623 (1986).

132. D. P. Giedroc and B. A. Johnson, unpublished observations.

133. J. R. Casas-Finet, N.-I Jhon, and A. H. Maki, *Biochemistry, 27,* 1172 (1988).

134. G. D. Rose, L. M. Gierasch, and J. A. Smith, *Adv. Protein Chem., 37,* 1 (1985).

135. J. F. Leszczynski and G. D. Rose, *Science, 234,* 849 (1986).

136. P. Schwartzberg, J. Collicelli, M. L. Gordon, and S. P. Goff, *J. Virol., 49,* 918 (1984).

137. C. Meric and P.-F. Spahr, *J. Virol., 60,* 450 (1986).

138. S. G. Devare, E. P. Reddy, J. D. Law, K. C. Robbins, and S. A. Aaronson, *Proc. Natl. Acad. Sci. USA, 80,* 731 (1983).

139. D. P. Bolognesi, R. C. Montelaro, H. Frank, and W. Schafer, *Science, 199,* 183 (1978).

140. A. H. Lu, M. M. Soong, and P. K. Y. Wong, *Virology, 93,* 269 (1979).

141. C. Meric, J.-L. Darlix, and P.-F. Spahr, *J. Mol. Biol., 173,* 531 (1984).

142. J.-L. Darlix and P.-F. Spahr, *J. Mol. Biol., 160,* 147 (1982).

143. W. J. Roberts, J. Elliott, and K. R. Williams, *Int. J. Peptide Protein Res.,* submitted.

144. B. Van Houten, H. Gamper, A. Sancar, and J. E. Hearst, *J. Biol. Chem., 262,* 13180 (1987).

145. A. D. Frankel, J. M. Berg, and C. O. Pabo, *Proc. Natl. Acad. Sci. USA, 84,* 4841 (1987).

146. J. T. Kadonaga, K. R. Carner, F. R. Masianz, and R. Tjian, *Cell, 51,* 1079 (1987).

147. K. Nagai, Y. Nakaseko, K. Nesmyth, and D. Rhodes, *Nature, 332,* 284 (1988).

148. M. Johnston and J. Dover, *Proc. Natl. Acad. Sci. USA, 84,* 2401 (1987).

149. D. E. Schwartz, R. Tizard, and W. Gilbert, *Cell, 32,* 853 (1983).

150. T. D. Copeland, M. A. Morgan, and S. Oroszlan, *Virology, 133,* 137 (1984).

151. M. Seiki, S. Hattori, Y. Hirayama, and M. Yoshida, *Proc. Natl. Acad. Sci. USA, 80,* 3618 (1983).

152. K. Shimotohno, Y. Takahashi, N. Shimizu, T. Gojobori, D. W. Golde, I. S. Y. Chen, M. Miwa, and T. Sigimura, *Proc. Natl. Acad. Sci. USA, 82,* 3101 (1985).

153. T. D. Copeland, M. A. Morgan, and S. Oroszalan, *FEBS Lett., 156,* 37 (1983).

154. P. Sonigo, M. Alizon, K. Staskus, D. Klatzmann, S. Cole, O. Danos, E. Retzel, P. Tiollais, A. Haase, and S. Wain-Hobson, *Cell, 42,* 369 (1985).

155. R. M. Stephens, J. W. Casey, and N. R. Rice, *Science, 231,* 589 (1986).

156. L. Ratner, W. Haseltine, R. Patarca, K. J. Livak, B. Starcich, S. F. Josephs, E. R. Doran, J. A. Rafalski, E. A. Whitehorn, K. Baumeister, L. Ivanoff, S. R. Petteway, Jr., M. L. Pearson, J. A. Lautenberger, T. S. Papas, J. Ghrayeb, N. T. Chang, R. C. Gallo, and F. Wong-Staal, *Nature, 313.* 277 (1985).

157. L. Chakrabarti, M. Guyader, M. Alizon, M. D. Daniel, R. C. Desrosiers, P. Tiollais, and P. Sonigo, *Nature, 328,* 543 (1987).

158. G. Franchini, C. Gurgo, H.-G. Guo, R. C. Gallo, E. Collatti, K. A. Fargnoli, L. F. Hall, F. Wong-Staal, and M. S. Reitz, *Nature, 328,* 539 (1987).

159. M. Guyader, M. Emerman, P. Sonigo, F. Clavel, L. Montagnier, and M. Alizon, *Nature, 326,* 662 (1987).

6

"Zinc Fingers": The Role of Zinc(II) in Transcription Factor IIIA and Related Proteins

Jeremy M. Berg
Department of Chemistry
The Johns Hopkins University
Remsen Hall-Dunning Hall
34th and Charles Streets
Baltimore, MD 21218

1. INTRODUCTION

1.1. Transcription Factor IIIA (TFIIIA)

A role for zinc ions in the function of the nucleic acid-binding protein transcription factor IIIA (hereafter TFIIIA) was discovered in 1983 [1]. This protein, isolated from the African clawed toad *Xenopus laevis,* binds specifically to both DNA and RNA [2]. It binds to the ≈50 base pair internal control region of 5S ribosomal RNA genes and plays a central role in activating the transcription of these genes. It also binds to the 5S RNA molecule itself to form 1:1 complexes, referred to as 7S storage particles, that accumulate in immature oocytes [3]. These substrates are shown schematically in Fig. 1. Hanas et al. [1] isolated 7S storage particles and showed that they contained 2-3 eq of Zn^{2+} ions per particle. TFIIIA isolated from these storage particles specifically bound to the internal control region of the 5S RNA gene. Dialysis against 1-5 mM EDTA resulted in loss of the specific binding activity. Treatment with 15 μM Zn^{2+} restored the specific binding activity but treatment with other metals (Mn^{2+}, Fe^{2+}, Co^{2+}, Ni^{2+}) at this concentration did not [1].

1.2. The "Zinc Finger" Hypothesis

An intriguing explanation for these observations became apparent when the amino acid sequence of TFIIIA, deduced from the nucleotide sequence of a cDNA clone [4], was analyzed. In 1985, two groups reported that the TFIIIA sequence contained tandem, imperfect repeats of a sequence of approximately 30 amino acids [5,6]. Each repeat has two cysteine residues and two histidine residues in conserved positions as well as three conserved hydrophobic residues. Furthermore, Miller et al. [5] showed that, when isolated under conditions which avoided potential chelating agents, 7S storage particles contained 7-11 Zn^{2+} ions. Based on these observations, it was hypothesized

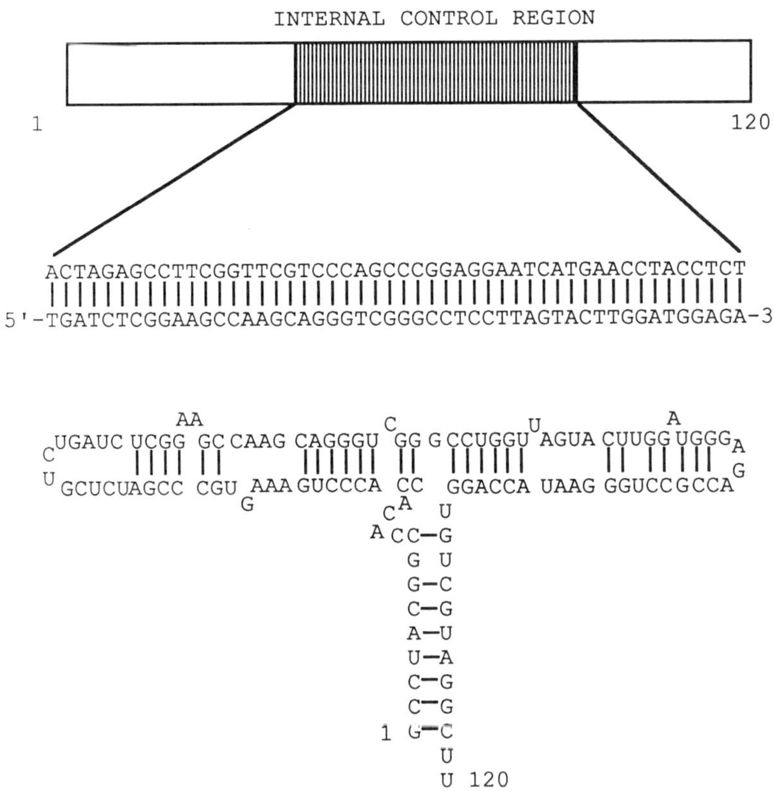

FIG. 1. The structure of the 5S RNA gene and the secondary structure of 5S RNA. Note the similarity between the internal control region of the gene and a potential long interrupted helix in the 5S RNA.

that each sequence repeat represents a structural domain organized around a Zn^{2+} ion coordinated to the conserved cysteinate and histidine residues. These units were termed "zinc fingers" [5]. A wide variety of subsequent results have provided support for this hypothesis. In addition, a large number of recently sequenced genes appear to encode proteins that contain strikingly similar sequences revealing that zinc finger domains are a frequently occurring feature of eukaryotic nucleic acid-binding proteins.

2. EVIDENCE FOR ZINC FINGERS IN TFIIIA

One line of support for the zinc finger hypothesis comes from early studies of the shape of the TFIIIA molecule. A single molecule with a molecular weight of 38 kD protects about 50 base pairs of DNA from enzymatic digestion indicating that the molecule can span approximately 50 • 3.4 Å = 170 Å [2,7,8]. Furthermore, sedimentation studies of free TFIIIA revealed that it is highly asymmetrical with an axial ratio of approximately 7.5 [8]. The observation that three-fourths of the molecule appears to consist of nine small structural domains provides an appealing explanation for the highly elongated structure of TFIIIA.

The hypothesis is also supported by limited proteolysis studies of the 7S storage particle [4]. Digestion with several different proteases resulted in limit digests that are suggestive of the presence of small domains that are relatively insensitive to proteolysis. For example, treatment with trypsin yielded a final mixture of fragments with molecular weights of approximately 3, 4, and 6 kD which were themselves degraded after sufficiently long times. Studies of a single zinc finger peptide (vide infra) confirm that such units are quite stable to proteolysis in the presence of suitable metal ions.

An approach that is based on the interactions between shortened versions of TFIIIA and DNA yielded strong support for the modular nature of TFIIIA and provided some details of the interaction between TFIIIA and a 5S RNA gene. Deletion mutants of TFIIIA were prepared by molecular biological methods and their interactions with the internal control region of a 5S RNA gene were examined using "footprinting" techniques [9]. Chemically generated hydroxyl radical footprinting [10] revealed that deletion of each zinc finger from the protein resulted in a localized decrease in the protection of the DNA molecule. Furthermore, these effects proceeded monotonically along the DNA molecule. These observations indicate that relatively independent structural domains of TFIIIA are aligned along the internal control region and allow correlation between individual portions of the TFIIIA amino acid sequence and contact regions on the DNA.

A completely different line of support comes from studies of the TFIIIA gene. Examination of the structure of this gene revealed that it consists of nine separate exons [11]. Remarkably, the first six putative zinc finger domains are encoded on individual exons. This suggests that TFIIIA has evolved by gene duplication of an exon that encoded a single domain and supports the idea that a single 30 amino acid sequence represents a structurally significant unit [12].

Finally, the nature of the zinc sites in the 7S particle has been probed using x-ray absorption spectroscopy [13]. The results revealed two important facts: (1) the overall intensity of the x-ray fluorescence observed is consistent with the presence of nine zinc ions per protein, and (2) the zinc sites appear to be quite similar to one another and the EXAFS (extended x-ray absorption fine structure) spectrum can be fit with two sulfur atoms at 2.30 Å and two histidine nitrogen atoms at 2.00 Å. Thus, these results are fully consistent with the structure of the zinc sites suggested by the zinc finger hypothesis.

3. OTHER PROTEINS CONTAINING ZINC FINGER SEQUENCES

The advent of gene isolation and DNA-sequencing techniques has greatly accelerated the rate at which protein sequences are determined. Often the deduced amino acid sequence of a protein is one of the first properties of a protein to be elucidated. Thus, the discovery of sequence templates characteristic of particular classes of proteins has become extremely important. The consensus sequence deduced from TFIIIA is:

$$\text{(Tyr,Phe)-X-Cys-X}_{2,4}\text{-Cys-X}_3\text{-Phe-X}_5\text{-Leu-X}_2\text{-His-X}_3\text{-His-X}_{2\text{-}5}$$

where X represents a relatively unconserved position. At the time the TFIIIA sequence was analyzed, no other protein sequences were known that contained this consensus sequence. Remarkably, a large

TABLE 1

Proteins that Contain Sequences that Match the Zinc Finger Consensus

Protein	Source	Zinc fingers	Ref.
TFIIIA	*Xenopus*	9	5,6
Xfin	*Xenopus*	7+6+8+7+3+6	14
Krüppel	*Drosophila*	5	15
Serendipity β	*Drosophila*	6	16
Serendipity δ	*Drosophila*	6+1	16
Hunchback	*Drosophila*	4+2	17
Snail	*Drosophila*	4	18
Terminus	*Drosophila*	1	19
Krh	*Drosophila*	2+	20
ADR1	Yeast	2	21
SWI5	Yeast	3	22
NGFI-A	Rat	3	23
mkr1	Mouse	7+	24
mkr2	Mouse	9+	24
TRS-1	Trypanosome	1+1+1+1+1	25
Sp1	Human	3	26
Testis-determining factor	Human	13+	27

Note: The number of zinc finger domains in each protein divided into sets according to the number of tandem domains in each set is listed. An entry ending with + indicates that a complete open-reading frame has not yet been sequenced.

number of protein sequences have subsequently been elucidated that contain one or more copies of the TFIIIA consensus. These proteins are summarized in Table 1 [5,6,14-27]. The majority of these proteins have been independent discoveries in that no attempt has been made to select genes for sequencing that encode any particular sequence. Instead, genes that have been identified on the basis of some other characteristic have been sequenced and the deduced amino acid sequences analyzed to reveal the presence of regions that match the zinc finger consensus. One of the proteins, ADR1 from yeast, is known to be a transcription factor, regulating alcohol dehydrogenase expression [21]. Interestingly, many of the other proteins identified are encoded by genes that appear to play roles in developmental processes. For example, the *Drosophila* proteins encoded by the *krüppel* [15] and *hunchback* [17] genes are involved in control of segmentation. An extremely interesting case is provided by testis-determining factor [27]. This gene appears to be the major sex-determining switch in humans. At this point, one exon has been identified and sequenced, and it is made up entirely of 13 zinc finger sequences.

The relatively conserved nature of the zinc finger sequences and, in some cases, the sequences linking them led several groups to try to clone genes on the basis of nucleic acid hybridization to zinc finger-encoding genes under conditions of low stringency [14, 20,24]. This approach has been quite successful in two regards. First, one complete open-reading frame (encoding a protein with 37 zinc finger sequences!) and several partial ones have been discovered that encode proteins containing sequences that match the zinc finger consensus. Second, these methods have allowed screening of a relatively large number of organisms for potential zinc finger-encoding sequences. Interestingly, every eukaryotic gene library that has been examined (ranging from yeast to man) contained members that hybridize to *krüppel*-derived probes detectably above background whereas an *Escherichia coli*-derived library did not. These results suggest that zinc finger sequences occur quite frequently in eukaryotes but not in prokaryotes, and raise interesting questions concerning the role of such proteins in evolution.

In all, 17 zinc finger-containing protein sequences have been published to date with a total of 122 individual zinc finger sequences. Many of these sequences occur in tandemly arrayed sets with short linkers of two to six amino acids between them (as in TFIIIA). However, in other proteins the sequences are isolated with large linkers

(>30 amino acids) separating them. This observation supports the idea that each sequence forms an independent structural unit. Given the rate at which these sequences have been discovered, the number of known zinc finger proteins should increase rapidly over the next years.

Genetic studies of ADR1 have provided further support for the importance of the zinc finger structures for gene regulation activity [28]. The ADR1 gene was randomly mutagenized with hydroxylamine and mutants that were deficient in ADR1 activity were selected. Mutants at nine positions in the amino acid sequence were isolated. Seven of these were in the two zinc finger domains including four of the conserved cysteine and histidine residues. Thus, as would be expected, disruption of these residues leads to significant loss of gene regulatory activity presumably by affecting the specific DNA-binding ability of the protein.

4. A PROPOSED THREE-DIMENSIONAL STRUCTURE

4.1. Construction of a Model Based on Known Protein Structures

The zinc finger is defined entirely in terms of its primary structure at present; no experimental evidence is yet available concerning its three-dimensional structure. Nonetheless, it has been possible to deduce a plausible proposal for a three-dimensional structure on the basis of crystallographically determined protein structures [29].

The prediction of protein tertiary structure from primary structure remains one of the most important unsolved problems in modern biology. Among the reasons for the difficulty of this problem are: (1) It is hard to access the importance of the nature of a particular residue for the stability of the three-dimensional structure, and (2) structures are only partially determined by local sequence and are strongly dependent on long-range interactions. Three aspects of the zinc finger sequences are favorable with regard to these concerns. First, the sequences are relatively short. Second, the availability of 122 sequences leads to a very well-defined consensus sequence with clearly identified conserved residues that presumably play important roles in structural stability. Finally, the context of the sequences is at least partially defined in that the conserved cysteine and histidine residues are almost certainly coordinated to zinc ions.

The consensus sequence for the zinc finger domains is:

{(Tyr,Phe)-X-Cys-$X_{2\ \mathrm{or}\ 4}$-Cys-X_3-Phe}-X_4-{X-Leu-X_2-His-X_3-His}-X_n

Cys-Cys loop Tip His-His loop Linker

It is useful to divide the consensus sequence as shown for the purpose of constructing the proposed three-dimensional structure. This model was developed as follows:

(1) The structures of crystallographically characterized metalloproteins [30] with spacings between the metal-binding residues that match those in the zinc finger sequences were examined. It was discovered that conserved three-dimensional structures occur that include the metal-binding amino acids and that also extend to the presence of hydrophobic residues in positions that match those in the zinc finger sequences. For the Cys-Cys loop the conserved structure is an antiparallel β sheet with the metal coordinated to the two cysteinate residues at a loop at its base. This structure is observed in the sequences shown below:

Leu-Lys-*Cys*-Lys-Tyr-*Cys*-Glu-Lys-Glu-*Phe*	aspartate transcarbamylase
Tyr-Thr-*Cys*-Thr-Val-*Cys*-Gly-Tyr-Ile-*Tyr*	rubredoxin
(*Tyr*,*Phe*)-X-*Cys*-X-X-*Cys*-X-X-X-*Phe*	zinc finger consensus

For the His-His loop the conserved structure is an α helix with the metal coordinated to the histidine residues through the ε nitrogens. This structure is observed in the sequences below (and in two sequences from hemocyanin [31] as well):

Asp-*Val*-Val-Ala-*His*-Glu-Leu-Thr-*His*-Ala	thermolysin
Gly-*Tyr*-Ala-Glu-*His*-Lys-Lys-Ala-*His*-Asp	hemerythrin
X-*Leu*-X-X-*His*-X-X-X-*His*-X	zinc finger consensus

(2) These substructures were placed against one another around a tetrahedrally coordinated metal ion bound to the cysteinate and histidine residues. One of the two possible arrangements packed all of the conserved hydrophobic residues into a compact core and left the carboxyl terminus of the Cys-Cys loop and the amino terminus of the His-His loop approximately 7.5 Å apart whereas the other appeared to be physically impossible. The two substructures were then joined with the Tip region. The construction of the model is summarized in Fig. 2.

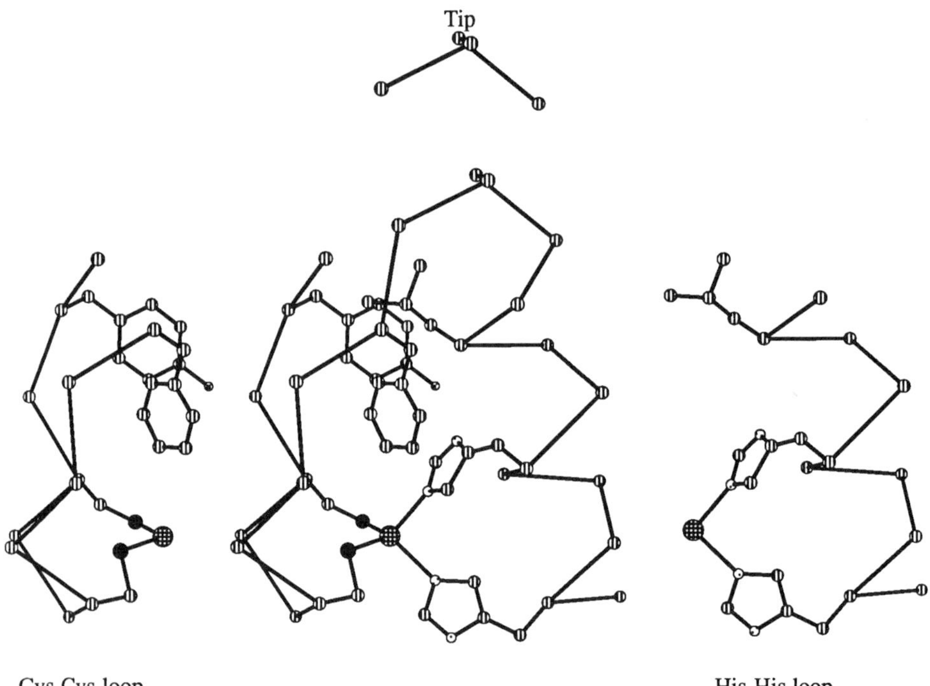

FIG. 2. The construction of a model for the three-dimensional structure of the zinc finger domains based on conserved substructures form crystallographically characterized metalloproteins. The α-carbon atoms are shown along with the side chains for the conserved metal-binding and hydrophobic residues. The substructures are shown with the complete structure in the center.

Thus, the proposed model consists of an antiparallel β sheet followed by an α helix. This model is supported by further analysis of the zinc finger sequences. First, secondary structure analysis averaged over a number of the zinc finger sequences suggests the presence of an α helix approximately in the region indicated [32]. Second, more detailed analysis of the sequences suggests roles for additional residues in stabilizing the proposed structure [33]. It will be extremely interesting to compare this structure with experimentally determined ones as they become available. It should be noted that the zinc ion is relatively buried and is almost certainly not in a position to coordinate additional ligands (such as potential ligands from DNA).

4.2. Implications for Nucleic Acid Binding

Based on a series of footprinting experiments with a number of chemical and enzymic reagents, Fairall et al. [34] proposed two models for the complex between TFIIIA and the 5S RNA gene. In the first model the protein wraps around the DNA following the major groove. This model has two implications with regard to the structure of the individual zinc-binding domains. First, the true repeat is one domain so that each domain can interact with the nucleic acid in approximately the same way. Second, since the protein moves along the major groove from one zinc finger to the next, the amino terminus and carboxyl terminus of each zinc finger domain should be at opposite ends of the structure. In the second model, the protein lies along one face of the DNA with the zinc fingers in the major groove alternating from one side of the DNA to the other. This model has quite different implications with respect to the structure of an individual zinc finger. The true repeat of this pattern is two fingers so that, by symmetry, there are two types of zinc finger-DNA interactions. Furthermore, since the protein lies along one face of the DNA, this model suggests that the amino and carboxyl termini of each domain should be close to one another in the three-dimensional structure.

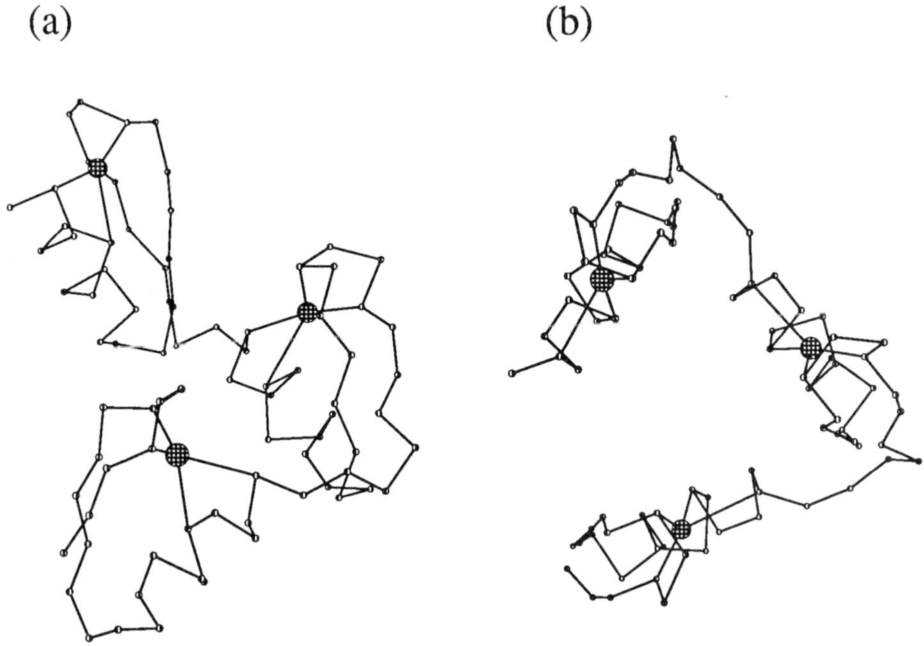

FIG. 3. Three zinc finger domains in an arrangement capable of wrapping around a DNA double helix. α-Carbon atoms are shown along with the zinc ions. (a) A view with the DNA helical axis vertical. (b) A view down the DNA helical axis.

Thus, these two models make quite different demands on the structural properties of the individual metal-binding domains.

Based on the observation of an apparent alternation in the strength of the protection ascribed to the individual zinc fingers, Fairall et al. [34] favored the second model. However, the proposed three-dimensional structure for the zinc finger discussed above clearly favors the first model. One possible arrangement of three zinc fingers in a structure capable of wrapping around DNA in the major groove is shown in Fig. 3.

Each zinc finger lies with its α helix in the major groove of the DNA with the second strand of the β sheet making contacts with the phosphodiester backbone. The placement of the α helix is based on crystal structures of other DNA-binding proteins and their complexes with oligonucleotides [35-37]. Further experimentation, including perhaps a crystal structure determination of a zinc finger protein-nucleic acid complex, will be required to determine which of the two models more closely represents the actual structure.

5. A ZINC FINGER DOMAIN PEPTIDE

Given that this family of proteins appears to be constructed from individual zinc finger modules, an appealing approach to determining the structural and metal-binding properties of these modules is the preparation of isolated single zinc finger peptides. Initial results of such a study have been reported [38]. The peptide produced has the sequence:

Pro-*Phe*-Pro-*Cys*-Lys-Glu-Glu-Gly-*Cys*-Glu-Lys-Gly-*Phe*-Thr-Ser-
Leu-His-His-*Leu*-Thr-Arg-*His*-Ser-Leu-Thr-*His*-Thr-Gly-Glu-Lys

where the proposed metal-binding residues and conserved hydrophobic amino acids are shown in italics. This peptide was prepared by synthesizing a gene that encodes the peptide with an additional amino terminal methionine residue, expressing this gene as part of a larger fusion protein that is expressed at high levels in *Escherichia coli*, and freeing the desired peptide via cyanogen bromide cleavage.

Studies of this peptide have revealed a number of interesting properties relevant to zinc fingers:

1. In the absence of appropriate metal ions such as Zn^{2+} or Co^{2+}, the peptide exists as a random coil in aqueous solution. This conclusion is based on ultraviolet circular dichroism spectroscopy and on the sensitivity of the free peptide to proteolytic digestion.
2. The peptide folds in the presence of Zn^{2+} or Co^{2+} to form a much more structured species based on substantial changes in the circular dichroism spectrum and on a great decrease in sensitivity to tryptic digestion.

3. Zn^{2+} binds significantly more tightly than does Co^{2+} to the peptide. This behavior seems to provide an explanation for the uniqueness of Zn^{2+} in reconstituting the specific DNA-binding activities of zinc finger proteins.
4. The ultraviolet-visible spectrum of the Co^{2+} complex strongly supports the proposed tetrahedral metal-binding site. A set of ligand field transitions are observed centered at 635 nm with extinction coefficients of $\approx 400\ M^{-1}\ cm^{-1}$. The correlation is made stronger by comparison with some recently prepared model complexes including $Co(2,4,6\text{-triisopropylbenzenethiolate})_2(N\text{-methylimidazole})_2$ [39].

Thus, these studies strongly support the zinc finger hypothesis. Furthermore, they clearly underscore the role of zinc in stabilizing small structural domains.

6. OTHER PROTEINS WITH POTENTIAL METAL-BINDING DOMAINS

The zinc finger domains in TFIIIA and the other proteins noted in Table 1 above are clearly recognized by (and, indeed, have been defined by) an amino acid sequence motif including especially a pattern of cysteine and histidine residues. Through systematic searches [40] and the discovery of new deduced amino acid sequences, several classes of proteins have been discovered that have other patterns that are different than those in the classical zinc finger proteins which nonetheless appear likely to form metal-binding domains. Of these, three classes will be discussed briefly here: (1) the yeast regulatory protein GAL4 and related proteins; (2) the steroid hormone receptor superfamily; and (3) the retroviral core nucleic acid-binding proteins.

GAL4 is a positive regulatory protein which binds to the GAL upstream activating sequences [41]. The DNA-binding activity has been localized to the first 74 amino acids [42]. Contained in this region is a sequence of the form Cys-X_2-Cys-X_6-Cys-X_6-Cys-X_2-Cys-X_6-Cys [43]. Sequences of this form have also been observed in other regulatory proteins [44]. Genetic experiments have demonstrated that the DNA-binding activity is zinc-dependent [45]. A number of mutants that were significantly defective in GAL4 activity under normal growth conditions were isolated [46]. One of these mutants (in which a conserved Pro residue in the cysteine-rich region was replaced with Leu) was unique [45]. In the absence of added zinc, yeast harboring this mutation were unable to grow on galactose as their sole carbon source. However, when a crystal of zinc sulfate was placed on the edge of a galactose plate inoculated with this strain, a halo of yeast grew around the crystal. No structural information is yet available on any member of this class of proteins. There is no reason to believe that the structures formed for these proteins is similar to that for the TFIIIA-like zinc finger proteins. The presence of six conserved cysteine residues suggests the possibility that these proteins may contain a structurally precedented [47] $Zn_2(S\text{-cys})_6$ unit rather than one or more mononuclear sites.

When a cDNA for the human glucocorticoid receptor was cloned and sequenced [48], a region of strong sequence similarity to the avian erythroblastosis virus v-erb-A oncogene was discovered [49]. This region encodes approximately 70 amino acids including nine conserved cysteine residues. This result foreshadowed the discovery of a large superfamily of nuclear receptors that is now known to include receptors for estrogen, progesterone, mineralocorticoid, thyroid hormones, and retinoic acid [50]. In addition to binding the substrates noted above, these proteins also bind to specific sites on DNA and couple hormone binding to transcriptional activation [51]. The cysteine-rich domain is responsible for the DNA-binding site specificity in a direct or indirect manner. This has been elegantly demonstrated by the generation of a chimeric receptor consisting of the human estrogen receptor with the glucocorticoid receptor cysteine-rich domain substituted for the native one [52]. This hybrid stimulates transcription from a normally glucocorticoid-responsive gene in an estrogen-dependent manner. Metal ions such as zinc(II) have been shown to affect the affinity of certain of these receptors for DNA [53] and for other proteins [53,54]. Again, no structural information is yet available for any member of this class of proteins and the nature of the structures formed from the cysteine-rich domains and metal ions remains to be discovered.

Finally, the low molecular weight nucleic acid-binding proteins from retroviruses also contain a short-sequence motif that is quite suggestive of a metal-binding domain [40]. Much like the TFIIIA-like zinc finger proteins, these proteins can contain multiple, tandem copies of the motif; these proteins have either one or two sequences of the form Cys-X_2-Cys-X_4-His-X_4-Cys [40,55]. These proteins appear to play an important role in RNA packaging. A two amino acid insertion in one of the cysteine-histidine motifs appears to significantly disrupt this function [56]. No detailed studies of the metal-binding or structural properties of these proteins have been reported.

7. FUNCTIONS OF METAL-BINDING DOMAINS

7.1. Structural Roles

A metal-binding domain is defined as a metal-binding site in a protein in which all of the ligands come from within one short stretch of amino acid sequence. The zinc finger metal-binding domains consist of approximately 25 amino acids. From the experimental work on the single zinc finger peptide, it is clear that these domains do not fold in the absence of appropriate metal ions but do in their presence. The absence of folding without a bound metal ion is consistent with the general observations that peptides of this length do not fold and that most structural domains in proteins contain at least 50 amino acids. The explanation for these observations appears to be that the folded forms of such small peptides do not have sufficient stabilization energy from hydrophobic interactions, van der Waals interactions, and hydrogen bonds to overcome the destabilizing influence of the high-conformation entropy of the denatured state. A major role of a bound metal ion in leading to folding is to destabilize the unfolded state by crosslinking it and therefore significantly decreasing the number of possible conformations.

It should also be noted that entropic terms are also of fundamental importance in the metal-binding process itself via the chelate effect. In aqueous solution, "free" metal ions are, of course, bound to six water molecules. Upon coordination to a metal-binding peptide or metal-binding domain in a protein, these water molecules are released. Thus, metal binding occurs despite the fact that six covalent bonds (to water) are broken and only four (albeit stronger ones) are formed.

7.2. Potential Regulatory Roles

As noted above, many of the zinc finger proteins whose functions are known are involved in developmental control processes. Obviously, this is also true for many members of the steroid hormone receptor superfamily as well. The fact that these two classes of proteins are zinc metalloproteins raises the possibility that the zinc ion may play regulatory roles related to its structural one. In principle, regulated changes of the concentration of zinc ion could exert considerable effects on gene expression through the action of these proteins. At this time, this idea is entirely speculative, but the rapid progress over the past few years in the discovery and sequencing of these proteins may be followed by equally exciting developments in elucidating their biological functions in the upcoming years.

REFERENCES

1. J. S. Hanas, D. J. Hazuda, D. F. Bogenhagen, F. Y.-H. Wu, and C.-W. Wu, *J. Biol. Chem., 258,* 14120 (1983).
2. H. R. B. Pelham and D. D. Brown, *Proc. Natl. Acad. Sci. USA, 77,* 4170 (1980).
3. B. Picard and M. Wegnez, *Proc. Natl. Acad. Sci. USA, 76,* 241 (1979).

4. A. M. Ginsberg, B. O. King, and R. G. Roeder, *Cell, 39,* 479 (1984).

5. J. Miller, A. D. McLachlan, and A. Klug, *EMBO J., 4,* 1609 (1985).

6. R. S. Brown, C. Sander, and P. Argos, *FEBS Lett., 186,* 271 (1985).

7. D. R. Smith, I. J. Jackson, and D. D. Brown, *Cell, 37,* 645 (1984).

8. J. J. Bieker and R. G. Roeder, *J. Biol. Chem., 259,* 6158 (1984).

9. K. E. Vrana, M. E. A. Churchill, T. D. Tullius, and D. D. Brown, *Mol. Cell. Biol., 8,* 1684 (1988).

10. T. D. Tullius and B. A. Dombroski, *Proc. Natl. Acad. Sci. USA, 83,* 5469 (1986).

11. J. Y. Tso, D. J. Van der Berg, and L. J. Korn, *Nucleic Acids Res., 14,* 2187 (1986).

12. W. Gilbert, *Nature, 271,* 501 (1978).

13. G. Diakun, L. Fairall, and A. Klug, *Nature, 324,* 698 (1986).

14. A. Ruiz i Altaba, H. Perry-O'Keefe, and D. A. Melton, *EMBO J., 6,* 3065 (1987).

15. U. B. Rosenberg, C. Schröder, A. Preiss, A. Kienlin, S. Côté, I. Riede, and H. Jäckle, *Nature, 319,* 336 (1986).

16. A. Vincent, H. V. Colot, and M. Rosbash, *J. Mol. Biol., 186,* 149 (1985).

17. D. Tautz, R. Lehmann, H. Schnürch, R. Schuh, E. Seifert, A. Kienlin, K. Jones, and H. Jäckle, *Nature, 327,* 383 (1987).

18. J. L. Boulay, C. Dennefeld, and A. Alberga, *Nature, 330,* 395 (1987).

19. R. M. Baldarelli, P. A. Mahoney, F. Salas, E. Gustavson, P. D. Boyer, M.-F. Chang, M. Roark, and J. A. Lengyel, *Dev. Biol., 125,* 85 (1988).

20. R. Schuh, W. Aicher, U. Gaul, S. Côté, A. Preiss, D. Maier, E. Seifert, U. Nauber, C. Schröder, R. Kemler, and H. Jäckle, *Cell, 47,* 1025 (1986).

21. T. A. Hartshorne, H. Blumberg, and E. T. Young, *Nature, 320,* 283 (1986).

22. A. Klug and D. Rhodes, *Trends Biochem. Sci., 12,* 464 (1987).

23. J. Milbrandt, *Science, 238,* 797 (1987).

24. K. Chowdhury, U. Deutsch, and P. Gruss, *Cell, 48,* 771 (1987).

25. E. Pays and N. B. Murphy, *J. Mol. Biol., 197,* 147 (1987).

26. J. T. Kadonaga, K. R. Carner, F. R. Masier, and R. Tjian, *Cell, 51,* 1079 (1987).

27. D. C. Page, R. Mosher, E. M. Simpson, E. M. C. Fisher, G. Mardon, J. Pollack, B. McGillivray, A. de la Chapelle, and L. G. Brown, *Cell, 51,* 1091 (1987).

28. H. Blumberg, A. Eisen, A. Sledziewsi, D. Bader, and E. T. Young, *Nature, 328,* 443 (1987).

29. J. M. Berg, *Proc. Natl. Acad. Sci. USA, 85,* 99 (1988).

30. F. C. Bernstein, T. F. Koetzle, G. J. B. Williams, E. F. Meyer, Jr., M. D. Brice, J. R. Rodgers, O. Kennard, T. Shimanouchi, and M. Tasumi, *J. Mol. Biol., 112,* 535 (1977).

31. W. P. J. Gaykema, A. Volbeda, and W. G. J. Hol, *J. Mol. Biol., 187,* 255 (1985).

32. R. S. Brown, and P. Argos, *Nature, 324,* 215 (1986).

33. J. M. Berg, in preparation

34. L. Fairall, D. Rhodes, and A. Klug, *J. Mol. Biol., 192,* 577 (1986).

35. C. O. Pabo and R. T. Sauer, *Ann. Rev. Biochem., 53,* 293 (1984).

36. J. E. Anderson, M. Ptashne, and S. C. Harrison, *Nature, 326,* 846 (1987).

37. J. A. McClarin, C. A. Frederick, B.-C. Wang, P. Greene, H. W. Boyer, J. Grable, and J. M. Rosenberg, *Science, 234,* 1526 (1986).

38. A. D. Frankel, J. M. Berg, and C. O. Pabo, *Proc. Natl. Acad. Sci. USA, 84,* 4841 (1987).

39. D. T. Corwin, Jr., R. Fikar, and S. A. Koch, *Inorg. Chem., 26,* 3079 (1987).

40. J. M. Berg, *Science, 232,* 485 (1986).

41. E. Giniger, S. M. Varnum, and M. Ptashne, *Cell, 40,* 767 (1985).

42. L. Keegan, G. Gill, and M. Ptashne, *Science, 231,* 699 (1986).

43. A. Laughon and R. Gesteland, *Proc. Natl. Acad. Sci. USA, 79,* 6827 (1982).

44. R. K. Beri, H. Whittington, C. F. Roberts, and A. R. Hawkins, *Nucleic Acids Res., 15,* 7991 (1987) and references therein.

45. M. Johnston, *Nature, 328,* 353 (1987).

46. M. Johnston and J. Dover, *Proc. Natl. Acad. Sci. USA, 84,* 2401 (1987).

47. A. D. Watson, Ch. Pulla Rao, J. R. Dorfman, and R. H. Holm, *Inorg. Chem., 24,* 2820 (1985).

48. S. M. Hollenberg, C. Weinberger, E. S. Ong, G. Cerelli, A. Oro, R. Lebo, E. B. Thompson, M. G. Rosenfeld, and R. M. Evans, *Nature, 318,* 635 (1985).

49. C. Weinberger, S. M. Hollenberg, M. G. Rosenfeld, and R. M. Evans, *Nature, 318,* 670 (1985).

50. M. Petkovich, N. J. Brand, A. Krust, and P. Chambon, *Nature, 330,* 444 (1987) and references therein.

51. K. R. Yamamoto, *Ann. Rev. Genet., 19,* 209 (1985).

52. S. Green and P. Chambon, *Nature, 325,* 75 (1987).

53. M. Sabbah, G. Redeuilh, C. Secco, and E.-E. Baulieu, *J. Biol. Chem., 262,* 8631 (1987).

54. E. M. Wilson, *J. Biol. Chem., 260,* 8683 (1985).

55. L. E. Henderson, T. D. Copeland, R. C. Sowder, G. W. Smythers, and S. Oroszlan, *J. Biol. Chem., 256,* 8400 (1981).

56. C. Méric and P.-F. Spahr, *J. Virol., 60,* 450 (1986).

7

Site-Directed Mutagenesis and Structure-Function Relations in EF-Hand Ca^{2+}-Binding Proteins

Sture Forsén, Torbjörn Drakenberg, Sara Linse, Peter Brodin,* Peter Sellers, Charlotta Johansson, Eva Thulin, and Thomas Grundström*
Physical Chemistry 2, Chemical Centre
Lund University, P.O. Box 124
S-221 00 Lund, Sweden
*Unit of Applied Cell and Molecular Biology
University of Umeå
S-901 87 Umeå, Sweden

1. INTRODUCTION

It is widely recognized that Ca^{2+} ions play a central role in a complex intracellular messenger system that mediates a wide variety of biological processes: muscle contraction, secretion, the metabolic processes of glycolysis and gluconeogenesis, ion transport, cell division, and growth. The detailed organization of this messenger system is presently the subject of considerable scientific activity [1]. A schematic outline of the system is presented in Fig. 1. An essential link in the Ca^{2+} messenger system is a family of highly homologous Ca^{2+}-binding proteins that undergo Ca^{2+}-dependent conformational changes and respond to transitory increases in intracellular Ca^{2+} concentrations. Well-known representatives of this family are calmodulin, parvalbumin, the S-100 proteins, and skeletal and heart muscle troponin C [2-8]. These proteins show a high degree of amino acid sequence homology, both internally and between different members. Furthermore, x-ray crystallographic studies on four of these proteins: carp parvalbumin [9,10], calf calbindin D_{9k} [11,12], turkey and chicken skeletal muscle troponin C [13-15], and rat and cow calmodulin [16,17] have shown that the calcium-binding sites are constructed in a similar fashion—colloquially termed "the EF hand" [9]. This structural unit is made up of two helices separated by a calcium-binding loop which is typically 12 amino acids long and is wrapped around the Ca^{2+} ion in such a way that the ion is approximately octahedrally coordinated with oxygen atoms. A striking property of the EF-hand-

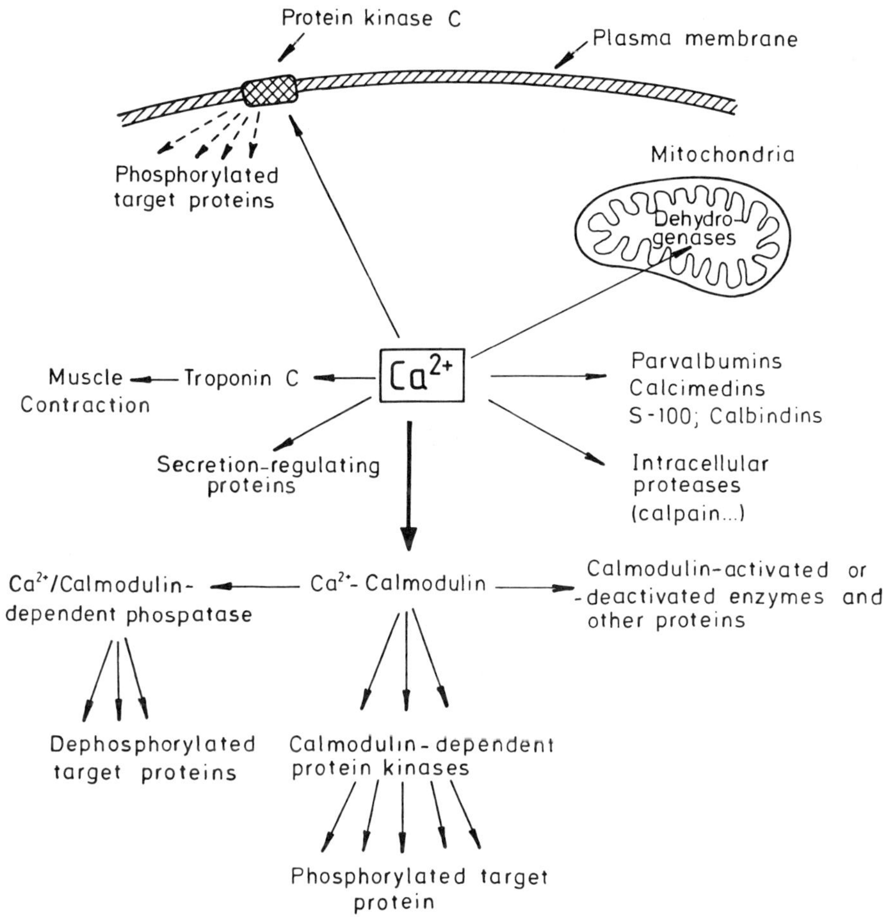

FIG. 1. A calcium-centered and schematic view of intracellular activities.

type proteins is that the Ca^{2+}-binding constants span a wide range from 10^5 M^{-1} or less to 10^9 M^{-1} with no simple and obvious correlation with the type and disposition of the Ca^{2+} ligands [11,13]. A second structural feature of this family of proteins is that *pairs* of EF hands rather than individual binding sites seem to be the functional unit.

As mentioned above, there is considerable experimental evidence that the regulatory Ca^{2+}-binding proteins undergo Ca^{2+}-dependent conformational changes that form the molecular basis for their biological function. The detailed nature of these conformational changes and the rate at which they occur is presently not well established but is of considerable importance. These and related questions have been addressed in a number of biophysical studies from many laboratories, our own included. These studies have primarily dealt with calmodulin (CaM), skeletal muscle troponin C (sTnC), and their respective tryptic fragments (for some recent reviews, cf. [18-22]). Needless to say these molecules are of utmost biological interest but are at the same time almost too complicated for detailed biophysical studies. The dynamics of Ca^{2+} binding serve to illustrate this. Both CaM and sTnC bind four Ca^{2+} ions and, although the affinities of the binding sites tend to fall into two distinct groups, the presence of four sites considerably complicates the interpretation of experimental stopped-flow and variable-temperature ^{43}Ca NMR results. In contrast *calbindin* D_{9k} constitutes an experimental system that is much more amenable to

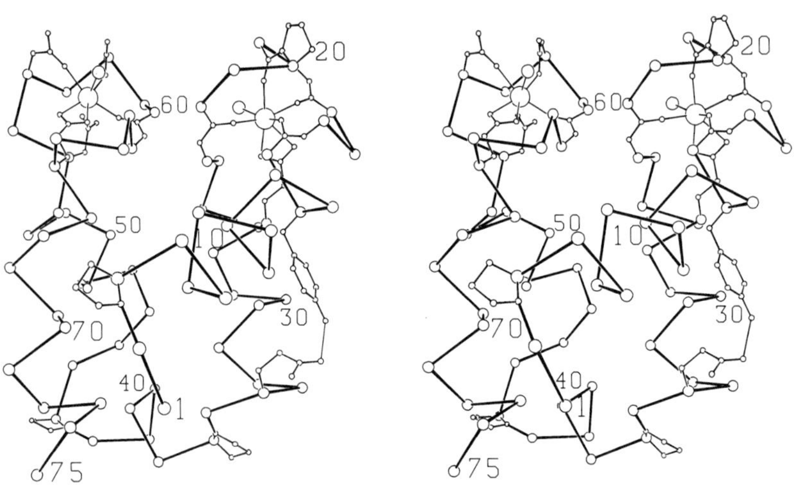

FIG. 2. Stereo view of the three-dimensional structure of calbindin D_{9k} according to x-ray studies by Szebenyi and Moffat [11]. Calcium ligands and Tyr-13 and Glu-35 side chains are indicated. (The figure has kindly been provided by Dr. D. Szebenyi, Cornell University.)

studies aimed at exploring the relationships between biological function and structure and dynamics among the EF-hand family of Ca^{2+}-binding proteins. This protein has only two Ca^{2+}-binding sites and its small size (75 amino acids; $M_r \approx 8{,}500$), stability, and high solubility in aqueous media make it very attractive for NMR studies. The x-ray structure of the "minor A" form of calbindin D_{9k} has recently been refined to a resolution of 2.3 Å [11]. A schematic stereo diagram of the structure is shown in Fig. 2. The overall features of the molecule are similar to those of the Ca^{2+}-binding domains of parvalbumin, CaM, and TnC. The Ca^{2+}-binding site in the C-terminal half (site II) of the molecule has the same fold as that in the archetypal EF hand, whereas the site in the N-terminal half (site I) constitutes a variant hand with two extra amino acids in the loop, one of which is a proline. A comparison of the amino acid sequences of the Ca^{2+}-binding loops of calbindin D_{9k} with those of other EF-hand-type proteins is shown in Fig. 3.

2. CHOICE OF MUTANTS FOR BIOPHYSICAL STUDIES

The power of site-directed mutagenesis in the study of enzyme mechanisms, protein stability, folding-unfolding pathways, and protein-protein or protein-nucleic acid interactions is by now well established and is amply illustrated through other contributions in this volume. How can this technique help in understanding structure-function relationships in regulatory Ca^{2+}-binding proteins? First let us briefly summarize some of the specific questions concerning this particular group of protein molecules.

- Why do EF-hand Ca^{2+} sites mainly appear in pairs?
- How is the Ca^{2+} affinity of the individual sites related to the nature and disposition of the ligands?
- How do the Ca^{2+} sites in EF-hand pairs interact and what is the free energy of interaction?
- What are the rates of Ca^{2+} binding and release?

In addition, other general biophysical questions, e.g., regarding the relationship between amino acid sequence and three-dimensional

Protein	X Y Z -Y -X -Z
PA CD	I-D-Q-D-K-S-G-F-I-E-E-D-E-L
EF	G-D-S-D-G-D-G-K-I-G-V-D-E-F
sTNC I	F-D-A-D-G-G-G-D-I-S-V-K-E-L
II	V-D-E-D-G-S-G-T-I-D-F-E-E-F-
III	F-D-R-N-A-D-G-Y-I-D-A-E-E-L
IV	G-D-K-D-N-D-G-R-I-D-F-D-E-F-
cTNC I	F-V G-A-E-D-G-C-I-S-T-K-E-L 'L'
CaM I	F-D-K-D-G-N-G-T-I-T-T-K-E-L
II	V-D-A-D-G-N-G-T-I-D-F-P-E-F
III	F-D-K-D-G-N-G-Y-I-S-A-A-E-L
IV	A-D-I-D-G-D-G-E-V-N-Y-E-E-F
Calbindin D_{9k} I	^{13}Y A-K-E-G-D-P Q-L-S-K-E-E-L^{28} 'A' 'N'
II	^{53}L-D-K-N-G-D-G-E-V-S-F-E-E-F^{66}

FIG. 3. A comparison of the calcium-binding regions of several calcium-binding proteins. PA = carp parvalbumin; sTNC = skeletal muscle troponin C; cTNC = cardiac muscle troponin C; CaM = calmodulin. The single-letter amino acid code is used. X, Y, Z, -Y, -X, and -Z are Ca^{2+} ligand coordinates in EF-hand sites as specified in Ref. 8.

structure, stability, and dynamic properties, are certainly as relevant to Ca^{2+}-binding proteins as to all other proteins.

Our initial set of mutants was aimed at perturbing and modifying the N-terminal, "pseudo-EF-hand", Ca^{2+}-binding site (site I) of calbindin. Only amino acids not directly involved as Ca^{2+} ligands, through either their side chains or their backbone carbonyl groups, were modified or deleted. A list of mutant proteins prepared to date is given in Table 1. In the following the wild-type minor A form of bovine calbindin D_{9k}, with an additional methionine residue at the N terminus (cf. Sec. 3) will be designated "MO". Mutant

TABLE 1

Definition of Mutant Calbindins D_{9k}

Mutant protein	Amino acid substitutions and deletions[a]
M1	Pro-20 → Gly
M2	Pro-20 → Gly; Asn-21 deleted
M3	Pro-20 deleted
M4	Tyr-13 → Phe
M5	Glu-17 → Gln
M6	Asp-19 → Asn
M7	Glu-17 → Gln; Asp-19 → Asn
M8	Glu-26 → Gln
M9	Glu-26 → Gln; Glu-17 → Glu
M10	Glu-26 → Gln; Asp-19 → Asn
M11	Glu-26 → Gln; Glu-17 → Gln; Asp-19 → Asn
M12	Pro-20 deleted; Ala-14 deleted
M13	Pro-20 → Gly; Ala-14 deleted
M14	Pro-20 → Gly; Ala-14 deleted; Asn-21 deleted
M15	Ala-14 deleted

[a]Note that the definitions of some mutants could be made in more than one way. Thus M2 could equally well be designated as "Pro-20 deleted; Asn-21 → Gly". Also, since the amino acids in positions 14 and 15 are both alanines it is evident that any mutant with "Ala-14 deleted" equally well could be regarded as "Ala-15 deleted". *Note:* For numbering of the amino acids, cf. text and Fig. 3.

proteins are denoted M1, M2, etc., as defined in Table 1. The numbering scheme of Fullmer and Wasserman [23] is used, which means that the N-terminal methionine is given number zero. This scheme has been used in Fig. 3.

Mutants M1-M3 were prepared in order to study the role of the proline at position 20 in the Ca^{2+}-binding loop of site I. According to the x-ray crystallographic study by Szebenyi and Moffat [11], the OH group of Tyr-13 is hydrogen-bonded to the carboxyl group of

Glu-35—thus linking the two helices flanking the N-terminal Ca^{2+} site (cf. Fig. 2). M4 was designed to study the role of this interhelix hydrogen bond. The series M5-M11 is aimed at investigating how the Ca^{2+}-binding properties of sites I and II are influenced by negatively charged amino acid side chains present at the protein surface near these sites. A schematic view of the Ca^{2+} sites with the positions of adjacent amino acids Asp and Glu indicated is shown in Fig. 4. Mutants involving "neutralization" also of Glu-60 are under preparation. The mutants M12-M15, and also M1, M2, and M3, have been prepared in order to stepwise modify the Ca^{2+}-binding loop of the "pseudo-EF site" (I) into a loop more resembling those observed in typical EF-hand sites, site II in calbindin included. In this series M12, which involves a shortening of the loop by two amino acids, and M14, for which in addition an ubiquitous Gly is

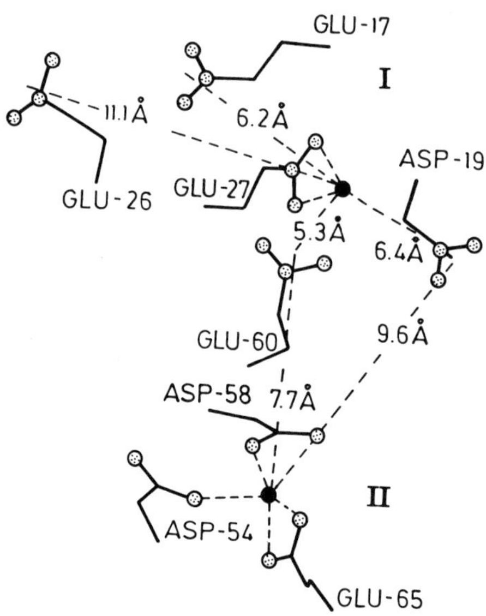

FIG. 4. The "cluster" of negatively charged amino acids around the two Ca^{2+} sites of calbindin D_{9k}. The view is from the top of Fig. 2. According to the x-ray studies, only Glu-27, Asp-54, Asp-58, and Glu-65 are part of the inner ligand sphere of sites I and II, respectively. Distances between the Ca^{2+} ions and the nonliganded carboxylate groups (midpoint between the oxygens) are indicated.

introduced between the "Z" and "-Y" ligands, may be regarded as the most EF-like mutants (cf. Fig. 3).

3. GENE SYNTHESIS AND EXPRESSION; PURIFICATION OF MUTANT PROTEINS

3.1. Gene Synthesis and Expression

To produce the minor A form of bovine calbindin in *E. coli*, a gene was designed [24] encoding the known amino acid sequence of the protein. Mainly codons found to be preferentially utilized in highly expressed *E. coli* genes were used, but a few moderately used codons were also utilized when restriction enzyme sites were to be introduced or avoided. A DNA sequence corresponding to the terminator of transcription of the *rrnC* gene was introduced behind the coding sequence to obtain independence from the existence of terminators in different expression vectors. The gene was constructed from 34 overlapping oligonucleotides synthesized by the very rapid segmented microscale method with cellulose disks as solid support [61,62]. The oligonucleotides were assembled into five DNA segments in different bacteriophage M13 derivatives [63] using the "shotgun ligation" technique [64]. In this method correctly ligated oligonucleotides are obtained through direct biological selection instead of using gel purifications. The method is very convenient for construction of large DNA segments from oligonucleotides. The DNA segments constructed were delimited by restriction enzyme sites placed at convenient distances in the gene and were cloned into the corresponding sites of different M13 bacteriophages. Progeny phages were isolated from all five ligations and the nucleotide sequences of the DNA segments were verified by dideoxy sequencing. The complete gene was then assembled from the DNA segments, and a phage with the wanted structure was isolated. The assembly of the gene from "DNA building blocks" bordered by restriction enzyme sites facilitates the construction of genes encoding calbindin molecules with amino acid substitutions because only the

DNA segment containing the substitutions desired has to be reassembled from oligonucleotides for each mutation. In order to further facilitate the construction of mutations in the N-terminal part of the protein, containing calcium-binding site I, a truncated version of the gene was constructed [25]. In this gene a 101 base pair segment corresponding to the first 33 amino acids in the protein was replaced by a short polylinker sequence. The restriction enzyme sites bordering this polylinker were subsequently used for reassembling the complete gene from the original set of oligonucleotides with replacements of the ones containing the codons to be changed.

Two different expression plasmids were developed to overproduce calbindin in *E. coli*. They both carry a strong inducible hybrid promoter, the "tac" promoter, in front of a polylinker containing several restriction enzyme sites, and a ribosome-binding site with high homology to the 3' end of 16S ribosomal RNA and with the most commonly found distance between this homology and the initiation codon. The first expression plasmid, pICB1 [63], is a vector based on the high copy number plasmid pBR 322. In order to keep the tac promoter repressed on such a multicopy plasmid in the absence of the inducer IPTG, the lac repressor has to be overproduced. Therefore, a 1.2 kb DNA fragment carrying $lacI^Q$, a lac repressor overproducing gene, was cloned into the plasmid. The second expression plasmid, pRCB1, is a derivative of the plasmid pBEU50 [65] where the EcoRI-HpaI restriction fragment containing a part of the *tet* gene has been replaced by a DNA segment from pICB1 containing the tac promoter, the synthetic calbindin gene, and the $lacI^Q$ gene. This plasmid is temperature-sensitive in the control of its copy number. At 30°C the vector is present in a moderate number of copies per cell, but when the temperature is raised above 35°C the replication control is lost and the number of copies per cell increases continuously, thereby increasing the gene dosage of the calbindin gene. Cell growth and protein synthesis continues normally for many generations at the higher temperature. By the use of this so-called runaway vector it has been possible to obtain an about fivefold higher level of overproduction of the mutated calbindins than with the pICB1 type of plasmids previously reported.

Stability problems have been encountered with foreign proteins in *E. coli*. A protease encoded from the *lon* gene has been shown to be responsible for a considerable part of this protein breakdown. For this reason, expression of calbindin from pICB1 was studied both in *lon*$^{+}$ and in *lon*$^{-}$ *E. coli* strains by SDS polyacrylamide gel electrophoresis of total protein extracts. The level of expression was found to be approximately three times higher from a *lon*$^{-}$ than from a *lon*$^{+}$ strain. The higher level of expression corresponds to 1-2 wt% of total *E. coli* protein [63]. We have found that the runaway expression system based on vector pRCB1 is not compatible with the *lon*$^{-}$ strain. However, this was not a problem because the high gene dosage which could be achieved by using the runaway vector pRCB1 leads to a high expression of calbindin in *lon*$^{+}$ *E. coli* reaching 5-10 wt% of total protein. Considering that calbindin has a more than five times lower molecular weight than an average *E. coli* protein, this represents, in molar terms, a very high level of expression.

3.2. Protein Purification

The purification of proteins in considerable quantities for biophysical measurements is always an important task and proteins expressed in *E. coli* are no exception. In this subsection we will briefly discuss some of the problems we encountered during the initial stages of our work and the way we solved them. It is our experience from talking to colleagues at meetings that similar problems have been encountered by others in studies of genetically engineered proteins but few written accounts have been published.

The wild-type protein, M0, and the mutant proteins, M1, M2, M3, and M4, were purified as described in detail elsewhere [24,25]. The same methods have been employed also for the more recently expressed mutants. The steps consisted of sonication of the cell and fractionation on DEAE-ion exchange and Sephadex G-50 columns. An urea step was found necessary to remove a low molecular weight contaminant that did not reveal itself on SDS-polyacrylamide or agarose gels but was observed in UV and ^{1}H NMR spectra. We have not identified the impurity with certainty but it appears to be a proteoglycan fragment from the *E. coli* cell wall [26]. It could not be removed from the protein using dialysis under normal conditions. However, urea treatment (8 M urea) prior to the Sephadex G-50 fractionation was effective as was a heat treatment at 95°C.

The purity of the mutant proteins was checked by SDS-polyacrylamide and agarose gel electrophoresis [27] in the presence of either 2 mM EDTA or 2 mM Ca^{2+}. This combination of gels is quite useful. Whereas all mutant proteins have the same mobility on the SDS gels, minor changes in charge show up on agarose gels. During the last year we have had access to a Pharmacia "Phast System" with thin gel electrofocusing. This technique has proved to be of great value in assessing the homogeneity of our protein preparations and has in some cases forced us to repurify samples regarded as pure on the basis of SDS and agarose gel electrophoresis. This is particularly important since deamidation ($R\text{-}CONH_2 \rightarrow R\text{-}COOH$), in particular of Gln, may occur during protein purification and aging [28]. In addition, ^{1}H NMR spectra were always obtained on the protein preparations during the final stages of purification. Not only did ^{1}H NMR permit the detection of contaminants due to the expression system but this method also allowed facile detection of EDTA used in the ion exchange steps. The binding of EDTA to calbindin is strong and is difficult to remove by dialysis using normal procedures. The binding is found to be increased in the mutants where a carboxylic acid has been replaced with an amide group.

Calbindin D_{9k} is not unique in its interaction with EDTA. In our experience many proteins will bind EDTA very strongly and this chelator is not easily removed. EDTA contamination in metalloprotein preparations appears in fact to be more common than is generally appreciated and may have been the cause of diverging results on metal-binding properties from different laboratories and on different preparations in the same laboratory. A case in point is the metallo-enzyme alkaline phosphatase. For several years the metal ion-binding properties of this enzyme were in dispute, with widely different stoichiometries and affinities reported by different research groups. It was finally convincingly shown, through the use of ^{14}C-labeled EDTA, that the enzyme had an extraordinary affinity for this chelator and that rather drastic measures had to be taken to remove it from allegedly metal-free preparations of the protein [29]. Once such measures were taken, consistent metal-binding results were obtained.

4. BIOPHYSICAL METHODS

In this section we will briefly outline the various biophysical methods used to study wild-type and mutant calbindins. Many of them may not be familiar to the general reader and in several instances they have been subject to considerable development and modifications.

4.1. Determination of Ca^{2+}-Binding Constants

The Ca^{2+}-binding constants were determined in Ca^{2+}-free 2 mM Tris-HCl buffer at pH 7.5 by three different methods: (1) titration of the protein with Ca^{2+} in the presence of the Ca^{2+} chelators Quin 2 or 5,5'-Br_2-BAPTA [58] as described elsewhere [25]; (2) titration of the protein with the Ca^{2+} concentration determined by means of a stabilized Ca^{2+}-selective electrode [25]; (3) measurements of NMR spectral changes as Ca^{2+} was successively added to the protein solutions. The first method gives very accurate data for Ca^{2+}-binding constants, a factor of 10 above or below the binding constant of the chelator used; for Quin 2 the range is 10^7-10^9 M^{-1} and for 5,5'-Br_2-BAPTA the range is 10^6 to 10^8 M^{-1}. The electrode method is good for binding constants in the range 10^4-10^7 M^{-1}. The NMR method gives qualitative results. Further details are given in Ref. 25.

4.2. NMR Studies

Hydrogen-1 and calcium-43 NMR spectra were obtained on ~7 and 1 mM solutions, respectively, using a Nicolet 360 WB spectrometer operating at 361.79 and 24.34 MHz, respectively. Isotope-enriched ^{43}Ca was used (enrichment ~50%, Technab Export, Moscow, USSR) and ^{43}Ca NMR spectra were obtained using a home-made high-sensitivity horizontal probe with a solenoidal coil. To eliminate the effects of acoustic ringing observed at lower rf frequencies, the RIDE sequence was used with a 90° pulse length of 40 μsec [30]. 10^3 data points were recorded with a spectral width of $\pm 10^4$ Hz. Up to 5×10^6 transients were accumulated during a total time of 15 hr.

Cadmium-113 NMR spectra were obtained at 56.55 MHz on a home-made spectrometer using an Oxford Instruments 6T wide-bore magnet. The probes were of the same horizontal solenoid type as was used for ^{43}Ca NMR. Normal one-pulse experiments were used and 2.5×10^4 to 1×10^5 transients accumulated, with a pulse length of 10 μsec (45°) and a repetition time of 0.6 sec. The isotope enrichment of ^{113}Cd was 95% (Oak Ridge, TN).

The rate of exchange of the Ca^{2+} ion from its protein-binding site(s) are obtained from studies of the ^{43}Ca NMR signal line shape at different temperatures. Rates of exchange in the range $10\text{-}10^4$ sec^{-1} are accessible with this method, which has been largely developed and applied in our laboratories and has recently been outlined in some detail in a paper by Tsai et al. [32].

4.3. Stopped-Flow Kinetic Studies

Stopped-flow experiments were performed on a modified Durrum-Gibson instrument as described elsewhere [33-35]. A solution of the Ca^{2+}-loaded protein was mixed with a Quin 2 solution. The mixture was excited at 366 nm and the fluorescent light intensity at wavelengths above 475 nm was measured. Typical concentrations before mixing were calbindin 15 μM, Ca^{2+} 30 μM, Quin 2 200 μM. Data analysis was performed by standard nonlinear least squares fitting to a one or two exponential function. In cases where photobleaching of Quin 2 was noticeable, the data were analyzed by incorporating a sloping baseline in the fitting function. Reaction amplitudes have been corrected, where necessary, for losses in the instrumental deadtime (1.1 msec).

4.4. Thermal and Urea Denaturation Studies

Thermal denaturation studies were made using a JASCO 500A spectropolarimeter to follow circular dichroism (CD) intensities at 223 nm. A commercial water-jacketed cuvette was used and the temperature measured with a digital thermometer with a nominal resolution of 0.1°C. The rate of temperature change was 1°C per min or less. CD spectra were recorded both on heating the samples from ambient temperature to 90°C and on cooling the samples from that temperature. All solutions were made in 10 mM Tris-HCl buffer at pH 7.5. The protein concentration was 10-15 μM. In studies of the Ca^{2+}-loaded forms of the proteins the solutions were made 50 μM in Ca^{2+} and in studies of the apo forms EDTA was added to a total concentration of 300 μM [36].

CD measurements of the urea-induced unfolding were made on a JASCO J-41-C spectropolarimeter equipped with an open-ended cuvette that simplified the change from one urea concentration to another [36]. Two different protein stock solutions were used—both in 20 mM Pipes buffer at pH 7.1 with a protein concentration of 10 μM and with EGTA added to a total concentration of about 1 mM. One of these solutions contained urea at a concentration of 8.0 M. By mixing these two solutions, any intermediate urea concentration could be obtained without changing the protein concentration.

In the thermal unfolding experiments the CD absorbance of the stock solutions at 25°C and 223 nm was taken as 100%. Similarly, in the urea unfolding studies the mean value of the CD absorbance at 222 nM over the first three to four low urea concentrations was taken as 100%. The small CD adsorbance in the pure buffer solutions was corrected for in both cases.

The urea-induced unfolding was treated using the simple two-state equilibrium model, N $\rightleftharpoons$ D, where N is the native and D the unfolded or denaturated form of the protein. The fractions of native (f_N) and unfolded (f_D) protein in the transition region were calculated from the normalized CD absorbance in the same region and the apparent free-energy difference; ΔG_D was calculated as $\Delta G_D = -RT \ln(f_N/f_D)$. As a first approximation ΔG_D was assumed to depend linearly on urea concentration:

$$\Delta G_D = \Delta G_D^{H_2O} + A[U] \tag{1}$$

where $\Delta G_D^{H_2O}$ is the free-energy difference between native and unfolded protein in the absence of urea. A is a parameter that describes the cooperativity of the unfolding transition and [U] is the urea concentration in mol/liters [36,37]. Since a slight curvature was noted in the ΔG_D vs. [U] curves, $\Delta G_D^{H_2O}$ values were also calculated from least-squares fit to a less truncated version of Eq. (1):

$$\Delta G_D = \Delta G_D^{H_2O} + A'[U] + B'[U]^2 \tag{2}$$

4.5. Microcalorimetry

Enthalpy changes upon the addition of Ca^{2+} to the apo forms of wild-type and mutant calbindins were made at 25°C using an LKB batch microcalorimeter equipped with an LKB titration syringe assembly mounted on the outside of the calorimeter block [38]. The calorimeter was calibrated electrically and also by using the standard procedure of protonation of Tris with HCl [39]. All measurements were made at 25°C in 20 mM Pipes buffer at pH 7.0. The protein concentration in the reaction cell (4-5 ml) was 0.2 to 0.8 mg/ml and the Ca^{2+} concentration in the titrating solution ($CaCl_2$ dissolved in Pipes buffer) was 10-90 mM depending on the magnitude of the observed heat effects. The calorimetric titration curves were analyzed using a computer program that allowed the simultaneous determination of equilibrium constants and enthalpy changes if the value of the highest constant is below 10 M^{-1}. The calculation of ΔH values in systems with higher binding constants requires access to binding constants obtained by other methods [25].

5. RESULTS

5.1. Ca^{2+}-Binding Constants

The two stoichiometric Ca^{2+} binding constants K_1 and K_2—also referred to as "macroscopic binding constants" and corresponding to the binding of one and two Ca^{2+} ions—for wild-type and some mutant calbindins are listed in Table 2.

The free energy of interaction ($\Delta\Delta G$) between the two sites in a system like calbindin D_{9K} may be expressed in terms of K_1, K_2, and the ratio, η, of the two site binding constants, K_I and K_{II}, as defined in Fig. 5 according to [25]:

$$\Delta\Delta G = -RT \ln\left\{\left(\frac{K_2}{K_1}\right) \frac{(1+\eta)^2}{\eta}\right\} \tag{3}$$

where $\eta = K_{II}/K_I$.

Knowledge of K_1, K_2, and η yields not only $\Delta\Delta G$ but also the four site binding constants K_I, K_{II}, $K_{I,II}$, and $K_{II,I}$ defined in Fig. 5. Note also that the value of $\Delta\Delta G$ is independent of the particular choice of site labels, I and II.

Needless to say, η cannot be deduced from the Quin 2 or ion-selective electrode measurements but must be determined by a method that distinguishes Ca^{2+} binding to the two different sites of the protein. As will be discussed below, ^1H NMR permits such studies. Even in the absence of a knowledge of η, a lower limit of the absolute value of $\Delta\Delta G$ can be obtained from the ratio K_2/K_1 since the function $(1 + \eta)^2/\eta$ has a minimum at $\eta = 1$. Thus we can already

TABLE 2

Macroscopic Ca^{2+}-Binding Constants (K_1 and K_2), Total Free-Energy Change for the Binding of Two Ca^{2+} Ions [$\Delta G_{tot}^{Ca^{2+}} = -RT \ln(K_1 \cdot K_2)$], and Lower Limit of the Free Energy of Site-Site Interaction ($|\Delta\Delta G|_{min}$) for Wild-Type and Mutant Calbindins at 25°C

Protein	K_1 (M^{-1})	K_2 (M^{-1})	$\Delta G_{tot}^{Ca^{2+}}$ (kJ/mol)	$|\Delta\Delta G|_{min}$[a] (kJ/mol)
M0	$2 \cdot 10^8$	$4 \cdot 10^8$	-97	-5.1 ± 0.4[b]
M1	$6.5 \cdot 10^7$	$(8 \pm 6) \cdot 10^6$	-84	
M2	$1.5 \cdot 10^8$	$(5 \pm 2) \cdot 10^5$	-80	
M3	$8 \cdot 10^7$	$(5 \pm 2) \cdot 10^5$	-78	
M4	$3.7 \cdot 10^8$	$2.5 \cdot 10^8$	-97	3.9
M5	$2 \cdot 10^7$	$1 \cdot 10^8$	-89	7.7
M6	$4 \cdot 10^7$	$1 \cdot 10^8$	-89	5.8
M7	$2 \cdot 10^7$	$1 \cdot 10^7$	-83	1.7
M8	$6 \cdot 10^7$	$2 \cdot 10^8$	-93	6.4
M9	$6 \cdot 10^6$	$6 \cdot 10^7$	-83	9.2
M10	$8 \cdot 10^6$	$4 \cdot 10^7$	-83	7.8
M11	$6 \cdot 10^6$	$5 \cdot 10^6$	-77	2.9
M12	$8 \cdot 10^7$	$\leq 5 \cdot 10^5$	≥ -77	
M13	$8 \cdot 10^7$	$\leq 5 \cdot 10^5$	≥ -77	
M14	$8 \cdot 10^7$	$\leq 5 \cdot 10^5$	≥ -77	
M15	$5 \cdot 10^7$	$2 \cdot 10^6$	-80	

[a]Calculated from the ratio K_2/K_1 using Eq. (3) with $\eta = 1$.
[b]The actual value of $\Delta\Delta G$ as obtained using additional NMR spectroscopic information on the population of individual Ca^{2+} sites (cf. text).

Note: The standard deviations are 10 to 20% of the value quoted unless otherwise indicated.

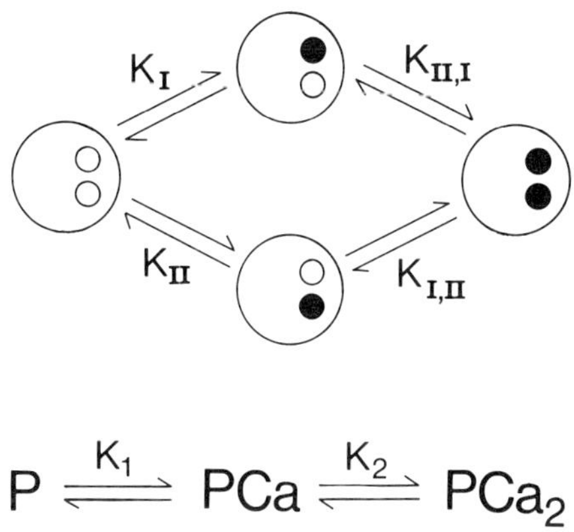

FIG. 5. Definition of binding constants: macroscopic (or stoichiometric) binding constants K_1 and K_2, and the microscopic (or site) binding constants K_I, K_{II}, $K_{I,II}$, and $K_{II,I}$. (Reproduced from Ref. 25 with permission of the American Chemical Society.)

infer from the macroscopic binding constants of Table 2 that *positive cooperativity* of Ca^{2+} binding (i.e., $\Delta\Delta G$ negative) is at hand for wild-type calbindin (M0) as well as for mutants M4 through M11. The lower limit of $-\Delta\Delta G$ is of the order of 3-5 kJ/mol. We will return to the question of the values of $\Delta\Delta G$ in connection with our presentation of the ^1H NMR spectral changes upon Ca^{2+} addition to the apo proteins. We may, however, also infer from the K_1 and K_2 values of Table 2 that in some mutants Ca^{2+} binding is clearly sequential with several powers of 10 separating the two stoichiometric binding constants. In such cases it will not be possible to deduce if cooperative binding still occurs—there is no direct way of determining the binding constant of the weaker of the two sites in the absence of a Ca^{2+} ion in the stronger site.

5.2. Denaturation Studies

5.2.1. Thermal Denaturation

The normalized CD absorbance at 223 nm as a function of temperature for wild-type and some mutant calbindins is shown in Fig. 6(a) (the

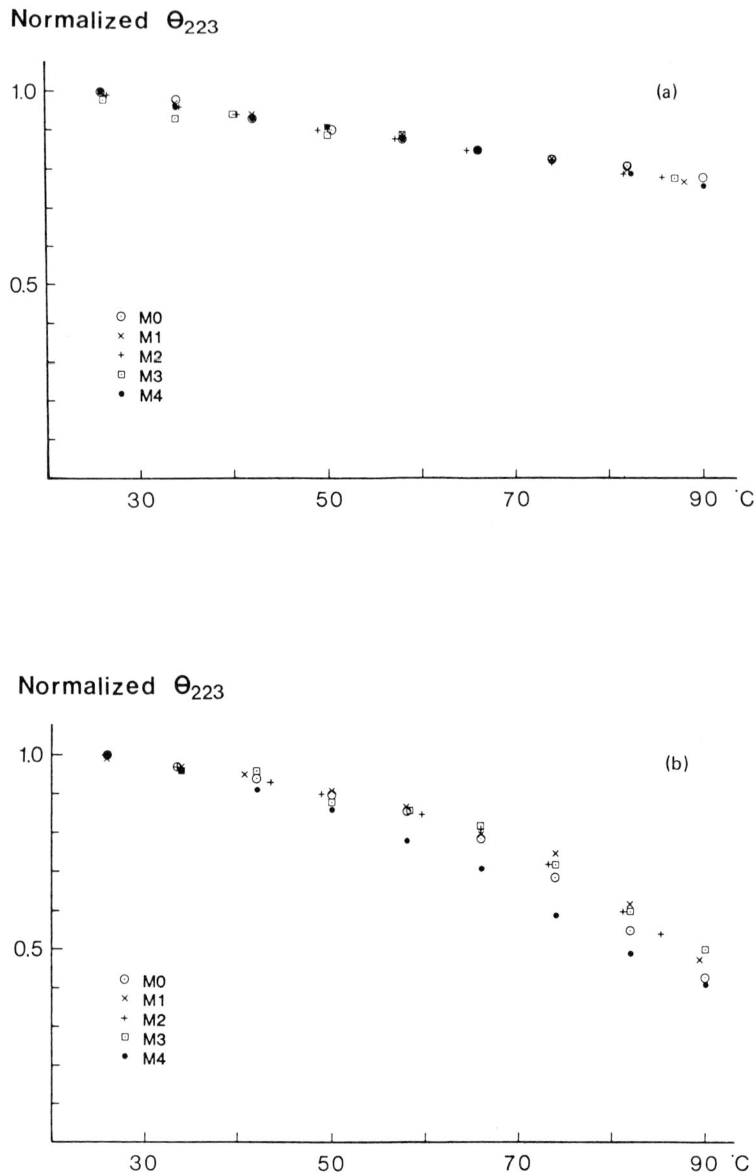

FIG. 6. Thermal denaturation of wild-type (M0) and some mutant calbindins followed by CD absorbance measurements at 223 nm. (a) Ca^{2+}-loaded proteins. (b) Proteins in the apo form. Experimental conditions are outlined in text and definition of mutants is given in Table 1.

Ca^{2+}-loaded forms) and 6(b) (the apo forms). Spectra recorded on heating and cooling the samples were essentially identical and no indication of irreversibility was observed.

5.2.2. *Urea Denaturation*

The normalized CD absorbance at 222 nm as a function of urea concentration on solutions of wild-type and mutant calbindins is shown in Fig. 7. Since measurements on the Ca^{2+}-loaded forms of proteins showed that they remained essentially unaffected by urea up to 8.0 M only the result for M0 is shown. Figure 8 illustrates how the experimental data for a representative protein (M0) conform to a simple two-state model and to the linear and quadratic extrapolation for $\Delta G_D^{H_2O}$

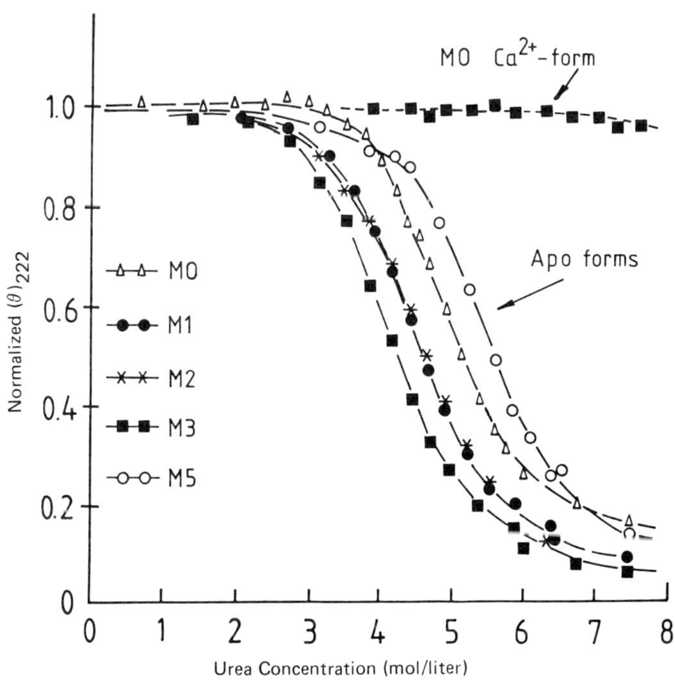

FIG. 7. Denaturation by urea of wild-type (M0) and some mutant calbindins followed by absorbance measurements at 222 nm. Experimental details are given in text. Stability data calculated from these curves are summarized in Table 3.

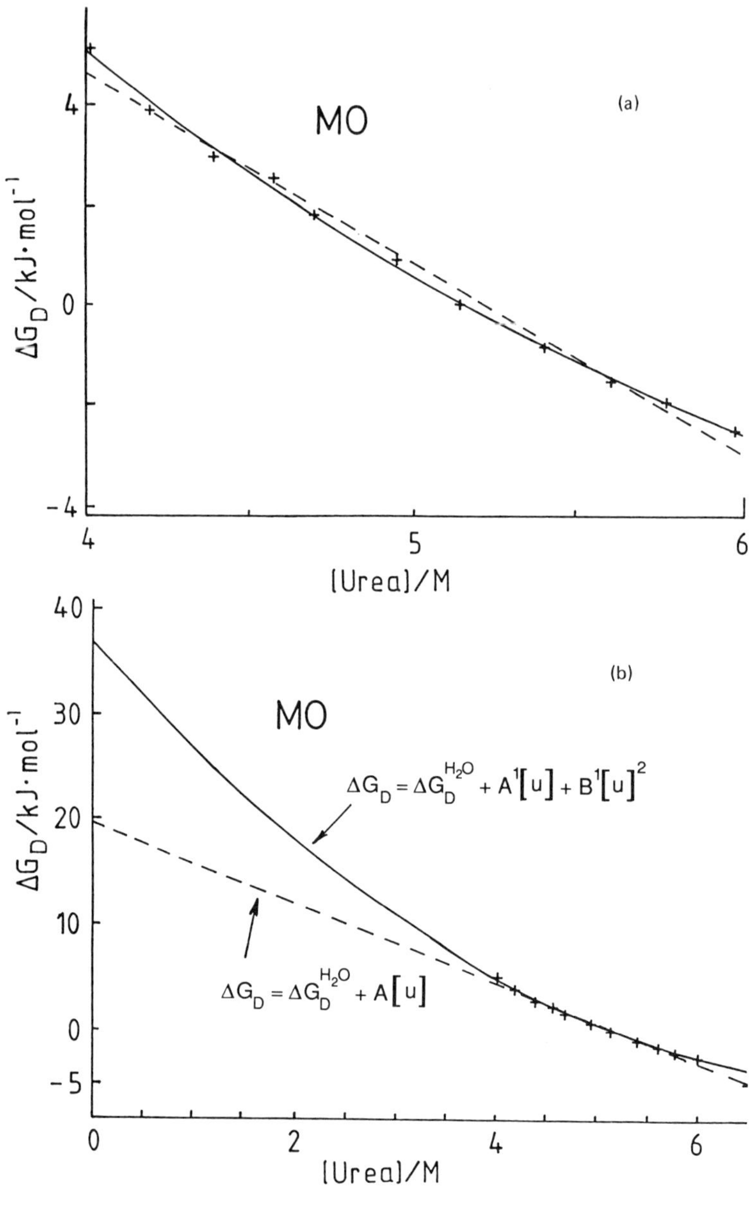

FIG. 8. Stabilization free energy ΔG_D of wild-type calbindin (M0) as a function of urea concentration. (a) The transition region and fits to the experimental points to Eq. (1) (- - -) and Eq. (2) (——); (b) Extrapolation to zero urea concentration using Eqs. (1) and (2), respectively.

TABLE 3

Stabilization Free Energies of Unfolding ($\Delta G_D^{H_2O}$) Urea Concentrations at Transition Midpoints (C_m) and Coefficients Multiplying the Linear and Quadratic Concentration Terms in Eqs. (1) and (2) for the Urea-Induced Unfolding of Wild-Type Calbindin 9k and Five Mutants in the Ca^{2+}-Free Form

Protein	$\Delta G_D^{H_2O}$ Eq. (1) (kJ/mol)	C_m (M, urea)	A (kJ/mol, M)	$\Delta G_D^{H_2O}$ Eq. (2) (kJ/mol)	A' (kJ/mol, M)	B' (kJ/mol, M^2)
M0	20 ± 1	5.2	-3.8	37 ± 2	-10.6	0.68
M1	15 ± 1	4.7	-3.3	26 ± 2	-7.9	0.50
M2	16 ± 1	4.7	-3.5	17 ± 1	-3.7	0.03
M3	15 ± 1	4.3	-3.6	26 ± 2	-8.8	0.61
M4	15 ± 1	4.1	-3.7	26 ± 2	-9.2	0.65
M5	22 ± 1	5.6	-3.8	35 ± 2	-9.0	0.48

according to Eqs. (1) and (2) of Sec. 4. The nonlinear equation provides an improved fit to the experimental data and results in higher values of $\Delta G_D^{H_2O}$ than those obtained with Eq. (1). The urea concentration at the midpoint of the urea-induced unfolding C_m is, however, not dependent on which equation is used. The C_m values therefore will be given emphasis in our comparison of the different mutant proteins. Our best set of values of $\Delta G_D^{H_2O}$, A, A', B', and C_m is given in Table 3.

5.3. Microcalorimetry

The microcalorimetric results on Ca^{2+} binding to wild-type and some mutant calbindins are summarized in Table 4. In the case of M0 and M4, ΔG° values were calculated using average values of the Ca^{2+}-binding constants listed in Table 2. For these proteins the assignment of ΔH° values to the individual sites is therefore arbitrary. For M0 and M4 the total enthalpy change was the same (-16 kJ/mol) within the experimental error. The whole titration curve is shown in Fig. 9. For the other mutants the enthalpy changes for the

TABLE 4

Thermodynamic Quantities for Binding of Ca^{2+} to Wild-Type and Mutant Calbindins at 25°C

Protein	Site	ΔG^θ (kJ/mol)[a]	ΔH^θ (kJ/mol)[b]	ΔS^θ (J/K•mol)
M0	I	-48	-8	134
	II	-48	-8	134
M1	I	-39	-11	94
	II	-45	-13	107
M2	I	-33	+5	128
	II	-47	-8	131
M3	I	-33	+5	128
	II	-45	-8	131
M4	I	-48	-8	134
	II	-48	-8	134

[a]Calculated from the binding constants in Table 2.
[b]The uncertainties are 1-2 kJ/mol.

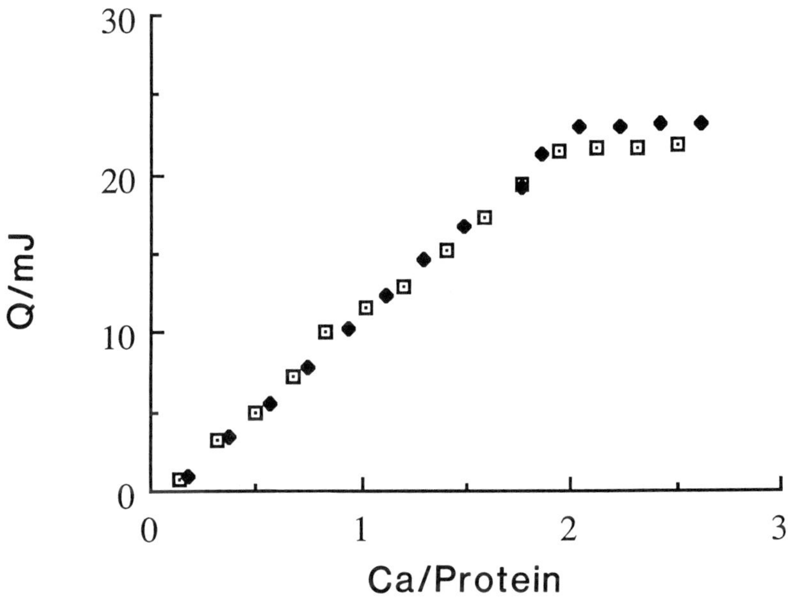

FIG. 9. Calorimetric titration curves showing the accumulated heat evolved as a function of $[Ca^{2+}]/[Protein]$ for the binding of Ca^{2+} to (♦) M0 and (▣) M4. The difference in plateau values is due to a slight diffcronce in protein concentration.

individual sites were calculated using an iterative computer program (cf. [25]) by fixing the binding constants to the values given in Table 2.

5.4. NMR Results

5.4.1. ^{113}Cd NMR

The results are summarized in the first two columns of Table 5. Shifts are reported relative to 0.1 M $Cd(ClO_4)_2$ at pH 7. Titrations of the apo proteins with $^{113}Cd^{2+}$ showed the binding of Cd^{2+} to be sequential for all mutants studied. The first Cd^{2+} entering the protein gave a signal at about -105 to -110 ppm. This is a shift range typical of EF-hand sites [22]. Upon further addition of Cd^{2+},

TABLE 5

^{113}Cd and ^{43}Ca NMR Chemical Shifts in ppm (δ) and Dissociation Rate Constants Obtained from Variable-Temperature ^{43}Ca NMR [k_{off}(NMR)]

Protein	Site	$\delta(^{113}Cd)$		$\delta(^{43}Ca)$	k_{off}(NMR) (sec^{-1})
		$(Cd)_1$	$(Cd)_2$		
M0	I		-156	-9	
	II	-105	-110	7	3[a]
M1	I		-88	—	430 ± 100
	II	-109	-110	7	<10
M2	I		—	—	5,000 ± 1,000
	II	-112	-115	8	<10
M3	I		—	—	5,000 ± 1,000
	II	-113	-115	8	<10
M4	I		-155	-9	3[a]
	II	-105	-110	7	
M5	I		~-150		
	II	-106	-110		
M9	I		~-154		
	II	-107	-109		
M11	I		-140		
	II	-107	-107		
M12	I		—	—	
	II	-110	~-110	7	7,000 ± 1,000
M13	I		—	—	≤10
	II	-108	-108	7	
M14	I		—		6,000 ± 1,000
	II	-110	-110		≤10
M15	I		—		800 ± 200
	II	-105	-105		≤10

[a]The value refers to the dissociation of two Ca^{2+} ions.

Note: $(Cd)_1$ and $(Cd)_2$ refer to proteins with one or two bound Cd^{2+}-ions, respectively.

a second signal appeared for some mutant calbindins with chemical shifts ranging from -156 ppm (M0) to -88 ppm (M1). Interestingly, the binding of the second Cd^{2+} to the protein is in the case of M0 through M4 accompanied by a change in chemical shift of the first Cd^{2+} ion bound. The displacement, which will be denoted $\Delta\delta^{II,I}$ and is of the order of 5 ppm, is indicative of site-site interactions.

The assignment of observed NMR signals to individual sites has previously been made for M0 [25,40]. For the other mutants our assignment is simply based on the assumption that the site modified will show the largest changes in NMR properties. Therefore, the resonance at about -110 ppm is assigned to Cd^{2+} bound to site II based on the shift as well as on the fact that this is the first signal to appear upon stepwise addition of Cd^{2+}.

5.4.2. *^{43}Ca NMR*

The results are summarized in the third and fourth column of Table 5. Shifts are reported relative to dilute aqueous solutions of $^{43}Ca^{2+}$. In the presence of two equivalents of Ca^{2+} the ^{43}Ca NMR spectra of M0 and several of the mutant calbindins show two broad, overlapping resonances that by curve fitting can be resolved into two Lorentzian lines with approximately equal integrated intensity. One of the resolved signals has the chemical shift and narrow line width typical of EF-hand sites (~7 ppm and 350 Hz, respectively). The other signal is generally much broader (~1,000 Hz) and with a shift of about -9 ppm, i.e., upfield from the reference position and quite distinct from EF site shifts. The assignment of the Ca^{2+} signals to the individual sites in the mutants follows that earlier made for M0 (25,40) as was the case for the ^{113}Cd NMR signals. The rates of exchange of Ca^{2+} given in the last column of Table 5 have been calculated from the observed line shapes at different temperatures (cf. [31]).

5.4.3. *^{1}H NMR*

The well-resolved aromatic regions of the ^{1}H NMR spectra of wild-type and some mutant calbindins in the presence and absence of Ca^{2+} are shown in Fig. 10.

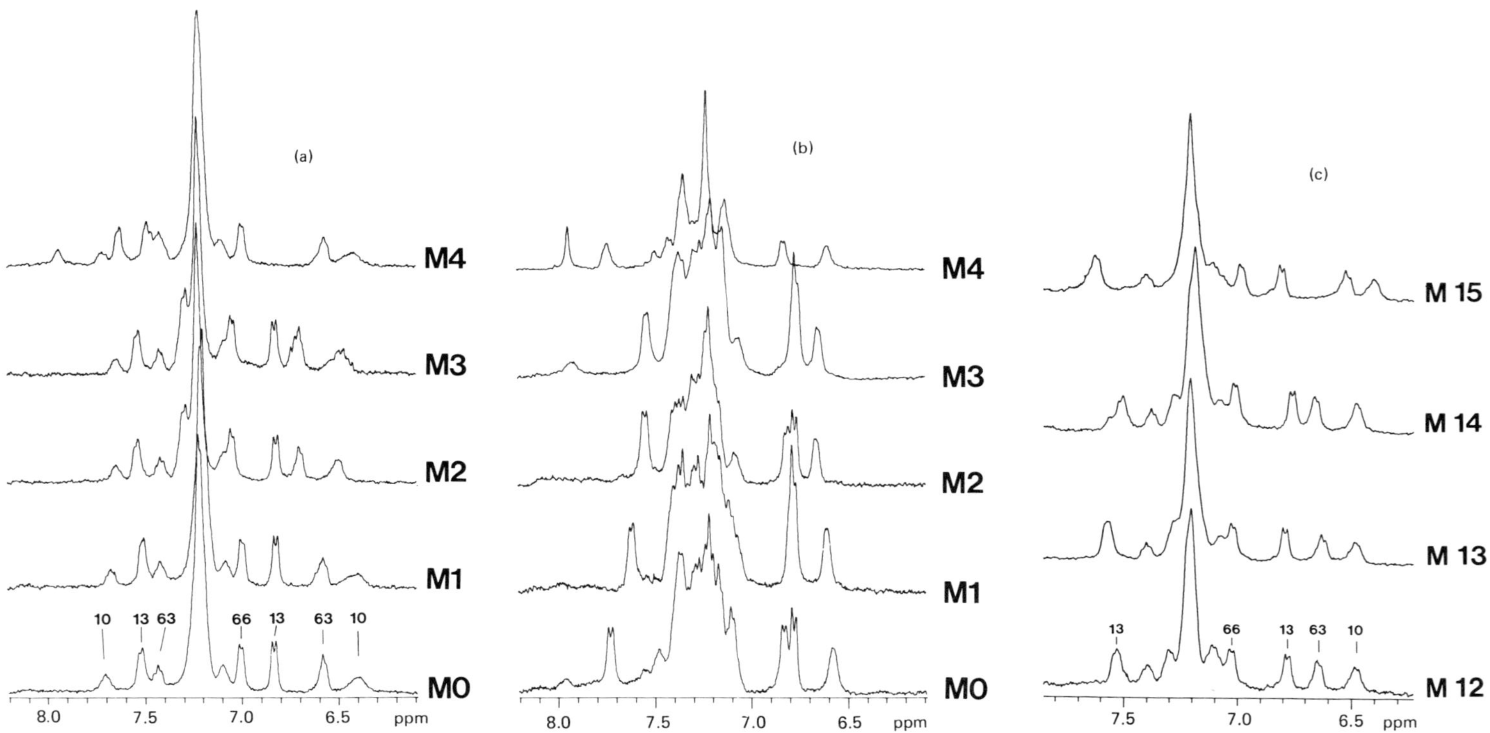

FIG. 10. 360-MHz ^{1}H NMR spectra of the well-resolved aromatic region of wild-type and mutant calbindins. Tentative assignments (W. Chazin and J. Kördel, personal communication) of some aromatic signals to individual amino acids are indicated (13 = Tyr-13; 10, 63, and 66 = Phe-10, 63, and 66, respectively). (a) and (c) Ca^{2+}-loaded forms; (b) apo proteins. (Parts (a) and (b) are reproduced from Ref. 25 with permission of the American Chemical Society.)

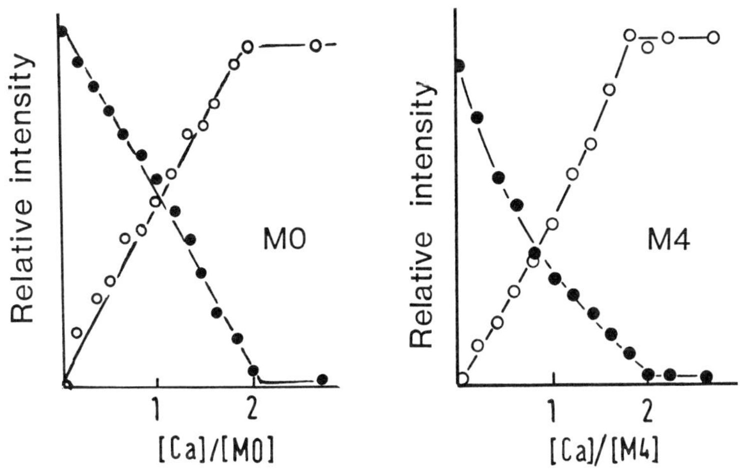

FIG. 11. Intensities of selected ^{1}H NMR signals of M0 and M4 as a function of the Ca^{2+}-to protein ratio. (left) M0: (○) Phe-66 resonance at 6.98 ppm; (●) Tyr-13 resonance at 6.77 ppm. (right) M4: (○) Phe-66 resonance at 6.98 ppm; (●) Phe-63 resonance at 6.82 ppm in the apo-protein. (Reproduced from Ref. 25 with permission of the American Chemical Society.)

At the time of writing a complete sequence-specific assignment of the ^{1}H NMR spectrum of wild-type calbindin in the Ca^{2+}-loaded form is nearly complete (J. Kördel and W. Chazin, private communication). A tentative assignment of a limited number of resonances in both the apo and Ca^{2+}-loaded forms was earlier made by Dalgarno et al. [41]. In Fig. 10 we have indicated the assignment of some selected well-resolved signals.

Although the ^{1}H NMR spectra of mutant and wild-type calbindins are very similar—in particular for the Ca^{2+}-loaded forms—distinct differences were observed during the Ca^{2+} titrations of the apo proteins. For M0 and M4 the Ca^{2+} ion exchange rates are slow on the NMR time scale and two overlapping spectra are seen during the titration. Several signals increase or decrease with added Ca^{2+} for [Ca^{2+}]/[protein] ratios up to 2. This is exemplified for a few resonances in Fig. 11.

In the case of mutants M1, M2, M3, M12, M13, M14, and M15, the addition of less than 1 eq of Ca^{2+} has the result that the intensities

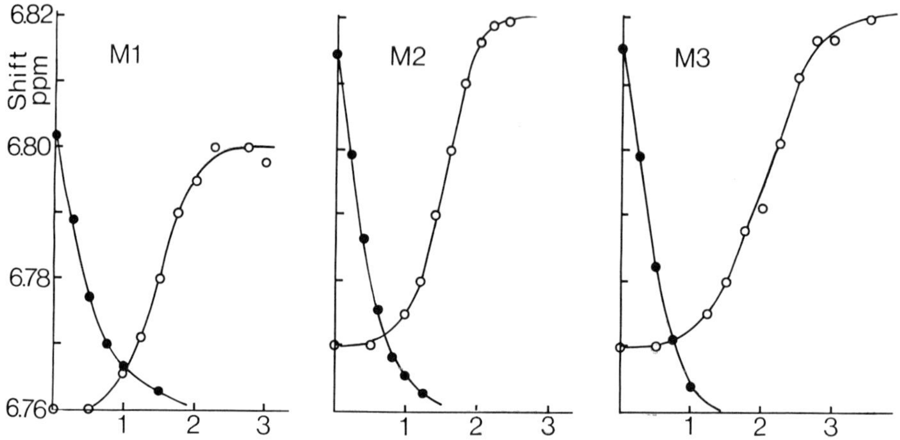

FIG. 12. Intensity and chemical shift changes of selected ^1H NMR signals in M1 to M3 as a function of the Ca^{2+}-to-protein ratio. (●) Intensity change of the resonance due to Phe-10 (2,4-hydrogens) with a chemical shift of 6.61-6.66 ppm; (○) chemical shift change in the Tyr-13 resonance (3,5-hydrogens). (left) M1; (center) M2; (right) M3. (Reproduced from Ref. 25 with permission of the American Chemical Society.)

of some signals are decreased while new signals appear. This type of behavior is expected when the first bound Ca^{2+} ion is in slow exchange with the bulk solution. When more than 1 eq of Ca^{2+} is added, some ^1H NMR signals are gradually shifted to a new position, as expected when the Ca^{2+} ions causing this change are in fast exchange. For some mutants like M1, certain resonances also become broadened during the Ca^{2+} titration. This shows that the exchange rate is not quite so fast as to completely average out the effects due to shift differences between the protein species with one and two Ca^{2+} ions bound. In Fig. 12 we show a few examples of changes in shifts and intensities for some resonances in the mutant calbindins M1, M2, and M3.

In the ^1H NMR shift range 5.1-5.6 ppm—a region generally attributed to $C_\alpha H$ protons in β-sheet structures—a well-resolved signal corresponding to a single proton is observed at low field (5.51 ppm) in the ^1H NMR spectrum of M0. This resonance is also found in the spectra of M1 and M4 through M11, but not in those of M2, M3, M12, M13, M14, and M15. This resonance was earlier assigned by Dalgarno et al. [41] to one of the α protons of Gly-18 in the wild-type protein. Recent two-dimensional NMR studies have shown, however, that this resonance stems from the α proton of Leu-23. This amino acid is one of the two members of a "mini β sheet" structure and is hydrogen-bonded to Val-61.

5.5. Stopped-Flow Kinetic Results

Table 6 summarizes rates and reaction amplitudes obtained for wild-type calbindin and mutants M1-M5 at 20°C. In the case of M0, M4, and M5 two Ca^{2+} ions dissociate with the same apparent rate constant in low ionic strength medium ("-KCl"). At higher ionic strengths ("+KCl"), however, two dissociation steps are observed—each corresponding to the release of one Ca^{2+} ion.

In the case of M1 two dissociation steps, each corresponding to the release of one Ca^{2+} ion, are clearly resolved in both low and high ionic strength media while for M2 and M3 only a single dissociation step is observed. For the latter two mutants it is apparent from variable-temperature ^{43}Ca NMR measurements that they also show a very fast dissociation step ($k_{off} \approx 5{,}000\ sec^{-1}$) which is complete within the instrumental dead time of the stopped-flow instrument.

6. DISCUSSION

In order to make our discussion of the different biophysical results more tractable, we will first cover different aspects in subsections and then try to summarize the results in a concluding section.

TABLE 6

Ca^{2+} Dissociation Rates (k) and Dissociation Amplitudes (A) Obtained from Stopped-Flow Experiments on Wild-Type and Genetically Modified Calbindins at 20°C[a]

	A(-KCl)				*B*(+KCl)			
Protein	k_s (sec^{-1})	A_s (mV)	k_f (sec^{-1})	A_f (mV)	k_s (sec^{-1})	A_s (mV)	k_f (sec^{-1})	A_f (mV)
M0	3.5 (0.5)[b]	1600 (200)[b]	—	—	8.6 (1.5)[b]	1010 (170)[b]	48 (7)[b]	1270 (26)[b]
M1	8.6 (1.5)	720 (140)	115 (20)	910 (160)	16.0 (2.8)	1060 (170)	210 (30)	1100 (260)
M2	10.0 (2.5)	590 (160)	—	—	38.0 (7.0)	780 (110)	—	—
M3	15.7 (2.5)	510 (150)	—	—	53.0 (8.0)	680 (160)	—	—
M4	2.8 (0.6)	1270 (250)	—	—	7.5 (1.6)	810 (180)	71 (10)	1080 (100)
M5	3.8 (0.8)	870 (140)	—	—	11.0 (1.6)	870 (110)	100 (18)	480 (160)
M8[c]	2.4	620	12	430				
M9[c]	3.9	470	18.6	410				
M10[c]	3.5	1020	—	—				

[a]Experimental conditions: *A*: 20 mM Pipes/KOH, pH 7.0; *B*: same as *A* but with buffer also 0.1 M in KCl.
[b]Values in parentheses are standard deviations.
[c]The kinetic data for these mutants are preliminary.

6.1. Stability of Mutants

One of the nightmares of an aspiring protein engineer is probably that the mutant proteins he or she so carefully designed and expressed will turn out to be very unstable and not really suitable for biophysical or biochemical studies. The thermal stability of the mutant calbindins so far prepared, however, appears to be very high both in Ca^{2+}-loaded and Ca^{2+}-free form as exemplified in Fig. 6(a) and (b). The thermal stability of wild-type calbindin (M0) in the Ca^{2+}-loaded form is similar to that of the homologous proteins calmodulin, troponin C and their proteolytic fragments [42]. In all these EF-hand proteins the denaturation temperature exceeds 90°C and is not experimentally accessible.

For the apo forms of these proteins the situation is, however, different. As seen in Fig. 6(b), the thermal unfolding of apo-M0 is not even half complete at 90°C. By contrast, the transition midpoint for apocalmodulin is about 55°C [42]. For the proteolytic fragments of calmodulin, TR_1C and TR_2C, the thermal transition midpoints occur at still lower temperatures, 49°C and 46°C, respectively. These two fragments represent essentially the N-terminal and C-terminal globular domains of calmodulin and have nearly the same overall size and three-dimensional structure as bovine calbindin D_{9k}. Evidently, apocalbindin is substantially more stable toward thermal denaturation than the corresponding fragments of calmodulin. The apo forms of skeletal muscle troponin C and the two analogous proteolytic fragments derived from this molecule have transition midpoints at 55°C, 60°C (N-terminal half), and ~20°C (C-terminal half), respectively [42]. We have presently no good explanation for these differences in the thermal stability of the apo forms.

The mutant calbindins M1-M5 show very much the same thermal stability as the wild-type protein (M0)—both in Ca^{2+}-loaded and Ca^{2+}-free form. The only significant change is observed with the apo form of M4 for which the normalized CD absorbance drops more rapidly with temperature than in the case of the other mutants studied. Also ^{1}H NMR spectral changes at elevated temperatures are more pronounced for M4 than for other mutants in the series M1-M5 [25].

The high resistance toward thermal unfolding of calbindin and mutants M1-M5 is reflected also in the resistance toward denaturation by urea as shown in Fig. 7. The stabilizing effect of Ca^{2+} on urea denaturation of calbindin is very similar to what is found for calmodulin where the transition midpoint (C_m) is shifted from >8 M urea (+ Ca^{2+}) to about 3.8 M (- Ca^{2+}) [43]. As indicated by Figs. 7 and 8, the urea-induced unfolding process follows closely the pattern expected for a two-state unfolding transition [44-46]. However, this should not be taken as evidence against the occurrence of intermediates, since the CD measurements primarily reflect structural changes affecting the α helices.

Values of the free energy of stabilization in aqueous solution, $\Delta G_D^{H_2O}$, generally depend on the method chosen for extrapolating ΔG_D to zero urea concentration. The linear extrapolation procedure generally seems to give lower estimates of $\Delta G_D^{H_2O}$ than alternative procedures [46,47] as is also found in our case as seen in Table 3. The value of $\Delta G_D^{H_2O}$ calculated for wild-type calbindin using the linear extrapolation procedure (i.e., 20 kJ/mol) is similar to that obtained for other small globular proteins. For example, $\Delta G_D^{H_2O}$ for α-lactalbumin ($M_r \approx 14{,}000$) has been determined as 18 kJ/mol at pH 7 [48], and for ribonuclease T_1 as 15 kJ/mol at pH 8.05 [47]. For the somewhat larger dihydrofolate reductase from *E. coli* ($M_r \approx 18{,}000$), $\Delta G_D^{H_2O}$ has recently been calculated as 24.7 kJ/mol at pH 7.8 [49].

With one exception, all genetically modified calbindins so far investigated by urea unfolding are less stable than the wild-type protein. The exception is M5 (Glu-17 → Gln) for which we find a clearly significant increase in C_m by some 0.4 M and a perhaps less significant increase in $\Delta G_D^{H_2O}$ by 2 kJ/mol [using Eq. (1)], by comparison with the wild-type.

Both enthalpy and entropy terms contribute to the free-energy difference between the folded and unfolded forms of a protein molecule. The contribution of any given residue X to the configurational part of the entropy of unfolding can be estimated as

$$\Delta S^{X}_{conf} = R \ln z \tag{4}$$

where R is the gas constant and z is the number of conformations available to that residue in the unfolded state—the number in the folded state assumed to be unity [50]. A detailed evaluation of ΔS^{X}_{conf} would require a statistical average over all accessible conformations available to amino acid residue X. As first suggested by Nemethy et al. [51], an approximate estimate of ΔS^{X}_{conf} can be obtained by considering the area available to residue X in a Ramachandran conformational map [52]. In particular, if we are interested in how much the entropy of unfolding would change if residue X was replaced with another residue Z ($\Delta S^{X\to Z}_{conf}$), we could simply write [51,53]:

$$\Delta S^{X\to Z}_{conf} = R \ln \frac{\gamma_Z}{\gamma_X} \tag{5}$$

where γ_Z and γ_X are the relative areas accessible to residues Z and X, respectively. Neglecting enthalpy changes, and equating $\Delta S^{X\to Z}_{conf}$ with the total entropy change upon substitution, we may thus estimate the accompanying change in $\Delta G_D^{H_2O}$. A substitution of Pro by Gly, as in M1, is in this manner calculated to result in a destabilization of the folded protein by 8 kJ/mol at 25°C. This value falls in between the experimental values of 5 ± 2 and 11 ± 4 kJ/mol found using the linear and quadratic extrapolation procedures, respectively (cf. Table 3).

Can the increased stability of mutant M5 (Glu-17 → Gln) be rationalized in a similar way? The areas accessible to Glu and Gln on a Ramachandran-type plot are identical and an explanation in terms of changes in ΔS_{conf} is not valid. It appears reasonable to assume that Glu-17 in the Ca^{2+}-free form will still be located at the surface of the protein, as it is in the Ca^{2+}-loaded form, and close to a number of other negatively charged amino acids (Asp-19, Glu-26, Glu-27 and possibly also Glu-60). Electrostatic interactions between these groups would tend to lower the stability of the native structure as compared with the unfolded form. Correspondingly, the removal of one negative charge, as in M5, should increase the stability of the native structure.

As is shown in Fig. 2, the x-ray studies of wild-type calbindin in its Ca^{2+}-loaded form show the OH group of Tyr-13 to be hydrogen-bonded to the carboxylate group of Glu-35. This hydrogen bond thus forms a link between the two α helices flanking the Ca^{2+}-binding loop of site I. Mutant M4 (Tyr-13 → Phe), in which this hydrogen bond is removed, has been investigated by a variety of biophysical techniques (cf. below). As compared with the wild-type protein M0, no significant changes have been observed in Ca^{2+}-binding properties or rate of Ca^{2+} dissociation. Yet, as we may note in Table 3, mutant M4 is determined to be 5 kJ/mol less stable toward unfolding than M0 in the apo form and the C_m value is lowered by 1.1 M. This result illustrates that changes in overall stability properties of a protein may not be reflected in structural or dynamic properties.

6.2. Structural Aspects

Although detailed structural studies on the mutants are still lacking (but are in progress), the results from ^{1}H and ^{113}Cd NMR are in a qualitative way very informative on this point. ^{1}H NMR spectra of the well-resolved aromatic region of the series of proteins M0-M4 and M12-M15 are overall very similar, in particular for the Ca^{2+}-loaded forms. It would appear, therefore, that no major changes in tertiary structure of the Ca^{2+}-loaded proteins have been caused by the modifications and/or deletions made so far. Logically, the next step should obviously be to determine the solution structure of the mutants by modern two-dimensional FT NMR techniques that have been pioneered by Wüthrich and others [54,55]. Such studies are currently in progress in our laboratory in collaboration with Dr. W. Chazin, Scripps Research Institute, La Jolla. Also x-ray crystallographic studies of the mutants are in progress.

For the apo forms of the above series of proteins, the ^{1}H NMR spectra show a larger variation than for the Ca^{2+}-loaded forms. The NMR results thus seem to agree with the results from the denaturation studies, which show that Ca^{2+} ions stabilize the native structure considerably.

Even if ^{1}H NMR spectra of in particular the Ca^{2+}-loaded forms of calbindins M0-M4 and M12-M15 indicate that no major global changes in tertiary structure have taken place, we know that the deletion and/or modifications made must have caused structural perturbations, at least in the loop of Ca^{2+} site I. Can we somehow assess these structural changes and probe to what extent they are carried over to site II? ^{113}Cd NMR provides qualitative information on this point. The Cd^{2+} ion has an ionic radius very similar to that of Ca^{2+} and a large body of spectroscopic and biochemical evidence on EF-hand-type Ca^{2+}-binding proteins has shown that Cd^{2+} is an excellent substitution probe for Ca^{2+} [40,56]. The ^{113}Cd NMR chemical shift covers some 1,000 ppm and is very sensitive to the nature and disposition of the surrounding ligands. We have earlier shown that the binding of Cd^{2+} to wild-type calbindin is sequential and that the first ^{113}Cd^{2+} enters into site II and has a chemical shift typical of other EF-hand sites (δ = -80 to -120 ppm in M0). Interestingly the observed chemical shift of ^{113}Cd^{2+} in site II is significantly different in proteins with no Cd^{2+} in site I (δ^{II} = -105 ppm) and in proteins in which this site is occupied by Cd^{2+} or Ca^{2+} (δ^{II} = -110 ppm). This change in chemical shift, $\Delta\delta^{II,I}$, is clearly an indication of interaction between the two Ca^{2+} sites. The effect is best seen in mutants where both sites have a high Ca^{2+} affinity, in mutants with a much reduced lower affinity of site I only site II will be occupied with Cd^{2+} under our experimental conditions.

The chemical shift of the ^{113}Cd^{2+} ion in site I (-156 ppm in M0) is quite extreme and only one example of a ^{113}Cd NMR signal at higher fields has been reported so far [57]. The NMR chemical shift of the ^{113}Cd^{2+} ion in site II in the mutant proteins—in the absence as well as presence of a second Cd^{2+} ion in site I—by and large remains close to its value in M0 (Table 3). Neither the mutation Tyr-13 → Phe (M4) nor the modification of Pro-20 into a Gly (M1) has any significant effect on δ^{II}. Neutralization of the negatively charged side chains around the Ca^{2+}-binding sites (mutants M5-M11) has only minor effects on the ^{113}Cd^{2+} shifts. Only in the case where three charges are neutralized (M11: Glu-17 → Gln, Asp-19 →

Asn, and Glu-26 → Gln) is the chemical shift of the $^{113}Cd^{2+}$ ion in site I significantly different from that in the wild-type protein. Other mutations in site I cause clearly measurable changes in δ^{II} but even the seemingly rather drastic deletion of two amino acids as in M12 and M14 result in only minor shift changes. Throughout the series of mutants studied so far, the value of δ^{II} remains within the range observed for EF-hand sites.

The general conclusion from our ^{113}Cd NMR studies, then, is that the peptide fold of site II is only slightly perturbed by the modifications made at site I. For the structural changes at site I a different picture emerges. The change of Pro-20 into Gly (M1) results in a change in the $^{113}Cd^{2+}$ chemical shift of site I (δ^{I}) by nearly 70 ppm, from -156 ppm to -88 ppm. Obviously this points to a rather substantial change in the disposition of the ligands at site I. Modifications of site I generally result in a lowering of the binding constants of Cd^{2+} as well as Ca^{2+} (cf. below), and for many mutants, such as M2, M3, M12, and M15, exchange broadening has prevented the determination of δ^{I}.

6.3. Equilibrium and Kinetic Properties

6.3.1. Equilibrium Properties

As mentioned briefly in Sec. 5, the binding constant measurements clearly indicate the binding of two Ca^{2+} ions in M0, M4, M5, and M7-M11 to be *cooperative*. From the ratios of the macroscopic (or stoichiometric) binding constants K_1 and K_2, only a lower limit of the absolute value of the interaction free energy $\Delta\Delta G$ can be obtained by putting η (= K_{II}/K_I) equal to unity. To proceed further we need information about the distribution of Ca^{2+} ions between the two sites. On this point the fact that the Ca^{2+} (or Cd^{2+}) binding is sequential in many of the mutants—since the mutations primarily reduce the Ca^{2+} affinity of the site modified—is of great help. ^{1}H NMR shows that some resonances primarily respond to the binding of Ca^{2+} (or Cd^{2+}) to one of two sites. Assuming that the intensity (or chemical shift if the exchange rate is high) of a particular signal is a measure of the concentration, say, of proteins with a cation in either site I or II, we may calculate theoretical curves showing how this intensity should vary with total Ca^{2+} concentration for different values of the parameters K_1, K_2, and η (= K_{II}/K_I). Since the macroscopic constants K_1 and K_2 are known from the Quin 2 (or 5,5'-Br_2-BAPTA [58]) titrations, the only unknown is η. For $\eta = 1$ the signal intensity is linearly dependent on $[Ca^{2+}]_{tot}$ while nonlinear behavior is observed for higher or lower values of η. By comparison of theoretical and experimental NMR intensities it is thus possible to delimit the values of η for a particular mutant. In this way it may be estimated that $\eta = 2 \pm 1$ for M0 and $\eta \gtrsim 5$ for M4. According to Eq. (3), this means that for M0 the free energy of interaction between the two sites, $\Delta\Delta G = -5.1 + 0.4$ *kJ/mol* while for M4 we have $\Delta\Delta G < -3.9$ *kJ/mol*. From this it appears that the cooperativity in M4 (Tyr-13 → Phe) may not be much different from that of the wild-type protein and whatever the molecular mechanism of site-site interaction is the hydrogen bond between Tyr-13 and Glu-35 certainly cannot play an important role.

Another highly interesting aspect of the finding that η (= K_{II}/K_I) is not very different from unity in M0 (perhaps also in M4) is the fact that despite the fact that the two Ca^{2+} sites are structurally very different, *their affinity for Ca^{2+} is almost equally high:* for M0 we calculate $K_I \approx 0.8 \cdot 10^8\ M^{-1}$ and $K_{II} \approx 1.5 \cdot 10^8\ M^{-1}$. It may further be inferred from the data in Table 2 that if we exclude mutants in which the net charge of the protein is affected (mutants M5-M11), most modifications of the loop of site I have resulted in a reduction of the Ca^{2+} affinity of this site while that of site II *is virtually unchanged.* This conclusion follows from the fact that when K_1 and K_2 become widely different, then $K_1 \approx K_{II}$ and $K_2 \approx K_{I,II}$.*

*The alternative interpretation ($K_1 \approx K_I$ and $K_2 \approx K_{II,I}$) can be ruled out from the NMR studies, and it would also be difficult to argue that only the affinity of site II should decrease when site I is perturbed.

While site-site interactions are present also in mutants M1-M3, according to the ^{113}Cd NMR ($\Delta\delta^{II,I} \neq 0$; cf. Table 5) the sequential Ca^{2+} binding makes it impossible to determine the true value of $\Delta\Delta G$, since there is no way to determine the value of the weaker of the two site-binding constants in the absence of a Ca^{2+} ion in the high-affinity site.

In mutants M5-M11 three negatively charged side chains that cluster around the Ca^{2+}-binding sites (cf. Fig. 4) are successively neutralized with a minimal change in side chain structure and volume requirements. The effects on the Ca^{2+} affinity of calbindin are highly interesting. As compared with the wild-type protein (M0), the Ca^{2+} affinity is gradually reduced as an increasing number of residues are neutralized. This trend is best reflected in the total free-energy change ΔG_{tot} associated with the binding of two Ca^{2+} ions (cf. Table 2):

$$\Delta G_{tot}^{Ca^{2+}} = -RT \ln (K_1 \cdot K_2) \tag{6}$$

For M0 this free-energy change is -97 kJ/mol. For mutants M5, M6, and M8, representing the neutralization of one negative surface charge, $\Delta G_{tot}^{Ca^{2+}}$ is about -90 kJ/mol, for mutants M7, M9, and M10, where two charges are neutralized, $\Delta G_{tot}^{Ca^{2+}}$ is around -83 kJ/mol while for M11, in which finally three charges are neutralized, the value of $\Delta G_{tot}^{Ca^{2+}}$ is -77 kJ/mol. On the average each reduction of a unit charge changes the value of $\Delta G_{tot}^{Ca^{2+}}$ by 6-7 kJ/mol. An estimate of the electrostatic contribution to the free energy, using an effective dielectric constant of 40 [59], shows that a unit charge 10 Å away from a Ca^{2+} ion would have an interaction free energy of that magnitude. This result does not seem unreasonable in view of the actual ion-ion distances observed in the x-ray structure of the Ca^{2+} form of calbindin (Fig. 4).

In the series M5-M10 the stepwise elimination of negative surface charges seems to affect K_1 more than K_2; in the mutants M9 and M10, where two charges are neutralized, the two macroscopic binding constants differ by about an order of magnitude. It is still too early to say which of the two site-binding constants, K_I and K_{II}, has been affected most. From the distances indicated in Fig. 4 one would expect that the Ca^{2+} affinity of site I would be influenced more than that of site II. If so K_1 should approximately be equal to K_{II} and K_2 equal to $K_{I,II}$ in mutants M5, M6, M8, M9, and M10; in M7 and M11 the situation is obviously unclear.

Accurate values of the free energy of interaction $\Delta\Delta G$ between the two Ca^{2+} sites in this series of mutants cannot be calculated until ^1H NMR titrations, like those performed for M0-M4 (cf. Figs. 11 and 12), hopefully enable us to determine the values of η ($= K_{II}/K_I$). We can nevertheless calculate a lower limit of $|\Delta\Delta G|$ from the values of K_1 and K_2 as outlined in Sec. 5.1.

The $|\Delta\Delta G|_{min}$ values listed in Table 2 have been calculated using Eq. (3) with $\eta = 1$. The result is truly unexpected. With the exception of M7 the cooperativity of Ca^{2+} binding is seen to *increase* as one or two negative surface charges are neutralized. While with a reduction of three charges (M11) the value $|\Delta\Delta G|_{min}$ drops below that of M0. Since we do not have access to the value of η for M7 and M11, the possibility remains that $\Delta\Delta G$ here still is lower than 5.1 ± 0.4 kJ/mol. The most pronounced cooperative binding appears to occur in M9 (Glu-26 → Gln; Glu-17 → Gln) where binding of the first Ca^{2+} ion increases the protein affinity for the second ion by a factor of $e^{9.2/RT} = 40$ at room temperature

How is this unexpected effect to be explained in molecular terms? As was first noted by Szebenyi and Moffat [11], calbindin has a rather assymetrical charge distribution with most negative charges present on the "upper" half of the molecule containing the Ca^{2+} sites (cf. Figs. 2 and 4). In the absence of Ca^{2+} ions in the two sites it appears reasonable to assume that the strong electrostatic repulsion between the negative charges would oppose a close approach of the two Ca^{2+}-binding loops. The first Ca^{2+} ion entering either of the two sites—presently we have no direct experimental information on the relative rates of binding to the two sites although the combined evidence from stopped-flow and binding constant studies on mutants M1, M2, and M3 indicates the on rates for the two Ca^{2+}

sites to be very similar (cf. discussion in the following subsection)—should reduce the loop-loop repulsion and possibly allow the two halves of the molecule to interact. If this "site-site" interaction poises the nonoccupied Ca^{2+} loop to adopt a conformation energically more favorable for Ca^{2+} binding, some of the free energy released upon binding of the first Ca^{2+} would be used in binding of the second ion, i.e., positive cooperativity of binding. When surface charges on the loops are neutralized, the reduced electrostatic interaction would facilitate a close approach of the two halves and less free energy spent on overcoming electrostatic repulsion. The result should be facilitated binding of the second Ca^{2+} ion, i.e., increased positive cooperativity. The net result would, however, be an overall reduced Ca^{2+} affinity of both Ca^{2+} sites. With three negative charges neutralized the schematic mechanism presented above seems no longer tenable; cooperativity appears to become less pronounced and the Ca^{2+} affinity is much reduced compared to that of the wild type. It is still possible, albeit not very likely, that the ratio η of the two site-binding constants ($\eta = K_{II}/K_{I}$) in mutant M11 is much larger than unity and thus $\Delta\Delta G$ higher than the lower limit calculated from the K_2/K_1 ratio.

In the series of mutants M12-M15 we note that K_1 always exceeds K_2 by more than an order of magnitude. Although this would seem to indicate that site-site interaction is absent and positive cooperativity is lost, this may not be the case. The extensive modifications of the loop of site I clearly have resulted in a much reduced value of K_I for this site. Thus K_1 will approach K_{II}, since the first Ca^{2+} will be bound at site II, and correspondingly K_2 approaches $K_{I,II}$. We may thus expect that $\eta = K_{II}/K_{I}$ is large and possibly larger than K_1/K_2 in which case $\Delta\Delta G$ will become negative.

The thermochemical results presented in Table 4 show several interesting features. The ΔH^{θ} values for the binding of Ca^{2+} to site II in each of the proteins, excluding M1, are all around -8 kJ/mol. This value is virtually the same as that found for the binding of Ca^{2+} to each site in the EF-hand-type protein calmodulin [21] under similar low ionic strength conditions. Thermochemical measurements have been made for several other calcium-binding proteins, including some with EF-hand sites such as parvalbumin and troponin C, but any direct comparisons with the results for the calbindins are complicated by the fact that most of these other measurements were made under high ionic strength conditions [66].

The most noticeable effects on the enthalpy changes are seen for M2 and M3, where the binding of Ca^{2+} to site II is exothermic but that to site I is endothermic. These positive ΔH^{θ} values for M2 and M3 are a dominating factor in the reduced Ca^{2+} affinity of the modified site I in M2 and M3 in comparison with M0, M1, and M4 and may well be connected with the shortening of the calcium-binding loop by one amino acid. In M1, on the other hand, the lower Ca^{2+} affinity of the modified site is a result of a reduction in ΔS^{θ} that more than compensates the more negative value of ΔH^{θ}.

Since the entropy of solvation of a gaseous Ca^{2+} ion is -254 JK^{-1} mol^{-1} [66], the loss of hydration water of the Ca^{2+} ion upon binding to the proteins will be expected to make a major contribution to the entropy changes. All the ΔS^{θ} values given here are positive but considerably lower than the dehydration values, which may be due to a loss of conformational entropy upon Ca^{2+} binding to the proteins. The exceptionally low ΔS^{θ} value for Ca^{2+} binding to M1 may be a result of more than one H_2O being liganded to Ca^{2+} in site I.

Further discussion will have to await the results of thermochemical studies now in progress on the series of mutants M5-M11. Considering the electrostatic nature of the binding of Ca^{2+} to the proteins, neutralization of one or more charges at the binding site will be expected to have significant effects on the thermodynamics of Ca^{2+} binding.

6.3.2. Kinetics

The ^{43}Ca NMR data of Table 5 and the stopped-flow data of Table 6 are in fair agreement where they overlap and are generally complementary; dissociation rate constants above 600 sec^{-1} are difficult to measure with stopped-flow techniques whereas the use of ^{43}Ca NMR enables rates up to about 5,000 sec^{-1} to be determined. The values of the reaction amplitudes in the stopped-flow studies as well as the variable-temperature NMR studies show that the dissociation of two Ca^{2+} ions in M0, M4, M5, and possibly also M10 occurs as one single exponential process. Computer simulations indicate that two processes, of equal amplitude, would appear as a single exponential if the rates differed by a factor of about 3 or less at the signal/noise ratio attainable (S. R. Martin, personal communication). For a number of mutants two rate processes, each corresponding to the release of one Ca^{2+} ion, are observed. In the majority of these cases the slow step has a rate constant of about 10 sec^{-1} or less. For the rest of the mutant proteins only a single rate process can be resolved in the stopped-flow measurements, although it is apparent that an initial very rapid Ca^{2+} release process is lost in the dead-time of the instrument. In the case of M2 and M3, ^{43}Ca NMR data show that this rapid process has a rate constant of about 5,000 sec^{-1}.

In the presence of 0.1 M KCl the Ca^{2+} dissociation rates are generally increased and there also appears to be a differential effect on the slow and fast rate processes. For proteins M0, M4, and M5 which disclose a single dissociation step in the absence of KCl two dissociation processes now become measurable. Similar salt effects on Ca^{2+} dissociation rates have also been observed for calmodulin [34]. The explanation we currently favor is that the transition state for Ca^{2+} release is highly ionic in nature, in which case increased ionic strength would be associated with a reduced activation energy. If this explanation is correct, the different effect of KCl on the two sites in M0, M4, and M5 may be taken as an indication that the relevant transition states for the two sites have significantly different ionic character. Since site II has three negatively charged side chains as ligands, it might be expected to possess the more ionic transition state and thus the fast process in 0.1 M KCl should be associated with dissociation from this site. The possible role of the unliganded negative surface charges around the two sites (Glu-17, Asp-19, Glu-26, and Glu-60) in the transition states remains, however, unsettled and our assignment of k_f to site II must be regarded as tentative.

The relation between equilibrium properties and kinetic data in wild-type and mutant calbindins may be discussed on the basis of the scheme of Fig. 13. The site-binding constants of this scheme are related to the macroscopic (or stoichiometric) binding constants through the relationships:

$$K_1 = K_I + K_{II} \tag{7a}$$

$$K_2 = \frac{K_{I,II} \cdot K_{II}}{(K_I + K_{II})} \tag{7b}$$

For the mutant proteins M1, M2, M3, M14, and M15 sequential Ca^{2+} binding and dissociation is well established. Since we are perturbing the loop of site I the most reasonable interpretation is that

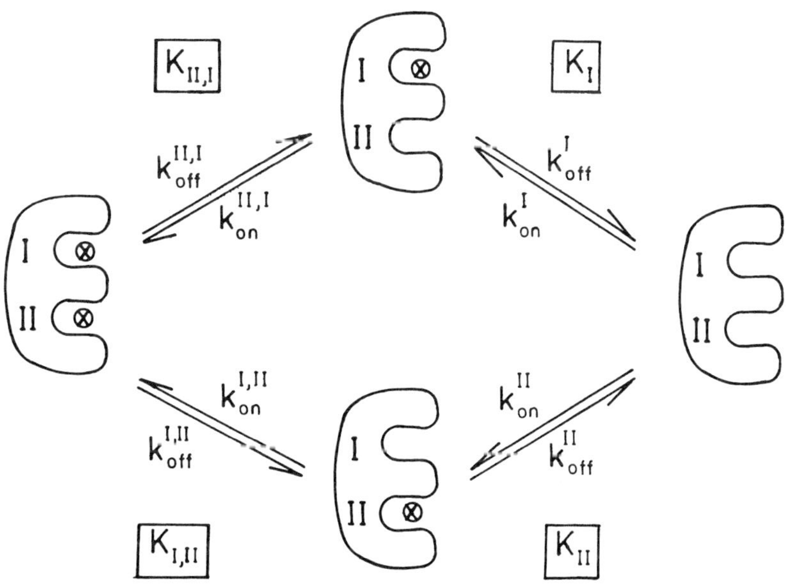

FIG. 13. Schematic diagram showing the different modes of Ca^{2+} dissociation and association in calbindin D_{9k}. The definition of the individual rate constants and their relation to the different site binding constants ($K = k_{on}/k_{off}$) is indicated.

TABLE 7

Rates of Ca^{2+} Binding to the Two Binding Sites in Mutant Calbindins for Which Sequential Binding and Dissociation Has Been Established[a]

Protein	$k_{on}^{I,II}$ (M^{-1}, sec^{-1})	k_{on}^{II} (M^{-1}, sec^{-1})
M1	$1 \cdot 10^9$	$6 \cdot 10^8$
M2	$2.5 \cdot 10^9$	$1.5 \cdot 10^9$
M3	$1.5 \cdot 10^9$	$1.2 \cdot 10^9$
M14	$\leq 1.8 \cdot 10^9$	$\sim 8 \cdot 10^8$
M15	$1.6 \cdot 10^9$	$\sim 5 \cdot 10^8$

[a]The rates of association are calculated from experimentally observed Ca^{2+}-binding constants and dissociation rates (obtained at low ionic strength) as outlined in the text. The rate constants $k_{on}^{I,II}$ and k_{on}^{II} are defined in Fig. 13.

the fast dissociation is due to Ca^{2+} release from site I and the slow process accordingly to Ca^{2+} release from site II; that is, we are observing the events in the lower pathway of the scheme in Fig. 13. Thus $K_1 = K_{II}$, $K_2 = K_{I,II}$, $k_s = k_{off}^{II}$, and $k_f = k_{off}^{I,II}$. For mutants M2 and M3 we use as k_f the value for the fast exchange process obtained by ^{43}Ca NMR (cf. Table 5). Since $K_{II} = k_{on}^{II}/k_{off}^{II}$ and $K_{I,II} = k_{on}^{I,II}/k_{off}^{I,II}$, the values of the rates of Ca^{2+} association, k_{on}^{II} and $k_{on}^{I,II}$, may now be calculated. These values are reported in Table 7. For mutants M8, M9, and M13 Ca^{2+} binding and release is certainly also sequential but at the time of writing we have insufficient NMR data to allow reliable conclusions about the pathway(s) of dissociation. In the case of proteins M0, M4, M5, and M10 the situation is much less clear. Dissociation of Ca^{2+} is observed as one single rate process and binding is not sequential. It is therefore possible that Ca^{2+} dissociation occurs by both pathways. For the moment we will leave open the question of the rates of association for these mutants.

As is inferred from Table 7 the calculated rate constants for the association of Ca^{2+} to mutants M1, M2, and M3 all fall within a narrow range, *$5 \cdot 10^8$ to $\sim 2 \cdot 10^9$ M^{-1} sec^{-1}*. In view of the "harmless" nature of the modifications made in these mutants—no charged amino acid residue is involved for one thing—it appears that these values should be valid for the wild-type protein as well. In fact using these on rates very reasonable dissociation rate constants close to those experimentally observed for M0 (and M4) are obtained from observed equilibrium constants and any reasonable model for the dissociation pathway. The observed on rates for Ca^{2+} binding are close to what is expected for a diffusion-controlled process. In view of the high concentration of negative charges around the protein-binding sites the theoretical limit of the on rates could, however, well be 10^{10} M^{-1} sec^{-1} or even higher [60], depending on the detailed model used for the charge distribution in the apo proteins. A striking result of the kinetic studies is also that the observed reduction in Ca^{2+} affinity of site I in mutants M1, M2, M3, M14, and M15 is almost entirely the result of an *increased off rate*.

This latter result may be contrasted with the situation in mutants M5-M11 where surface charges are successively neutralized. Even if a detailed kinetic pathway is still lacking we may note (cf. Table 6) that observed rates of dissociation differ only by a factor of at most 3-5 from the values observed in M0 and M4 (or from the slow process in M1, M2, and M3) while Ca^{2+} affinities are reduced by considerably larger factors—between one and two orders of magnitude in some cases. Thus it appears that in this series of mutants the reduction in Ca^{2+} affinity is mainly due to a *reduced rate of Ca^{2+} association*. This indicates that the negative surface charges, which do not have a side chain directly liganded to Ca^{2+}, around the two sites play a very prominent role in the binding process, perhaps by enhancing the affinity of an initial encounter complex.

7. SUMMARY

The 9k calbindin molecule has a considerable tolerance toward the perturbations induced by the modifications accomplished through site-directed mutagenesis. Although folding-unfolding studies so far have been performed only on a limited number of mutants, it appears that the stability toward unfolding has been affected to only a small degree. Structural changes caused by substitutions and/or deletions in the Ca^{2+}-binding loop of site I appear primarily restricted to this region. Only minor structural effects are inferred for site II or for the global structure of the protein.

In the wild-type protein the loops of Ca^{2+} sites I and II differ both in amino acid sequence and three-dimensional structure. Site II has the archetypal EF-hand structure found in parvalbumin, calmodulin, and troponin C whereas site I—two amino acids longer than the site II loop—has an unusual "pseudo-EF-hand" fold. The fold of the loop of site I in the mutant calbindins has been monitored by NMR spectroscopy, in particular using $^{113}Cd^{2+}$ as a substitution probe for Ca^{2+}. When this loop is shortened by two amino acids and further modified so as to become more homologous to the site II loop, spectroscopic evidence suggests that the fold may change to that of a typical EF-hand site.

The structural stability of Ca^{2+} site II with respect to modifications at site I is also reflected in a near constancy of the dynamic and equilibrium properties pertinent to site II.

Cooperative Ca^{2+} binding has been established in wild-type calbindin (M0) and, through spectroscopic information obtained from studies of some mutants in which sequential Ca^{2+} binding has been induced, it has been possible to determine the free energy of interaction $\Delta\Delta G$ in M0 as *-5.2 ± 0.4 kJ/mol*. Cooperative Ca^{2+} binding has also been established for a number of other mutant calbindins, though here only a lower limit of $|\Delta\Delta G|$ could be calculated from the stoichiometric Ca^{2+}-binding constants K_1 and K_2. An unexpected discovery is that the neutralization of one or two negative amino acid side chains of the surface around the Ca^{2+} sites—side chains not directly liganded to the Ca^{2+} ions—results in an *increased cooperativity* in Ca^{2+} binding. In the extreme case (M9: Glu-17 → Gln; Glu-26 → Gln) $\Delta\Delta G$ is -9.2 kJ/mol or lower. This result is taken to indicate that a reduction of the charges at the Ca^{2+} sites, which are about 11 Å from each other—facilitates a closer interaction of the two halves of the protein. The net result, however, is an overall reduced Ca^{2+} affinity of both sites.

The neutralization of each surface charge around the Ca^{2+} sites results in a near constant reduction in the total free energy of Ca^{2+} binding, $\Delta G_{tot}^{Ca^{2+}}$ (= $-RT \ln K_1K_2$), of about 6-7 kJ/mol per charge. In the wild-type protein (M0) the stoichiometric binding constants are $K_1 = 2 \cdot 10^8\ M^{-1}$ and $K_2 = 4 \cdot 10^8\ M^{-1}$ and, in mutant M11 in which three surface charges (Glu-17, Asp-19, and Glu-26) are neutralized, the values become reduced to $K_1 = 6 \cdot 10^6\ M^{-1}$ and $K_2 = 5 \cdot 10^6\ M^{-1}$. These results show unambiguously that amino acids not directly liganded to the Ca^{2+} ions have a large influence on the Ca^{2+} affinity.

Kinetic data also show that the reduced Ca^{2+} affinity caused by the neutralization of surface charges primarily is due to a *lowered rate of* Ca^{2+} *association*. By contrast other deletions and substitutions not affecting charged groups in the loop of site I but that also cause a reduction in the Ca^{2+} affinity, result in an *increased rate of dissociation*. This rate increase in fact explains most of the observed reduction in Ca^{2+} affinity and the rate of Ca^{2+} association appears almost unaffected. In the five mutants where the on rate could be calculated with some confidence (M1, M2, M3, M14, and M15), it is shown to be very high: $(1.3 \pm 0.8) \cdot 10^9\ M^{-1}\ sec^{-1}$.

The combined application of biophysical methods and site-directed mutagenesis has proved to be a very fruitful approach for the study of structure-function relationships in EF-hand-type Ca^{2+}-binding proteins.

ACKNOWLEDGMENT

The work described here was supported by grants from the Swedish Natural Science Research Council (K-KU 2541-116) and the National Swedish Board for Technical Development (512-87-350 and 732-87-575). The permission to quote some as yet unpublished stopped-flow results obtained by Stephen R. Martin and Peter M. Bayley, National Institute for Medical Research, Mill Hill, is gratefully acknowledged.

REFERENCES

1. H. Rasmussen, *N. Engl. J. Med., 314,* 1094-1101, 1164-1170 (1986).
2. C. B. Klee and T. C. Vanaman, *Adv. Protein Chem., 35,* 214-321 (1982).
3. A. S. Manalan and C. B. Klee, *Adv. Cyclic Nucleotide Protein Phosphor. Res., 18,* 227-278 (1984).
4. W. D. McCubbin and C. M. Kay, *Acc. Chem. Res., 13,* 185-192 (1980).
5. J. D. Potter and J. D. Johnson, in "Calcium and Cell Function", Vol. 1 (W. Y. Cheung, ed.), Academic Press, New York, 1982, pp. 145-173.
6. S. Forsén, "Calcium in biological systems" in "Introduction to Inorganic Biochemistry" (I. Bertini, H. B. Gray, and S. J. Lippard, eds.), Univ. Sci. Books, Mill Valley, CA, in press.
7. K. B. Seamon and R. H. Kretsinger, in "Metal Ions in Biology", Vol. 6 (T. Spiro, ed.), Wiley-Interscience, New York, 1983, pp. 1-52.
8. R. H. Kretsinger, *CRC Crit. Rev. Biochem., 8,* 119-174 (1980).
9. R. H. Kretsinger and C. E. Nockolds, *J. Biol. Chem., 248,* 3313-3326 (1973).
10. P. C. Moews and R. H. Kretsinger, *J. Mol. Biol., 91,* 201-228 (1975).
11. D. M. E. Szebenyi and K. Moffat, *J. Biol. Chem., 261,* 8761-8777 (1986).
12. D. M. E. Szebenyi, S. F. Obendorf, and K. Moffat, *Nature, 294,* 327-332 (1981).
13. O. Herzberg and M. N. G. James, *Biochemistry, 24,* 5298-5302 (1985).
14. O. Herzberg and M. N. G. James, *Nature, 313,* 653-659 (1985).

15. M. Sundaralingam, R. Bergstrom, G. Strasburg, S. T. Rao, P. Roychowdhury, M. Greaser, and B. C. Wang, *Science, 227,* 945-948 (1985).

16. Y. S. Babu, J. S. Sack, T. G. Greenhough, C. E. Bugg, A. R. Means, and W. J. Cook, *Nature, 315,* 37-40 (1985).

17. R. H. Kretsinger and L. J. Weissman, *J. Inorg, Biochem., 28,* 289-302 (1986).

18. "Calcium and the Cell", CIBA Foundation Symposium No. 12, Ciba Foundation, Wiley and Sons, Chichester, 1986.

19. H. Sigel, ed., "Metal Ions in Biological Systems", Vol. 17, Marcel Dekker, New York, 1984.

20. B. A. Levine and R. J. P. Williams, "The chemistry of the calcium ion and its biological significance", in "Role of Calcium in Biological Systems" (L. J. Anghileri and A. M. T. Anghileri, eds.), CRC Press, Boca Raton, FL, 1982, pp. 3-26.

21. S. Forsén, H. J. Vogel, and T. Drakenberg, "Biophysical studies of calmodulin", in "Calcium and Cell Function", Vol. 6 (W. Y. Cheung, ed.), Academic Press, New York, 1987, pp. 249-309.

22. H. J. Vogel and S. Forsén, "NMR of calcium binding proteins", in "Biological Magnetic Resonance" (L. Berliner and J. Reuben, eds.), Plenum Press, New York, Vol. 7, 1986, pp. 113-157.

23. C. S. Fullmer and R. H. Wasserman, *J. Biol. Chem., 256,* 5669-5674 (1981).

24. P. Brodin, T. Grundström, T. Hofmann, T. Drakenberg, E. Thulin, and S. Forsén, *Biochemistry, 25,* 5371-5377 (1986).

25. S. Linse, P. Brodin, T. Drakenberg, E. Thulin, P. Sellers, K. Elmdén, T. Grundström, and S. Forsén, *Biochemistry, 26,* 6723-6735 (1987).

26. T. Hofmann, private communication.

27. B. G. Johansson, *J. Clin. Lab. Invest., 29,* Suppl. No. 124, 214-321 (1972).

28. A. M. Klibanov and T. J. Ahern, in "Protein Engineering" (D. L. Oxender and C. F. Fox, eds.), Alan R. Liss, New York, 1987, pp. 213-218.

29. H. Csopak, K. E. Falk, and H. Szajn, *Biochem Biophys. Acta, 258,* 466-473 (1972).

30. P. D. Ellis, Univ. of South Carolina, private communication (1980); P. S. Belton, I. Y. Cox, and R. K. Harris, *J. Chem. Soc. Faraday Trans., 281,* 63-75 (1985).

31. T. Drakenberg, S. Forsén, and H. Lilja, *J. Magn. Reson., 53,* 412-422 (1983).

32. M.-D. Tsai, T. Drakenberg, E. Thulin, and S. Forsén, *Biochemistry, 26,* 3635-3643 (1987).

33. P. M. Bayley, P. Ahlström, S. R. Martin, and S. Forsén, *Biochem. Biophys. Res. Commun.*, *120*, 185-191 (1984).

34. S. R. Martin, A. Andersson Teleman, P. M. Bayley, T. Drakenberg, and S. Forsén, *Eur. J. Biochem.*, *151*, 543-550 (1985).

35. S. Forsén, B. Wendt, S. Linse, P. M. Bayley, and S. R. Martin, manuscript in preparation.

36. B. Wendt, S. R. Martin, P. M. Bayley, P. Brodin, T. Grundström, E. Thulin, S. Linse, and S. Forsén, *Eur. J. Biochem.*, in press.

37. J. A. Schellman, *Biopolymers*, *17*, 1305-1316 (1978).

38. A. T. Chen and I. Wadsö, *J. Biochem. Biophys. Methods*, *6*, 307-316 (1982).

39. G. Öjelund and I. Wadsö, *Acta Chem. Scand.*, *22*, 2691-2699 (1968).

40. H. J. Vogel, T. Drakenberg, S. Forsén, J. D. J. O'Neil, and T. Hofmann, *Biochemistry*, *24*, 3870-3876 (1985).

41. D. C. Dalgarno, B. A. Levine, R. J. P. Williams, C. S. Fullmer, and R. H. Wasserman, *Eur. J. Biochem.*, *137*, 523-532 (1983).

42. H. Brzeska, S. V. Venyaminov, Z. Grabarek, and W. Drabikowski, *FEBS Lett.*, *153*, 169-173 (1983).

43. S. R. Martin and P. M. Bayley, *Biochem. J.*, *238*, 485-490 (1985).

44. C. Tanford, *Adv. Prot. Chem.*, *23*, 121-223 (1968).

45. C. Tanford, *Adv. Prot. Chem.*, *24*, 1-77 (1970).

46. C. N. Pace, *CRC Crit. Rev. Biochem.*, *3*, 1-43 (1975).

47. C. N. Pace, "Methods in Enzymology", Vol. 131 (C. H. W. Hirs and S. N. Timasheft, eds.), Academic Press, New York, 1986, pp. 266-280.

48. F. Ahmad and C. C. Bigelow, *J. Biol. Chem.*, *257*, 12935-12938 (1982).

49. K. M. Perry, J. J. Onuffer, N. A. Touchette, C. S. Herndon, M. S. Gittelman, C. R. Matthews, J.-T. Chen, R. J. Mayer, K. Taira, S. J. Benkovic, E. E. Howell, and J. Kraut, *Biochemistry*, *26*, 2674-2682 (1987).

50. J. A. Schellman, *C.R. Trav. Lab. Carlsberg Ser. Chim.*, *29*, 230-259 (1955).

51. G. Nemethy, S. J. Leach, and H. J. Scheraga, *J. Phys. Chem.*, *70*, 998-1004 (1966).

52. G. N. Ramachandran, C. Ramakrishnan, and V. Sasisekharan, *J. Mol. Biol.*, *7*, 95-99 (1966).

53. B. W. Matthews, H. Nicholson, and W. J. Becktel, *Proc. Natl. Acad. Sci. USA*, *84*, 6663-6667 (1978).

54. K. Wüthrich, "NMR of Proteins and Nucleic Acids", John Wiley and Sons, New York, 1986.

55. W. R. Croasmun and R. M. K. Carlson (eds.), "Two-Dimensional NMR Spectroscopy: Applications for Chemists and Biochemists", Verlag Chemie, New York, 1987.

56. S. Forsén, T. Drakenberg, and H. Wennerström, *Quart. Rev. Biophysics, 19,* 83-114 (1987).

57. I. Morishima, M. Kurono, and Y. Shiro, *J. Biol. Chem., 261,* 9391-9399 (1986).

58. R. Y. Tsien, *Biochemistry, 19*, 2396-2404 (1980).

59. A. Warshel and S. T. Russell, *Quart. Rev. Biophys., 17,* 283-422 (1984).

60. P. Debye, *Electrochem. Soc., 82,* 265 (1942).

61. H. W. D. Mathes, W. M. Zenke, T. Grundström, A. Staub, M. Wintzerith, and P. Chambon, *EMBO J., 3,* 801-885 (1984).

62. H. W. D. Mathes, A. Staub, and P. Chambon, Chemical synthesis and mutagenesis by rapid microscale "shot gun gene synthesis", in "Methods in Enzymology" (R. Wu, ed.), Academic Press, New York, in press.

63. P. Brodin, T. Grundström, T. Hofmann, T. Drakenberg, E. Thulin, and S. Forsén, *Biochemistry, 25,* 5371-5377 (1986).

64. T. Grundström, W. M. Zenke, M. Wintzerith, H. W. D. Matthes, A. Staub, and P. Chambon, *Nucleic Acids Res., 13,* 3305-3316 (1985).

65. B. E. Uhlin, V. Schweickart, and A. J. Clark, *Gene, 22,* 255-265 (1983).

66. J. D. Buchanan, R. J. T. Corbett, and R. S. Roche, *Biophys. Chem., 23,* 183-199 (1986).

8

Genetic Alteration of Active Site Residues of Staphylococcal Nuclease: Insights into the Enzyme Mechanism

Albert S. Mildvan and Engin H. Serpersu
Department of Biological Chemistry
The Johns Hopkins University School of Medicine
725 N. Wolfe Street
Baltimore, MD 21205

1. INTRODUCTION

The enzyme staphylococcal nuclease is a monomeric protein, 149 amino acids in length with a molecular weight of 16,800. It requires Ca^{2+} for activity and catalyzes the hydrolysis of phosphodiester bonds found in DNA and RNA, yielding 3'-mono- and dinucleotides (Fig. 1) [1-3]. It is a powerful enzyme, accelerating the rate of phosphodiester hydrolysis by a factor of at least $1.7 \cdot 10^{15}$ fold as deduced by comparing k_{cat} (95 sec^{-1}) with the pseudo-first-order rate constant for nonenzymatic phosphodiester hydrolysis, extrapolated to 24°C, and to pH 7.4 ($5.7 \cdot 10^{-14}$ sec^{-1}) [4,5]. A somewhat larger rate enhance-

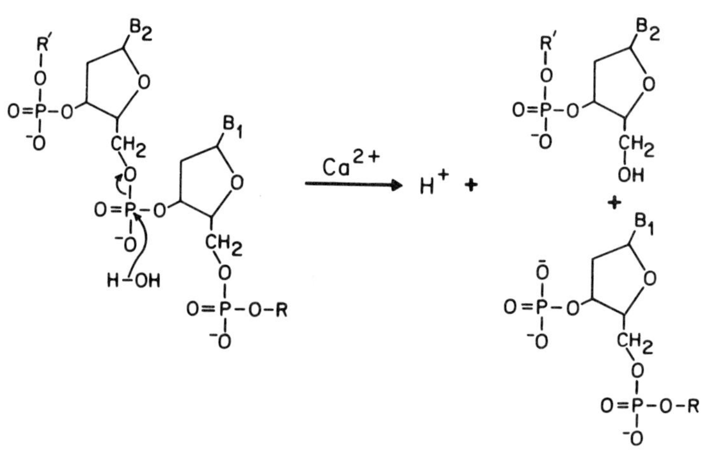

FIG. 1. Reaction catalyzed by staphylococcal nuclease.

ment of $\approx 10^{16}$-fold may be more accurate, since k_{cat} may be limited by steps other than bond breaking [6-8].

A chemical mechanism for staphylococcal nuclease was proposed [9] based on extensive kinetic studies of the intact and semisynthetic enzymes [1,10] and on the 1.5-Å x-ray structure of the ternary complex of the enzyme, Ca^{2+}, and the competitive substrate analog 3',5'-pdTp, the 5'-phosphate of which is directly coordinated to the metal (Fig. 2) [9]. In this mechanism (Fig. 3), the essential Ca^{2+} activator is in an octahedral complex, receiving three cis ligands from the protein, Asp-21, Asp-40, and the amide carbonyl group of Thr-41, as well as a phosphodiester oxygen of the polynucleotide substrate. Two cis water ligands complete the octahedral coordination, and second-sphere water molecules were also detected. Both the inner sphere water ligand adjacent to Asp-21 and the nearby second-sphere water molecule are close enough to the reaction center phosphorus to attack it in an associative mechanism (4.0 and 3.5 Å, respectively). While the second-sphere water molecule is better

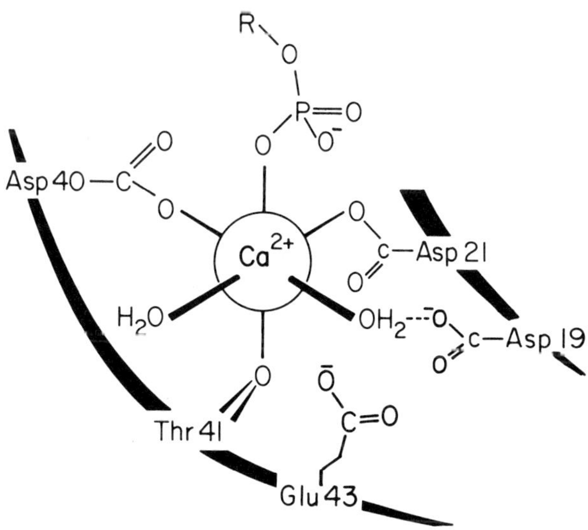

FIG. 2. Ligands of Ca^{2+} in the ternary staphylococcal nuclease-Ca^{2+}-3',5'-pdTp complex based on the 1.5-Å resolution x-ray structure [9].

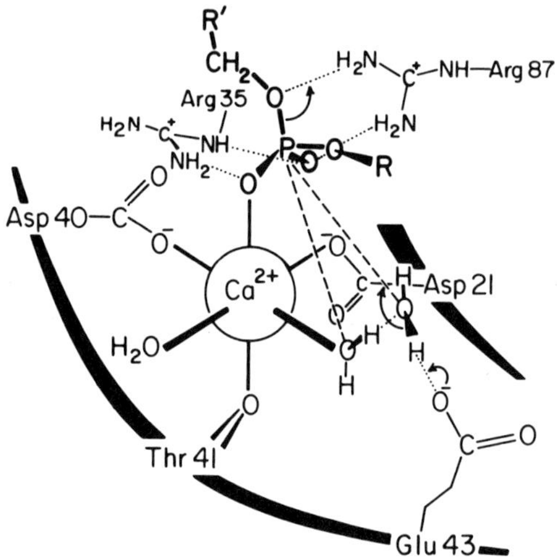

FIG. 3. Mechanism of staphylococcal nuclease based on x-ray structure [9] and on kinetic studies [1].

positioned to donate a hydrogen bond to Glu-43, a residue which probably functions as the general base [9-13], Glu-43 is flexible and could therefore deprotonate either the second-sphere water molecule, as shown, or the inner sphere water ligand, preparing either to attack the phosphodiester. Support for the direct attack of water on the phosphodiester with no phosphoenzyme intermediate was provided by the detection of stereochemical inversion at phosphorus in the staphylococcal nuclease reaction [14]. From the x-ray structure, the phosphodiester group of the substrate was proposed as being held by the metal, and also by Arg-35 and Arg-87 with each Arg acting as a bifunctional hydrogen bond donor (Fig. 3) [9]. By studying the effects of mutations at these positions, we have shown that Arg-87 does not contribute significantly to the binding of pdTp yet is nevertheless important in catalysis, suggesting that this residue interacts with the phosphorane transition state rather than with the phosphodiester ground state [5].

The availability of nearly 100 mutants of this 149-residue protein, made possible by random mutagenesis and precise genetic mapping of the amino acid changes [15,16], has permitted a detailed test of the mechanism of Fig. 3 and an evaluation of the quantitative contributions of each of the active site components to binding and catalysis.

2. METHODS

2.1. Enzyme Preparation

The isolation and cloning of the structural genes of the wild-type and mutant forms of staphylococcal nuclease have been described by Shortle and Lin [15,16]. Methods for purification and kinetic analysis of these enzymes from overproducing strains of *E. coli* have been described by Serpersu et al. [4,5].

2.2. Kinetic Studies

The initial rates of the reaction, typically measured with five concentrations of the activator Ca^{2+} at each of five concentrations of the substrate DNA, were analyzed to obtain five parameters: V_{max}, K_M of Ca^{2+} extrapolated to zero DNA (K_A^{Ca}) and to infinite DNA (K_M^{Ca}) and the K_M of DNA extrapolated to zero Ca^{2+} (K_S^{DNA}) and to infinite Ca^{2+} (K_M^{DNA}). In addition, the K_I values of Mn^{2+} and of Co^{2+} were determined. These cations do not activate staphylococcal nuclease but are linear competitive inhibitors of the enzyme with respect to Ca^{2+}. Representative kinetic data are summarized in Table 1.

2.3. Metal and Nucleotide Binding

The binding of Mn^{2+} to the enzyme and to enzyme-nucleotide complexes was studied by electron paramagnetic resonance (EPR), which measures free Mn^{2+}, and by the enhancement of $1/T_1$ of water protons, which

TABLE 1

Kinetic Parameters[a] of Wild-Type and Active Site Mutants of Staphylococcal Nuclease at pH 7.4

Enzyme	Relative k_{cat}	K_A^{Ca} (μM)	K_M^{Ca} (μM)	K_S^{DNA} (μM)[c]	K_M^{DNA} (μM)[c]	K_I^{Mn} (μM)	K_I^{Co} (μM)
Wild-type	1.0[b]	460	110	12.8	2.5	6.8	350
D40G	$3.3 \cdot 10^{-2}$	3440	1130	19.9	3.3	105	380
D40E	$8.1 \cdot 10^{-2}$	3830	1750	11.4	3.6	1.6	55
T41P	$2.7 \cdot 10^{-2}$	1490	930	14.3	8.1	13.0	—
D21E	$6.7 \cdot 10^{-4}$	1100	480	12.9	3.8	33	390
D21Y	$\leq 3.5 \cdot 10^{-5}$	—	—	—	—	—	—
E43S	$3.7 \cdot 10^{-4}$	2170	950	28.6	5.8	51	570
R35G	$\leq 2.8 \cdot 10^{-5}$	—	—	—	—	—	—
R87G	$\leq 2.8 \cdot 10^{-5}$	—	—	—	—	—	—
R87C	$\leq 9.8 \cdot 10^{-5}$	—	—	—	—	—	—

[a]Parameters are defined in the text. Errors are ≤42%.
[b]k_{cat} of wild type is 95 sec^{-1} at pH 7.4.
[c]Assuming the substrate to be a tetranucleotide of molecular weight 1,400. The factor 1.40 converts these to μg DNA/ml.
Source: Data from Refs. 4, 5, 22, and 30.

detects bound Mn^{2+}. The binding of other metals, such as Ca^{2+} or Co^{2+}, was studied by these same methods, in competition with Mn^{2+} [4,5,17]. The binding of nucleotides to the enzyme was monitored by competition with 1,N^6-ethenoadenosine 3',5'-bisphosphate based on the quenching of the fluorescence of this ligand by the enzyme [4,5], and the binding of nucleotides to enzyme-Mn^{2+} complexes was studied by the $1/T_1$ method [4,5,17]. The binding data were analyzed to yield the six dissociation constants which characterize the ternary complex of an enzyme (E), a metal (M), and a ligand (L). These are the dissociation constants of the three binary complexes,

$$K_1 = \frac{[M][L]}{[ML]} \tag{1}$$

$$K_D = \frac{[E][M]}{[EM]} \tag{2}$$

$$K_S = \frac{[E][L]}{[EL]} \tag{3}$$

and the dissociation constants characterizing the three ways that the ternary EML complex can dissociate.

$$K'_A = \frac{[EL][M]}{[EML]} \tag{4}$$

$$K_2 = \frac{[E][ML]}{[EML]} \tag{5}$$

$$K_3 = \frac{[EM][L]}{[EML]} \tag{6}$$

From the definitions of these dissociation constants it is easily shown that

$$K_1K_2 = K_3K_D = K'_AK_S \tag{7}$$

Dissociation constants were measured for ternary complexes of the wild-type enzyme and of active site mutants with Mn^{2+}, Ca^{2+}, and Co^{2+} and the ligands 5'-TMP and 3',5'-pdTp. Examples are given in Table 2 [4,5].

2.4. Pulsed EPR and NMR Studies

The structural environment of the active site metal in ternary enzyme-Mn^{2+}-nucleotide complexes was studied by three methods. The frequency dependence of $1/T_1$ of water protons provided information on the symmetry and mobility of the ligands of Mn^{2+} and also permitted quantitation of the number of fast exchanging water ligands, i.e., those that dissociate from Mn^{2+} with a rate constant exceeding 10^6 sec^{-1} [4,5]. The pulsed EPR or electron spin echo envelope modulation (ESEEM) method, which provides information on the nature of the ligands of enzyme-bound metals [18,19], was also used to quantitate the number of water ligands on enzyme-bound Mn^{2+}, without

TABLE 2

Dissociation Constants[a] of Metal Ions and 3',5'-pdTp and Enhancement Factors[c,d] of Binary and Ternary Complexes of Staphylococcal Nuclease and Mutants

Metal:	Ca^{2+}					Mn^{2+}						Co^{2+}
	K_D	K_S[b]	K'_A	K_2	K_3	K_D	K'_A	K_2	K_3	ε_b[c]	ε_T[d]	K'_A
Enzyme			(μM)					(μM)				(μM)
Wild-type	510	95	71	5.6	13.1	438	14	2.2	2.5	8.4	24.8	47
D40G	1660	61	259	13.0	9.5	1130	24	3.2	1.2	3.8	8.1	69
D40E	6495	51	220	9.4	1.7	686	7.1	0.8	0.56	3.3	13.0	—
T41P	3000	67	207	11.6	4.6	324	9.3	1.3	2.4	3.6	13.3	—
D21E	997	7.0	184	1.4	1.7	397	21	0.4	0.50	6.2	11.1	48
D21Y	3480	17.4	1390	20.2	7.0	2365	45	1.5	0.31	2.4	19.5	144
E43S	5400	66.2	228	12.6	2.8	3400	11.3	1.6	0.22	5.8	22.0	14
R35G	3940	900	3780	4760	1450	802	788	1600	800	7.0	—	—
R87G	4590	52.5	1420	62.1	16.2	904	146	13	7.2	6.0	7.0	541
R87C	2150	654	2790	1520	849	451	85	117	123	5.6	6.5	—
R87C-S-CH_2-CH_2-NH_3^+	2300	589	1620	795	415	856	120	149	83	7.4	9.7	—
R87C-S-CH_2-CO-NH_2	2050	>1310	2680	>2920	>1710	451	85	>330	>347	7.8	—	—

[a]Dissociation constants are defined in Eqs. (1) to (6) in the text. Errors in dissociation constants are ≤50%. $K_1(Ca^{2+}$-3',5'-pdTp) = 1,200 μM; $K_1(Mn^{2+}$-3',5'-pdTp) = 474 μM: $K_1(Co^{2+}$-3',5'-pdTp) = 209 μM.
[b]By definition, K_S is independent of metal ion.
[c]Enhancement of paramagnetic effect of Mn^{2+} on $1/T_1$ of water protons in binary E-Mn^{2+} complex (±10%).
[d]Enhancement in ternary E-Mn^{2+}-pdTp complex (±10%).
Source: Data from Refs. 4, 5, 22, and 30.

restriction to those which exchange rapidly [20], and to detect a strong interaction of enzyme-bound Mn^{2+} with a phosphate of pdTp [21].

Metal nucleotide interactions on Co^{2+} complexes of staphylococcal nuclease were also studied by the paramagnetic effects of enzyme-bound Co^{2+} on $1/T_1$ of the phosphorus and protons of 3',5'-pdTp and of 5'TMP [13,21,22]. These data permitted the calculation of Co^{2+} to phosphorus distances and Co^{2+} to proton distances [23], which were used to determine the conformation of enzyme-bound pdTp. An independent test of this conformation was provided by measurements of interproton distances on enzyme-bound pdTp [22] using the nuclear Overhauser effect (NOE) [24].

3. KINETIC, THERMODYNAMIC, AND STRUCTURAL PROPERTIES OF COMPLEXES OF WILD-TYPE STAPHYLOCOCCAL NUCLEASE

3.1. Kinetic Scheme

A kinetic analysis of the enzyme-catalyzed hydrolysis of DNA at pH 7.4 confirmed the essentiality of a ternary enzyme-Ca^{2+}-DNA complex for catalysis, which could form via either a binary E-Ca^{2+} or a binary E-DNA complex in a random kinetic scheme (Fig. 4). Evidence for only a single class of Ca^{2+}- and DNA-binding sites was obtained. The mutual tightening of Ca^{2+} binding by DNA ($K_M^{Ca} < K_A^{Ca}$) and of DNA binding by Ca^{2+} ($K_M^{DNA} < K_S^{DNA}$), as detected kinetically, are consistent with the existence of a

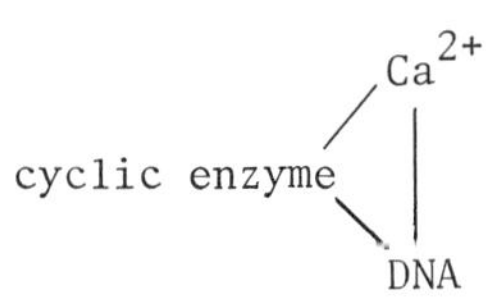

bridge complex.

3.2. Dissociation Constants of Metal and Nucleotide Complexes

Independent evidence for a cyclic metal bridge complex was provided by direct binding studies of metals and of nucleotides to the enzyme.

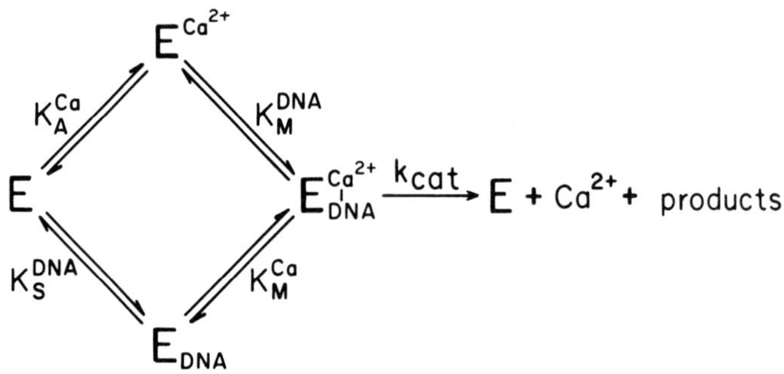

FIG. 4. Kinetic scheme of staphylococcal nuclease, based on kinetic and binding studies [4,5].

Binary enzyme-metal complexes were detected with Mn^{2+}, Ca^{2+}, and Co^{2+}, and binary enzyme-nucleotide complexes were detected with the substrate analogs 5'-TMP and 3',5'-pdTp. The stoichiometries were 1:1 and the dissociation constants of the E-Ca^{2+} and E-nucleotide complexes found in the binding studies were consistent with those obtained in the kinetic analysis, supporting a rapid equilibrium random kinetic scheme with a subsequent rate-limiting step, such as the formation or dissociation of products (Fig. 4). With all of the metals, ternary 1:1:1 complexes of enzyme, metal, and nucleotide were detected in the binding studies, indicating a single binding site for the essential Ca^{2+}. The dissociation constants of the ternary Ca^{2+} complexes were in reasonable agreement with the K_M^{Ca} and K_M^{DNA} obtained in the kinetic studies, considering the structural differences between mononucleotides and polynucleotides, thus providing further support for the rapid-equilibrium random kinetic scheme.

As found in the kinetic studies, the synergistic binding of metals and nucleotides were detected in the binding studies, supporting the formation of

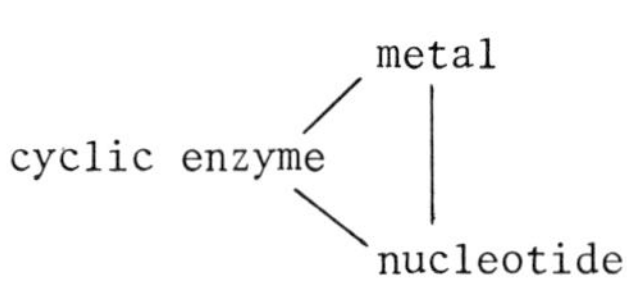

bridge complexes.

Proof of the existence of such metal bridge complexes in solution was obtained by pulsed EPR studies of the enzyme-Mn^{2+}-pdTp complex and by ^{31}P NMR relaxation studies of the enzyme-Co^{2+}-pdTp complex and the enzyme-Co^{2+}-TMP complex.

Both Mn^{2+} and Co^{2+} bind at the Ca^{2+} site of staphylococcal nuclease, as shown by competition in kinetic and binding studies, by synergy in binding of these metals and the substrate DNA or substrate analogs, and by comparable effects of mutations of liganding residues of the enzyme on the dissociation constants of all three cations (vide infra). However, the paramagnetic metals Mn^{2+} and Co^{2+} do not activate the enzyme. Indeed, they are less effective than Ca^{2+} by a factor of at least $10^{4.6}$ in promoting catalysis, despite the fact that they occupy the Ca^{2+} site. These observations have been interpreted to indicate that the proper metal, Ca^{2+}, contributes a factor $\geq 10^{4.6}$ to catalysis by staphylococcal nuclease [5].

3.3. Metal-Phosphate Interactions on the Enzyme

The reason for the inability of Mn^{2+} or Co^{2+} to activate staphylococcal nuclease is not clear, but may be related to the fact that these ions are significantly smaller than Ca^{2+}, with metal-oxygen bond lengths shorter by ≈0.4 Å [25]. Nevertheless Mn^{2+} and Co^{2+} are valuable paramagnetic probes, providing detailed information on the structure at the Ca^{2+}-binding site.

Pulsed EPR studies of frozen solutions of the ternary enzyme-Mn^{2+}-pdTp complex in H_2O revealed both high- and low-frequency modulations in the electron spin echo envelope, reflecting two kinds of magnetic nuclei near the bound Mn^{2+}. On Fourier transformation, in addition to the expected singlet resonance at 12.8 MHz due to exchangeable and nonexchangeable protons near Mn^{2+}, a doublet was detected, centered at 5.3 MHz, the Larmor frequency of phosphorus, with a Fermi contact splitting of 3.9 MHz or 1.4 gauss. Such Fermi contact splitting by a ^{31}P nucleus indicates strong Mn^{2+}-phosphate bonding in which the unpaired electrons of Mn^{2+} are partially delocalized into an s-type orbital of phosphorus [21].

Attempts to determine which phosphate of pdTp was coordinated by the enzyme-bound Mn^{2+}, using the NMR method of paramagnetic effects on T_1 of phosphorus, were unsuccessful due to the slow dissociation rate of pdTp out of the ternary Mn^{2+} complex compared to its relaxation rate in the ternary complex. When such slow exchange occurs, it is often helpful to use a metal with a lower effective magnetic moment, such as Co^{2+}, thereby decreasing the relaxation rate in the ternary complex to a value lower than the dissociation rate of the ligand. Since Co^{2+}, like Mn^{2+}, occupies the Ca^{2+} site of staphylococcal nuclease, paramagnetic effects of enzyme-bound Co^{2+} on the ^{31}P relaxation rates of pdTp and TMP were measured. The data, now in fast exchange, permitted the calculation of distances from enzyme-bound Co^{2+} to the 3'P and 5'P of pdTp of 4.4 ± 0.7 Å and 2.7 ± 0.4 Å, respectively, and 3.3 ± 0.5 Å to the 5'P of TMP. These distances establish direct coordination of the 5'-phosphate of the nucleotides by the enzyme-bound metal [21]. The distance from Co^{2+} to the 3'P of pdTp agrees with those found by x-ray studies of second sphere Co^{2+}-phosphate complexes with an intervening water ligand (4.75 ± 0.03 Å) [26]. While the distance from enzyme-bound Co^{2+} to the 5'P of TMP agrees with those found by x-ray studies of monodentate Co^{2+}-phosphate complexes (3.31 ± 0.04 Å) [26], the corresponding distance from Co^{2+} to the 5'P of the better substrate analog pdTp is somewhat shorter and is appropriate for a bidentate complex in which two phosphate oxygens are simultaneously coordinated to the metal. From model building, a Co^{2+}-to-phosphorus distance of 2.7 ± 0.1 Å is expected in a bidentate phosphate complex. Such bidentate coordination of Mn^{2+} on staphylococcal nuclease would explain the Fermi contact interaction detected by pulsed EPR. While monodentate coordination of Ca^{2+} is observed by x-ray analysis of the enzyme-Ca^{2+}-pdTp complex, bidentate coordination of Ca^{2+} in the catalytically active system could facilitate catalysis by stabilization of the trigonal, bipyramidal, phosphorane transition state. This possibility is suggested by the existence of many bidentate metal complexes of the planar trigonal nitrate anion [27], by the 10^9-fold rate acceleration of phosphate ester hydrolysis in a model reaction in which the phosphate was constrained to be a bidentate ligand of Co^{3+} [28], and by 30- and 12-fold lower activities of the D40G and D40E mutant staphylococcal nucleases [4,5] (Table 1), in which these strong metal-phosphate interactions are not observed [21]. These data suggest that bidentate coordination could, by transition state stabilization, contribute an order of magnitude to the rate acceleration of phosphodiester hydrolysis by this enzyme. Transition state stabilization by bidentate coordination could therefore explain at most only $\approx 10^{1.5}$ of the $\geq 10^{4.6}$-fold activation of staphylococcal nuclease by Ca^{2+}. The additional catalytic effects of the metal could be electrostatic and conformational.

3.4. Conformation of the Enzyme-Bound Substrate Analog 3',5'-pdTp

Distances from enzyme-bound Co^{2+} to the protons and phosphorus of pdTp, determined by paramagnetic effects on T_1, and interproton distances in pdTp in the enzyme-Ca^{2+}-pdTp complex, determined by nuclear Overhauser effects, yielded mutually consistent low-anti, C2' endo conformations ($\chi = 9 \pm 15°$; $\delta = 138 \pm 10°$) for the enzyme-bound nucleotide (Fig. 5) [22]. This conformation is very similar to the low-anti, C2' endo conformation ($\chi = 23°$, $\delta = 134°$) found by x-ray diffraction in the crystalline enzyme-Ca^{2+}-pdTp complex [9,29].

3.5. Metal-Water Interactions in Enzyme Complexes

The other substrate of the staphylococcal nuclease reaction is a water molecule (Fig. 1), which is either in the inner [29] or second coordination sphere of the enzyme-bound metal (Fig. 3) [9] as suggested by x-ray studies of the crystalline enzyme-Ca^{2+}-pdTp complex. In agreement with the x-ray data, two fast-exchanging water ligands on Mn^{2+} in the ternary enzyme-Mn^{2+}-pdTp complex in solution were detected by the frequency dependence of $1/T_1$ of water protons [4,5]. If no significant change in ligand geometry takes place upon coor-

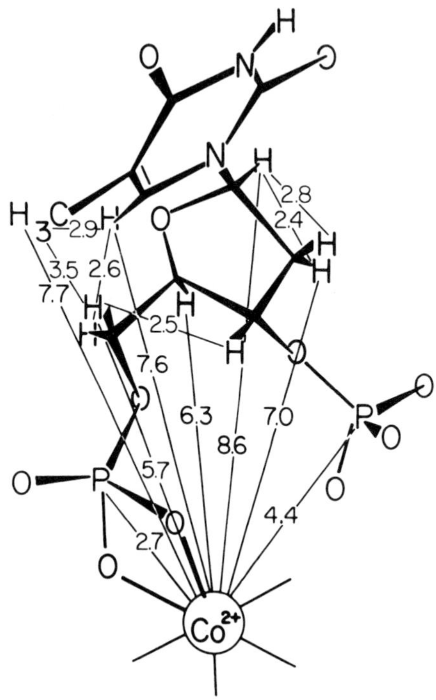

FIG. 5. Conformation of 3',5'-pdTp bound in the ternary enzyme-metal-pdTp complex based on distances from Co^{2+} to protons and phosphorus measured by paramagnetic effects on T_1 [21,22], and on interproton distances in the Ca^{2+} complex measured by the nuclear Overhauser effect [22].

dination of the nucleotide, the binary enzyme-Mn^{2+} complex, for which no corresponding x-ray structure exists, should have three water ligands coordinated to Mn^{2+}. However, $1/T_1$ measurements of water protons showed only one fast-exchanging water ligand on Mn^{2+} [4,5], due either to slow exchange of two of the three water ligands, or to the donation of two additional ligands to Mn^{2+} by the protein. The latter alternative is indicated by pulsed EPR studies of the binary and ternary complexes, which are not restricted to fast-exchanging ligands. Fourier-transformed difference spectra in ^2H_2O revealed one less ^2H_2O ligand of Mn^{2+} in the binary enzyme-Mn^{2+} complex than in the ternary enzyme-Mn^{2+}-pdTp complex [21]. Likely candidates for the two additional ligands to Mn^{2+} from the protein, based on the x-ray structure (Fig. 2) [9,29] and on the lack of other obvious candidates, are Glu-43 and Asp-19, which are in the second coordination sphere of the metal. The direct coordination of Glu-43 by the metal, which would require facile rotations of the side chain of this residue, is suggested by the 5- and 14-fold weaker binding of Ca^{2+} and Mn^{2+}, respectively, to the E43S mutant than to the wild-type enzyme [13]. However, the direct coordination of Asp-19 would require the movement of a portion of the backbone of the enzyme, which would make this less favorable. Of course, after the binding of the DNA substrate, the dissociation of Glu-43 from the metal would be essential to permit this residue to function as the general base.

4. STUDIES OF THE ROLE OF GLU-43 IN CATALYSIS BY THE USE OF MUTANTS

The earliest convincing evidence for the importance of Glu-43 in the reaction catalyzed by staphylococcal nuclease was the ≥100-fold lower activity of a semisynthetic enzyme in which Glu-43 was replaced by Asp [10]. Modern, site-specific mutagenesis by Gerlt and coworkers [11,12] revealed a 190-fold lower k_{cat} in the E43D mutant, and 10^3- to 10^4-fold lower maximal activities of the E43N, E43Q, E43A, and E43S mutants, suggesting that Glu-43 may contribute a factor as large as 10^4 to catalysis [5].

The use of such decreases in k_{cat} to estimate the quantitative contribution of Glu-43 to catalysis has been criticized [12] based on the detection of small conformational differences in the E43 mutants as reflected by their increased thermal stability and, in the case of the E43D and E43S mutants, by small changes in chemical shifts and NOEs ascribable to 0.4- to 0.6-A movements of Val-74 relative to Phe-34 and Phe-36, which are remote (18 and 28 Å) from Glu-43. However, the effects of these small and remote structural changes on the conformation around the active site and on catalytic activity are not clear [12]. In the crystalline state, preliminary x-ray studies of the ternary E43D-Ca^{2+}-pdTp complex reveal that Asp-40 has dissociated from the metal [29]. Such an effect would be expected to lower enzymatic activity by only one order of magnitude, as found with the D40G mutant which, like the E43S mutant, shows monodentate phosphate coordination by the metal [21] (see below).

Only small structural changes at the active site in solution are reflected by ≤ninefold weaker metal binding in ternary Ca^{2+} complexes of E43S which may be ascribed to the loss of the electrostatic effect of the carboxylate, a twofold weakening of DNA binding, and a fivefold tightening of pdTp binding (Tables 1 and 2). In the ternary E43S-Co^{2+}-pdTp complex, the distance from Co^{2+} to the 5'-phosphorus (3.3 ± 0.5 Å) is consistent with monodentate rather than bidentate phosphate coordination [13], which can explain at most a 30-fold loss in maximum catalytic activity (see above). Moreover, with wild-type staphylococcal nuclease, k_{cat} increases with pH, leveling off at pH 9.2, while k_{cat} of the E43S mutant continues to increase in proportion to [OH^-] up to pH 10, above which both activities fall irreversibly. These data establish that OH^- can at least partially replace Glu-43, supporting the proposed role of Glu-43 as a general base [13].

5. STUDIES OF METAL-LIGANDING MUTANTS

5.1. Mutants of Asp-40

In accord with the x-ray structure of the ternary Ca^{2+} complex of staphylococcal nuclease (Fig. 2), conversion of the metal ligand Asp-40 to Gly in the D40G mutant causes an order of magnitude weaker binding of Ca^{2+}, Mn^{2+}, and, to a lesser degree, Co^{2+} as detected by increases in K_A^{Ca}, K_M^{Ca}, K_I^{Mn} (Table 1), and in K_D and K_A', of these metals (Table 2). The mutation causes little or no effect on the binding of DNA, as indicated by essentially unaltered values of K_S^{DNA} and K_M^{DNA} (Table 1) or of pdTp as indicated by unchanged K_3 values (Table 2). Enlargement of Asp-40 to Glu in the D40E mutant also weakens the binding of Ca^{2+}, but tightens the binding of the smaller Mn^{2+} and Co^{2+} ions (Table 1). While DNA binding is essentially unaltered (Table 1), pdTp binding is tightened (Table 2).

The largest effects of mutation of Asp-40 are on k_{cat}, which shows a 30-fold decrease below wild-type in D40G and a 12-fold decrease in D40E. These inhibitory effects are reasonably ascribed to small structural differences in the immediate environment of the metal, as detected by three independent physical techniques. Analysis of the frequency dependence of $1/T_1$ of water protons in ternary enzyme-Mn^{2+}-pdTp complexes reveals greater asymmetry and mobility of the metal ligands in both mutants, compared to the wild-type enzyme, as reflected by lower enhancement factors (Table 2), and requires an additional ligand from the protein to Mn^{2+} in the D40G mutant, possibly Asp-19 [4,5]. The latter point is confirmed by pulsed EPR studies which show the same number of ^2H_2O ligands on Mn^{2+} in the ternary complex of the wild-type enzyme and the D40G mutant [20]. On both the D40G and the D40E mutants, weaker metal-phosphate bonding than on the wild-type enzyme is indicated by the absence of Fermi contact interaction between Mn^{2+} and phosphorus [21]. This point is confirmed by ^{31}P relaxation studies of the ternary D40G-Co^{2+}-pdTp complex which yield a distance from Co^{2+} to the 5'P of the nucleotide (3.6 ± 0.5 Å) consistent with monodentate phosphate coordination (3.31 ± 0.04 Å), rather than bidentate coordination (2.7 ± 0.1 Å) which was found with the wild-type enzyme [21]. Monodentate phosphate coordination may be responsible for the greater ligand mobility detected in the D40G and the D40E mutants.

5.2. Mutant of Thr-41

Because the backbone amide carbonyl group of Thr-41, rather than the side chain, provides a ligand to the metal (Fig. 1), a proline mutant at position 41 was studied, since proline would be most likely to alter the backbone. The T41P mutant showed a 37-fold lower k_{cat}, which may result from small changes in the arrangement of ligands on the metal. Such changes are suggested by an order-of-magnitude weaker binding of Ca^{2+} (Tables 1 and 2) and by the lower enhancement factors obtained with Mn^{2+} complexes (Table 2), reflecting a shorter electron spin relaxation time [5]. The affinities of T41P for Mn^{2+}, DNA, and 3',5'-pdTp are very similar to those found for the wild-type enzyme (Tables 1 and 2). No further structural information is available on the T41P mutant.

5.3. Mutants of Asp-21

While the enlargement of the Asp ligand to Glu at position 40, in the D40E mutant, results in only a 12-fold decrease in k_{cat}, the identical enlargement of Asp-21 in the D21E mutant results in a 1,500-fold decrease in k_{cat}, and further enlargement to Tyr in the D21Y mutant reduces k_{cat} by a factor $\geq 10^{4.5}$ (Table 1). Structural changes at the active sites of the D21E and D21Y mutants, such as the weaker binding of Ca^{2+} and Mn^{2+} (Tables 1 and 2), and decreases in enhancement factors (Table 2) resulting from a decreased ligand mobility on Mn^{2+} [5] are insufficient to explain these marked decreases in k_{cat}, since comparable effects are seen in the D40 and T41 mutants, which have lost factors of only $10^{1.5}$ in activity. Moreover, as found with the wild-type enzyme, intact enzyme-metal-pdTp bridge complexes of the D21E and D21Y mutants are detected by pulsed EPR, revealing Fermi contact interaction between Mn^{2+} and phosphorus, and by NMR measurements, yielding a short distance from Co^{2+} to the 5'P of pdTp (2.6 ± 0.5 Å) in the D21E-Co^{2+}-pdTp complex, suggesting strong bidentate phosphate coordination [21]. An alternative explanation is therefore necessary for the $\geq 10^3$-fold greater effect on k_{cat} of enlarging Asp-21 than that of enlarging Asp-40. This additional effect may reasonably be ascribed to occlusion of the nearby binding site for the attacking water molecule, and the resulting loss of catalysis by approximation (Figs. 3 and 6) [5]. Model-building studies indicate that, depending on the orientation of its side chain, Glu-21 can partially overlap and Tyr-21 can totally overlap with either the adjacent inner sphere or second-sphere water ligand of Ca^{2+} [5].

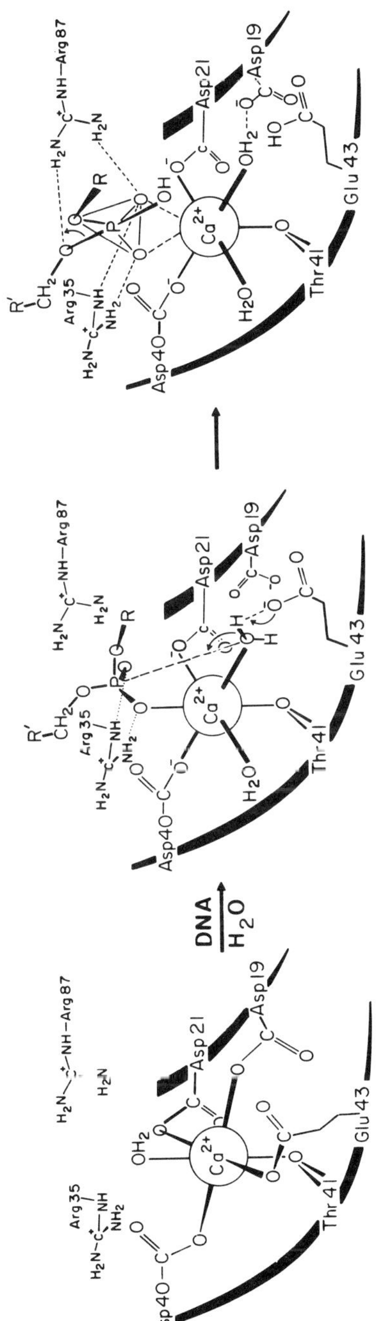

FIG. 6. Modified mechanism of substrate binding and phosphodiester hydrolysis catalyzed by staphylococcal nuclease, based on pulsed EPR [20,21], NMR [13,21], and kinetic studies of mutant enzymes [5,13].

6. STUDIES OF THE ROLES OF ACTIVE SITE ARGININE RESIDUES BY THE USE OF MUTANTS

6.1. Mutant of Arg-35

Based on the x-ray structure of the ternary enzyme-Ca^{2+}-pdTp complex (Fig. 2), Arg-35 and Arg-87 were proposed to be bifunctional hydrogen bond donors to the bound phosphodiester substrate, with Arg-87 also functioning as the general acid, protonating the 5' oxygen of the leaving polynucleotide product (Fig. 3). Accordingly, the replacement of Arg-35 by Gly in the R35G mutant profoundly decreases k_{cat} by a factor $\geq 10^{4.6}$ (Table 1) and weakens the binding of the substrate analog 3',5'-pdTp by factors approaching 10^3, but weakens the binding of metals by factors less than 10 (Table 2).

Hence the $\geq 10^{4.6}$-fold loss of catalytic activity may be ascribed predominantly ($\geq 10^3$-fold) to the inability of the R35G mutant to properly bind the DNA substrate, resulting in the loss of catalysis by approximation. A smaller contribution to the loss in activity of, at most, $10^{1.5}$-fold may be assigned to structural changes around the metal, based on the kinetic studies of the D40 and T41 mutants. From the data on the R35G mutant, and those on the D21 mutants, which block the binding of the other substrate, water, catalysis by approximation contributes a factor of at least 10^3 to the 10^{16}-fold overall rate acceleration produced by staphylococcal nuclease.

6.2. Mutants of Arg-87

As found with Arg-35, the mutation of Arg-87 to Gly also decreases k_{cat} by a large factor ($\geq 10^{4.6}$) (Table 1) and weakens metal binding by factors less than 10 (Table 2). However, unlike R35G, the R87G mutant shows normal affinity for the substrate analog pdTp, in binary and ternary Ca^{2+} and Mn^{2+} complexes, comparable to that found with the wild-type enzyme (Table 2). This finding is contrary to the conclusions of the x-ray analysis (Fig. 2) [9] and to the mechanism of Fig. 3 in which Arg-87 donates a bifunctional hydrogen bond to the tetrahedral phosphodiester group of the substrate. Clearly, in solution, this hydrogen bond is either not made or, if formed, contributes little to the binding of pdTp [5]. Nevertheless, Arg-87 is clearly essential for catalysis, since mutants at this position are reduced in k_{cat} by a factor $\geq 10^{4.6}$. It was therefore reasonably proposed that Arg-87 interacts with the trigonal bipyramidal transition state rather than with the tetrahedral ground state of the bound DNA substrate [5], as shown in the modified mechanism of Fig. 6. The kinetic data (Table 1) suggest that such transition state stabilization and proton donation by Arg-87 contribute a factor $\geq 10^{4.6}$ to catalysis [5].

6.3. Chemical Modification of the Arg-87-Cys Mutant

An independent test of the role of Arg-87 was made by a combination of mutagenesis and chemical modification [30]. Since staphylococcal nuclease contains no Cys residues, the R87C mutant contains a single, modifiable Cys residue at position 87. The R87C mutant is less active than the wild-type enzyme by a factor $\geq 10^4$ with DNA as substrate, by a factor $\geq 10^{4.2}$ with the more convenient substrate, p-nitrophenyl-5'-pdT, and shows somewhat weaker binding of divalent cations and pdTp (Tables 1 and 2). Chemical modification of Cys-87 by disulfide interchange with cystamine ($^{+}H_3N\text{-}CH_2\text{-}CH_2\text{-}S\text{-}S\text{-}CH_2CH_2\text{-}NH_3^{+}$) yields a pseudo-lysine derivative, R87C-S-CH_2-CH_2-NH_3^+ or R87"K", in which the length of the extended side chain (7.5 Å) better approximates that of Arg (7.3 Å) than does that of lysine (6.3 Å). Such modification restores a factor of $\geq 10^{1.8}$ to k_{cat}, and the activity of R87"K" is abolished by reduction of the disulfide with $NaBH_4$ [30]. The native structure around the metal ion is also partially restored in R87"K", as indicated by the increased enhancement factors of the binary and ternary Mn^{2+} complexes, compared to those of R87C (Table 2). While k_{cat} of the wild-type enzyme increases by a factor of 6.0 between pH 8.0 and 8.8 in proportion to $[OH^-]$, k_{cat} of R87"K" increases by only 1.25-fold over this pH range, probably because of the lower pK_A of pseudo-lysine than of arginine. These results indicate the need for a cationic proton donor at position 87, ideally arginine which, because of its bifunctionality, may better stabilize the transition state (Fig. 6).

No activation of R87C results from the covalent modification of Cys-87 with iodoacetamide to form the pseudo-glutamine or R87"Q" derivative, with an extended side chain 6.8 Å in length [30]. While the metal-binding properties of R87"Q" differ little from those of R87C and R87"K", pdTp binding is significantly weakened in this derivative. Hence the negligible activity of R87"Q" may result from the loss of both catalysis by approximation and the cationic proton donor which stabilizes the transition state.

6.4. Spin Labeling of the Arg-87-Cys Mutant

Spin labeling of Cys-87 with 3-(2-iodoacetamido)proxyl does not greatly alter the kinetic, metal-, or pdTp-binding properties of R87C [30]. From its EPR spectrum and its 4.0-fold enhanced effect on $1/T_1$ of water protons, the nitroxide is immobilized and exposed to the solvent. The presence of the nitroxide decreases the paramagnetic effects of enzyme-bound Mn^{2+} on $1/T_1$ of water protons, reducing ε_b from 5.6 in R87C to 4.3, indicating a weak dipolar spin-spin interaction between Mn^{2+} and the nitroxide. The paramagnetic effect of Mn^{2+} on the amplitude of the nitroxide EPR spectrum, analyzed as described by Leigh [31], yields a Mn^{2+}-to-nitroxide distance of 22 ± 2 Å, placing it far from the active site, possibly in a cleft near Met-32 [30]. Because of its remoteness from the active site, not surprisingly, reduction of the nitroxide to a cationic secondary ammonium, with ascorbate, does not restore activity to the modified R87C mutant.

7. MECHANISTIC CONCLUSIONS

With the important exception of the mutants at Arg-87, the kinetic, thermodynamic, and structural properties of active site mutants of staphylococcal nuclease are consistent with the mechanism of Fig. 3. However, the modified mechanism of Fig. 6, necessitated by studies of the R87G and R87C mutants, is preferable to that of Fig. 3 for three additional reasons. First, delaying the interaction of Arg-87 with the phosphodiester until the transition state forms preserves electroneutrality at the reaction center throughout the course of the reaction. Second, the mechanism of Fig. 6 explains how a poor proton donor, guanidinium, can protonate a poor proton acceptor, the 5'-bridge oxygen of the leaving group, by awaiting the development of a negative charge on this oxygen as it begins to leave. Third, the mechanism of Fig. 6 has enabled us to quantitatively dissect the factors which contribute to the $\approx 10^{16}$-fold rate acceleration of phosphodiester hydrolysis (k_{cat}) produced by this enzyme [5].

The preservation of electroneutrality results from the divalent cation Ca^{2+} being neutralized by Asp-21 and Asp-40. The phosphodiester, a monoanion in the ground state, is neutralized by the guanidinium cation of Arg-35 only. As the entering hydroxide bonds to phosphorus, converting the reaction center from a phosphodiester monoanion to a phosphorane dianion, a second guanidinium cation, that of Arg-87, becomes necessary to neutralize the additional negative charge which develops in the transition state (Fig. 6). Such development of additional negative charge on the equatorial oxygens of the transferred nucleotidyl group is characteristic of an associative nucleophilic substitution at phosphorus. The other major criteria of an associative mechanism, general base catalysis, and a reaction coordinate distance ≤4.9 Å [32], are also met by the mechanism of Fig. 6, thus providing a qualitative understanding of the reaction catalyzed by staphylococcal nuclease.

The mechanism of Fig. 6 may also provide a quantitative understanding of the $\approx 10^{16}$-fold rate acceleration produced by this enzyme. This may be shown by using the kinetic properties of the active site mutants (Table 1) to separate the factors contributing to k_{cat}. As discussed in detail previously [5], such an analysis requires that the catalytic and structural effects of each mutation be limited to the residue altered, and that each specific contribution to the rate enhancement be independent and counted only once. While structural changes beyond the residues mutated have been detected in the active site mutants we have studied, these effects are generally small, as reflected in changes in dissociation constants within one order of magnitude of those obtained with the wild-type enzyme (Table 2). The kinetic effects of these mutations are much greater, in some cases, decreasing k_{cat} by three to five orders of magnitude (Table 1). A preliminary x-ray structure of the ternary complex of the E43D mutant of staphylococcal nuclease reveals the dissociation of Asp-40 from Ca^{2+} in the ternary pdTp complex [29] which, by analogy with the D40G mutant, may be responsible for the loss of at most one order of magnitude in activity. Even smaller structural changes have been obtained with mutant forms of other enzymes [33,34]. To avoid counting an effect twice, we have made comparisons between complexes with active and inactive metal ions, and between appropriate pairs of mutants such as D40E with D21E, and R35G with R87G.

If the indirect effects of mutations on catalysis are indeed small, and if no catalytic effect has been counted twice, then the product of the contributions of metal catalysis by Ca^{2+} ($\geq 10^{4.6}$), general base catalysis by Glu-43 ($\approx 10^{4}$), catalysis by approximation ($\geq 10^{3}$), and transition state stabilization and proton donation by Arg-87 ($\geq 10^{4.6}$) are sufficient to explain the overall rate acceleration of $\approx 10^{16}$-fold produced by staphylococcal nuclease [5].

ACKNOWLEDGMENTS

We are grateful to D. Shortle, M. D. Pease, J. McCracken, J. Peisach, and J. A. Gerlt for valuable collaborations. This work was supported, in part, by National Institutes of Health grant DK28616.

REFERENCES

1. C. B. Anfinsen, P. Cuatrecasas, and H. Taniuchi, "Enzymes", 3rd ed., Academic Press, New York, *4*, 177 (1971).

2. F. A. Cotton and E. E. Hazen, Jr., "Enzymes", 3rd ed., Academic Press, New York, *4*, 153 (1971).

3. P. W. Tucker, E. E. Hazen, Jr., and F. A. Cotton, *Mol. Cell. Biochem.*, *23*, 67 (1979).

4. E. H. Serpersu, D. Shortle, and A. S. Mildvan, *Biochemistry*, *25*, 68 (1986).

5. E. H. Serpersu, D. Shortle, and A. S. Mildvan, *Biochemistry*, *26*, 1289 (1987).

6. E. Sulkowski and M. Laskowski, Sr., *J. Biol. Chem.*, *243*, 4917 (1968).

7. B. M. Dunn, C. DiBello, and C. B. Anfinsen, *J. Biol. Chem.*, *248*, 4769 (1973).

8. P. W. Tucker, E. E. Hazen, Jr., and F. A. Cotton, *Mol. Cell. Biochem.*, *22*, 67 (1978).

9. F. A. Cotton, E. E. Hazen, Jr., and M. J. Legg, *Proc. Natl. Acad. Sci. USA*, *76*, 2551 (1979).

10. I. M. Chaiken and G. R. Sanchez, *J. Biol. Chem.*, *247*, 6743 (1972).

11. D. W. Hibler and J. A. Gerlt, *Fed. Proc.*, *FASEB*, *45*, 1875 (1986).

12. D. W. Hibler, N. J. Stolowich, M. A. Reynolds, J. A. Gerlt, J. A. Wilde, and P. H. Bolton, *Biochemistry*, *26*, 6278 (1987).

13. A. S. Mildvan, E. H. Serpersu, D. W. Hibler, and J. A. Gerlt, *FASEB J.*, *2*, #1698, A588 (1988).

14. S. Mehdi and J. A. Gerlt, *J. Am. Chem. Soc.*, *104*, 3223 (1982).

15. D. Shortle, *Gene*, *22*, 181 (1983).

16. D. Shortle and B. Lin, *Genetics*, *110*, 539 (1985).

17. A. S. Mildvan and J. L. Engle, *Methods Enzymol.*, *26C*, 654 (1972).

18. W. B. Mims and J. Peisach, in "Biological Magnetic Resonance", Vol. 3 (L. J. Berliner and J. Reuben, eds), Plenum Press, New York, 1981, p. 213.

19. J. Peisach, W. B. Mims, and J. L. Davis, *J. Biol. Chem.*, *254*, 12379 (1979).

20. J. McCracken, E. H. Serpersu, J. Peisach, and A. S. Mildvan, *FASEB J.*, *2*, #1697, A588 (1988).

21. E. H. Serpersu, J. McCracken, J. Peisach, and A. S. Mildvan, *FASEB J.*, *2*, #1696, A588 (1988).

22. E. H. Serpersu and A. S. Mildvan, unpublished observations, 1988.

23. A. S. Mildvan and R. K. Gupta, *Methods Enzymol.*, *49G*, 322 (1978).

24. A. S. Mildvan, P. R. Rosevear, J. Granot, C. A. O'Brian, H. N. Bramson, and E. T. Kaiser, *Methods Enzymol.*, *99*, 93 (1983).

25. C. H. MacGillavry and G. D. Rieck (eds.), "International Tables for X-Ray Crystallography", Vol. 3, Kynoch Press, Birmingham, UK, 1962, pp. 260, 269.

26. Cambridge Structural Database System, Cambridge, Crystallographic Data Center, University Chemical Laboratory, Cambridge, England, Version 2, 1985.

27. A. F. Wells, in "Structural Inorganic Chemistry", Clarendon Press, Oxford, UK, 1984, p. 822.

28. B. Anderson, R. M. Milburn, J. MacB. Harrowfield, G. B. Robertson, and A. M. Sargeson, *J. Am. Chem. Soc.*, *99*, 2652 (1977).

29. P. Loll and E. Lattman, private communication, 1988.

30. M. D. Pease, E. H. Serpersu, and A. S. Mildvan, *Fed. Proc., FASEB*, *46*, 1932 (1987).

31. J. S. Leigh, Jr., *J. Chem. Phys.*, *40*, 2686 (1970).

32. A. S. Mildvan and D. C. Fry, *Advances Enzymol.*, *59*, 241 (1987).

33. T. Alber, S. Dao-pin, K. Wilson, J. A. Wozniak, S. P. Cook, and B. W. Matthews, *Nature*, *330*, 6143 (1987).

34. J. E. Villafranca, E. E. Howell, S. J. Oatley, N. Xuang, and J. Kraut, *Biochemistry*, *26*, 2182 (1987).

9

Alcohol Dehydrogenase: Structure, Catalysis, and Site-Directed Mutagenesis

Y. Pocker
Department of Chemistry, BG-10
University of Washington
Seattle, WA 98195

1. INTRODUCTION

Nicotinamide-nucleotide-linked dehydrogenases were among the earliest two-substrate enzymes to be subjected to a detailed analysis of structure and activity. It is the purpose of this chapter to bring together biochemical, metalloorganic, and physicochemical studies in an attempt to survey the structure and function of alcohol dehydrogenase (EC 1.1.1.1). This enzyme is an NAD^+-dependent oxidoreductase that acts on primary (and some secondary) alcohols by a mechanism involving both a hydride transfer to NAD^+ and a proton transfer from substrate to solution [Eq. (1)]:

$$RCH_2OH + NAD^+ \rightleftharpoons RCHO + NADH + H^+ \qquad (1)$$

Our understanding of reaction mechanisms and of transition state structures in horse liver and yeast alcohol dehydrogenases has been significantly advanced by the confluence of complete amino acid sequences [1-3], high-resolution x-ray crystallography [4-7], spectroscopy [8,9], metal ion substitution [10,11], and kinetic studies [12, 13]. Structural, chemical, and mutagenic modifications are presented both to provide a frame of reference for the interpretation of kinetic data and for gaining insights into the genetic and molecular basis of isozyme patterns. The maturity of research on alcohol dehydrogenases is attested to by recent comprehensive reviews [14-18] and international workshops [19,20].

2. STRUCTURE

Horse liver alcohol dehydrogenase is a dimer composed of two identical 40,000 mol wt subunits. Each subunit contains two firmly bound zinc ions and is divided into two unequal domains. The sequence of 374 amino acid residues in the single polypeptide chain constituting the subunit of the "ethanol-active" isozyme was determined by Jörnvall [1], and detailed information is now available from x-ray crystallographic data of Brändén, Eklund, and coworkers [4-7]. Native horse liver alcohol dehydrogenase crystallizes in an orthorhombic space group. This enzyme has been determined at a resolution of 2.4 Å and used to deduce the structures of a number of isomorphous enzyme-inhibitor complexes by difference Fourier techniques. Coenzyme-containing complexes usually give monoclinic or triclinic crystals depending on the preparation conditions. Structures of several ternary enzymic complexes crystallizing in triclinic space group have been determined to 2.9-Å resolution. They correspond to a more densely packed conformational state (closed form) [21-24] than that of orthorhombically crystallizing enzyme species (open form) [4-7]. The general fold is very similar in both the closed and open forms. The small and large domains of the enzyme subunit are separated by a cleft with a deep pocket. This pocket accommodates the substrate and the nicotinamide part of the coenzyme. The smaller domain, composed of residues 178-318, contains most of the groups participating in coenzyme binding. The larger domain provides ligands for the protein-bound zinc ions and contains most of the residues controlling substrate binding and catalytic activity. The coenzyme-binding domains are joined to form a central core in the enzyme dimer. A transition from open to closed form involves a rigid 10° body rotation of each catalytic domain relative to the central core with almost complete preservation of the internal structures in the subunit domains.

One of the two zinc ions in the enzyme subunit is essential for catalysis. This active site zinc ion is positioned at the bottom of the domain-separating cleft about 20 Å from the protein surface. It is bound in a distorted tetrahedral coordination sphere by three protein ligands (Cys-46, His-67, and Cys 174), with the fourth coordination position being accessible to water and other ligands from solution.

The second zinc ion is tetrahedrally coordinated by the sulfur atoms from cysteine residues 97, 100, 103, and 111. It is located about 20 Å from the catalytic metal ion site in a polypeptide lobe that projects out of the catalytic domain. The function of this zinc ion remains unknown, although recent results have been interpreted in support of a structural role [25].

3. BINDING DOMAINS

3.1. Coenzyme Binding

The adenine portion of the coenzyme is accommodated by the hydrophobic cleft. The adenine-ribose segment forms hydrogen bonds to the side chains of Asp-223 and Lys-228. These interactions involve both ribose hydroxyl groups [4].

The coenzyme pyrophosphate group is bound close to the rim of the domain-separating cleft and makes a bend over the edge of the pleated-sheet structure composing this region. This conformation appears to be stabilized by hydrogen bonds between the phosphate oxygens and peptide nitrogens as well as through charge interactions with the side chain residues of Arg-47, Arg-369, and Lys-228 [7]. The nicotinamide ribose is tightly bound in the narrow cleft between the two domains of the subunit. One of the ribose hydroxyl groups forms hydrogen bonds to the side chains of Ser-48 and His-51 [7]. It has been proposed that this system of hydrogen bonds acts as a shuttle for proton transfer between the active site and the solution [24]. The nicotinamide ring is positioned in the pocket near the bottom of the domain-separating cleft and interacts with residue side chains. The side chains on one side of the pocket are predominantly hydrophobic, while those on the other side are predominantly hydrophilic and consist of the catalytic zinc ion and its ligands. Orientation of the nicotinamide ring appears to be determined mainly by the hydrogen bonds from its carboxamide group to the residue side chains at the bottom of the pocket.

3.2. Substrate Recognition

Almost all structures reported by Brändén, Eklund, and coworkers [17,23] refer to enzyme complexes containing either a substrate or a competitive inhibitor. These molecules bind directly to, or in the vicinity of, the catalytic zinc in a 5-10-Å-wide by ~20-Å-long pocket that represents a continuation of the domain-separating cleft leading from exterior solution to the active site metal ion. Auramine O, a fluorescent dye, interacts specifically with the substrate-binding pocket and has been used as a reporter ligand in studies of complex formation at this site.

The bonding interaction between substrate and protein has been studied at 2.9 Å resolution in the ternary enzyme-H_2NADH-dimethylaminocinnamaldehyde and enzyme-NAD^+-bromobenzyl alcohol complexes, which crystallize in the closed conformation [21-24]. The oxygen atom of these substrates is bound directly to the catalytic zinc; in the case of bromobenzyl alcohol, it is positioned within a single hydrogen-bonding distance to the hydroxyl group of Ser-48. The crystallographic data are neither in accord with an outer sphere coordinated substrate nor with a water molecule bound to a penta-coordinated zinc. The substrate molecules as seen in ternary complexes by the averaging nature of x-ray structures are not optimally oriented for direct hydride ion transfer to or from bound coenzyme, but such an orientation can be readily achieved through rotation around the oxygen-metal bond.

From microspectrophotometric measurements on crystal complexes, Bignetti et al. [26] concluded that x-ray data on putative enzyme-NAD^+-alcohol complexes may actually refer to abortive complexes containing NADH instead of NAD^+. The criticism appears justified in the light of results recently presented for the Cd(II)-substituted enzyme [27]. These authors also found that substrate diffusion into crystals of the enzyme-NADH complex leads to reversible oxidation of the bound coenzyme without disruption of the crystal lattice. Based on the latter observation, the principle of microscopic reversibility, and related structural data, it has been suggested that the same coenzyme-binding pattern prevails for both abortive (enzyme-NADH-alcohol) and productive (enzyme-NAD^+-alcohol) complexes.

3.3. Sequential Active Site Dehydration

Here one may wish to raise a question about sequential and reversible displacement of water molecules from the active site cavity. Refinement of high-resolution data for different liver alcohol dehydrogenase complexes has provided some detailed information on well-ordered water molecules located inside the enzyme in the crystal structures [7]. When the coenzyme is absent, approximately 40 of these water molecules are in the domain-separating cleft and the substrate-binding pocket. About 30 water molecules are displaced on binding the coenzyme and substrate, in the enzyme-H_2NADH-methylpentanediol complex which crystallizes in the open enzyme conformation, leaving approximately 10 in the active site [7,21]. The transition from open to closed conformation narrows the domain-separating cleft, brings the catalytic zinc ion about 1 Å closer to the coenzyme binding site, and results in a complete dehydration of the active site region. This latter observation indicates that hydrophobic stabilization may represent a major driving force for the open to closed conformational transition. Occupation of the nicotinamide-nucleotide-binding site appears to be necessary for stabilization of the closed form [16]. The conformational transition is sensitive to ligand structure, as is illustrated by an open conformation for the enzyme-H_2NADH-methylpentanediol complex, but closed conformations for the enzyme-NADH-methylpentanediol and enzyme-H_2NADH-dimethylaminocinnamaldehyde complexes [21,24]. Results recently obtained with carboxymethylated enzyme indicated that, in solution, there may be an equilibrium distribution of species in open, closed, and partially closed states [25]. It is instructive to speculate on the reversible conformational changes in the active site which induce this successive dehydration followed by sequential rehydration.

4. CHEMICALLY MODIFIED ENZYMES

4.1. Coordination Chemistry

A useful tool for evaluating the role of catalytic zinc ion in metalloenzymes is its substitution by other metals such as Co(II), Cu(II),

Cd(II), or Ni(II). Anaerobic dialysis of LADH against divalent metal ions leads within a few hours to the formation of hybrid enzymes according to Eq. (2):

$$\begin{aligned} &2M^{2+} + Zn_4LADH \rightarrow M_2Zn_2LADH + 2Zn^{2+} \\ &M^{2+} \equiv {}^{65}Zn^{2+},\ Co^{2+},\ Cd^{2+} \end{aligned} \qquad (2)$$

However, the second pair of metal ions is exchanged much more slowly, e.g., the complete exchange of Zn^{2+} by Co^{2+} takes 72 hr [28].

The catalytic properties of the hybrid enzymes M_2Zn_2LADH prepared by the above dialysis techniques are virtually unchanged, clearly demonstrating that the noncatalytic zinc ions are replaced first [29]. These noncatalytic sites are close to the surface of the enzyme molecule and should therefore be easily accessible; thus, it seems plausible that exchange at the noncatalytic site proceeds via binuclear intermediates [Eq. (3)]:

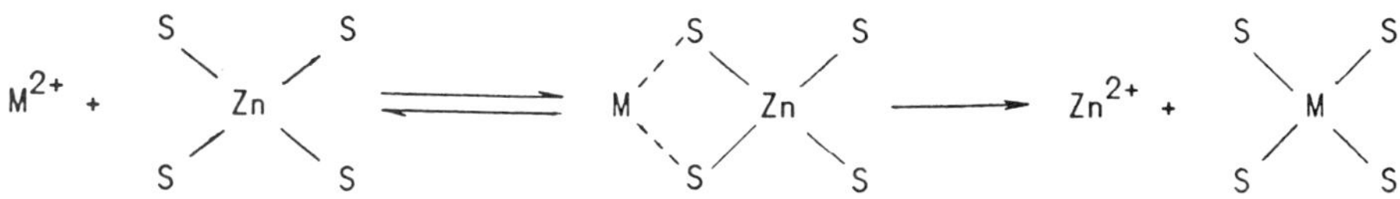

(3)

LADH can be fully depleted of protein-bound zinc ions by dialysis at low pH against chelating reagents [14]. The metal-free enzyme is catalytically inactive and it loses SH-groups on prolonged exposure to air. This apoenzyme can be reconstituted by treatment with Zn^{2+} in the presence of reducing agents [30].

Zeppezauer and coworkers were able to selectively deplete the catalytic zinc ions by treating crystalline suspensions of the enzyme with chelating agents under anaerobic conditions [10]. The inactive enzyme thus obtained lacks catalytic zinc but otherwise retains the crystal structure of the native enzyme except for minor changes at the metal sites [31]. Selectively metal ion-depleted enzyme can be fully reconstituted by treatment with Zn^{2+}, Co^{2+}, Cd^{2+}, Cu^{2+}, Ni^{2+}, Fe^{2+}, Pb^{2+}, or Fe^{3+}. Under the usual conditions of assay [32], the Zn(II), Co(II), Cd(II), and Ni(II) enzymes are catalytically active. Crystal structures of the reconstituted enzymes have been determined and found to be similar to those of the native enzyme [33].

Ultraviolet-visible absorption spectra of Co(II)-, Ni(II)-, Cd(II)-, and Cu(II)-substituted enzyme show transitions attributable to metal-sulfur charge transfer complexes that are consistent with a tetrahedral coordination geometry at the catalytic as well as the noncatalytic metal ion site. Tetrahedral coordination at the catalytic site is also supported by circular dichroic, resonance Raman, and electron paramagnetic resonance (EPR) spectral evidence. Particular attention has been drawn to the spectral properties of the selectively substituted Cu(II) derivative [11,34], which are similar to those of type 1 Cu^{2+} in "blue" copper proteins.

4.2. Chemical Modification

Pronounced effects on enzyme activity were observed with a variety of reagents that react more or less specifically with histidine, lysine, arginine, and cysteine residues [14]. Histidine modification with diethyl pyrocarbonate leads to complete loss of activity [35]; however, see Sec. 8 on mutagenic studies. Alkylation of lysine residues leads to an increased rate of coenzyme dissociation caused specifically by modification of Lys-228 [14]. Interactions that normally exist between the amino acid residues and the AMP portion of bound coenzyme appear to be affected upon alkylation of Lys-228; however, additional factors may contribute to impaired binding [36].

Reaction of the zinc ligand, Cys-46, with iodoacetate leads to a carboxymethylated enzyme that is catalytically active [25,37]. Alkylation is promoted by imidazole which binds to the catalytic zinc ion, an effect which can be attributed to the metal-mediated electron delocalization from the imidazole ligand to the cysteinyl ligand [38]. McKinley-McKee and coworkers have elegantly demonstrated that a particularly rapid and specific modification of Cys-46 is achieved with α-bromo-β-(5-imidazolyl)propanoic acid [39,40], a reagent which may interact simultaneously with catalytic zinc ion and Arg-47 at the pyrophosphate subsite [41].

Kinetic data for enzyme inactivation by a series of halogenated carboxylate ions were reported by Dahl and McKinely-McKee [42]. Chadka and Plapp [43] found that the enzyme inactivation caused by 3-bromopropanoate derives from the modification of Cys-174. The availability of adjacent sites for prior binding of the alkylating reagent and the metal coordination of residues Cys-46 and Cys-174 may be the cause of the particular reactivity of these residues. Butanedione or phenylglyoxal treatment of LADH results in modification of two arginine residues with concomitant loss of enzymic activity [44]. One of these residues has been identified as Arg-84 [45]; the other may be Arg-47, which plays a role in tight coenzyme (pyrophosphate moiety) binding [46].

5. INHIBITOR BINDING

Strong indications that the catalytic zinc ion may have an essential mechanistic function came from inhibition studies involving imidazole, pyrazole, 1,10-phenanthroline, and 2,2'-bipyridine [14]. Spectroscopic results obtained with metal-substituted enzyme confirm the presence of inner sphere complexes [33]. Resonance Raman spectra of complexes between Cu(II) enzyme and pyrazole are in accord with the tetrahedral structure of the catalytic metal ion [47], whereas perturbed angular correlation spectroscopy on Cd(II) enzyme led to the conclusion that a five-coordinate complex is formed [48]. Solvent magnetic resonance data for Cu(II) enzyme suggest that the metal ion accepts pyrazoles as a fifth ligand in the absence of coenzyme, whereas it retains the four coordinate geometry in the presence of bound NAD^+ [49]. The converse inference was drawn from the EPR spectral parameters determined with Co(II) enzyme [50]. Such disparity may reflect, in part, the difficulties associated with establishing metal coordination numbers, and in part the diverse coordinative capacity of various metal ion-substituted derivatives and the temperature dependence of two equilibrating enzyme conformations. Pyrazole, 1,10-phenanthroline, and 2,2'-bipyridine have been extensively used as reporter ligands in spectroscopic studies. Such studies have shown that catalytic zinc ions provide a binding site for a variety of compounds including such anionic ligands as the higher homologs of carboxylate ions [14,51]. While acetate associates ~15 times more tightly with the pyrophosphate-binding subsite than with the catalytic zinc, propanoate and higher homologs appear to bind preferentially to the metal ion site [51]. In view of these findings, it is perhaps not surprising that iodoacetate and 3-bromopropanoate lead to the modification of different cysteine residues.

Stopped-flow kinetic measurements demonstrate that the association of 2,2'-bipyridine and pyrazole with LADH free of coenzyme require a protonated group in the enzyme of pK_a 9.2 [52-54]. Coenzyme binding causes a substantial shift from the apparent pK_a of 9.2. Thus saturating the enzyme with NAD^+ shifts the effect of pH on the binding of bipyridine, pyrazole, or decanoate to a pK_a of 7.6 [52,54,55], whereas NADH shifts the pK_a of the group that controls binding of imidazole, pyrazole, or decanoate to ~11.2 [54,55]. These results support the hypothesis that zinc-bound water undergoes substantial and opposite electrostatic pK_a shifts on binding oxidized (NAD^+) vs. reduced (NADH) coenzyme [56]. It is attractive to infer that the observed differential effect of pH on NAD^+ and NADH binding derives from the ionization of zinc-bound water (Scheme 1). Spectroscopic studies of Co(II) enzyme have demonstrated inner sphere coordination of acetate and chloride ions to the metal ion site in the presence of NAD^+ [57,58].

$$\mathrm{E{\bullet}Zn-OH_2} \underset{+\,\mathrm{H_2O}\ (-\mathrm{Ligand})}{\overset{+\,\mathrm{Ligand}\ (-\mathrm{H_2O})}{\rightleftharpoons}} \mathrm{E{\bullet}Zn-Ligand}$$

$$\mathrm{E{\bullet}ZnOH_2} \overset{K_a}{\rightleftharpoons} \mathrm{E{\bullet}ZnOH} + \mathrm{H^+}$$

$$\text{Enzyme} - \mathrm{NAD^+} : pK_a = 7.6$$

$$\text{Free enzyme} : pK_a = 9.2$$

$$\text{Enzyme} - \mathrm{NADH} : pK_a = 11.2$$

Scheme 1. The binding of nonionizing ligands to catalytic zinc in LADH.

6. KINETIC ASPECTS

6.1. Kinetic Schemes

Liver alcohol dehydrogenase operates by a ternary complex mechanism. Over the concentration ranges where Michaelis-Menten kinetics are obeyed, the compulsory order mechanism, depicted in Scheme 2, accounts for the steady-state rate behavior of the enzyme with most primary alcohols and aldehydes tested.

Previous kinetic data have been analyzed in terms of an ordered mechanism, with preferential utilization of the pathway involving prior binding of the coenzyme [14,15]. Several attempts have been made to explain why this binding order is preferred [14,17,59]. A detailed analysis indicates that the apparent steady-state rate equation will be of Michaelis-Menten type corresponding to Scheme 2, irrespective of the pathway that is catalytically competent. Evidence that the reaction may not be strictly ordered as assumed comes from isotope exchange measurements, and the kinetic significance of species such as the abortive enzyme-NADH-alcohol and enzyme-NAD^+-aldehyde complexes. Thus equilibrium-binding studies which show that substrates and coenzymes form binary as well as abortive ternary complexes indicate that a generalized random order mechanism may actually be operative (Scheme 3). Nevertheless, in most cases examined, product desorption is essentially an ordered process with coenzyme release occurring last [55,60]. In addition, parameter values obtained with regular substrates in standard kinetic experiments provide information only on rate constants for reaction steps in Scheme 2 [60]. Hence, discussions of the kinetic behavior of the enzyme in terms of an ordered mechanism (Scheme 2) remain valid.

Scheme 2. Ordered kinetic scheme for the catalytic interconversion of primary alcohols (Alc) and aldehydes (Ald). The catalytic step involving proton transfer is not shown.

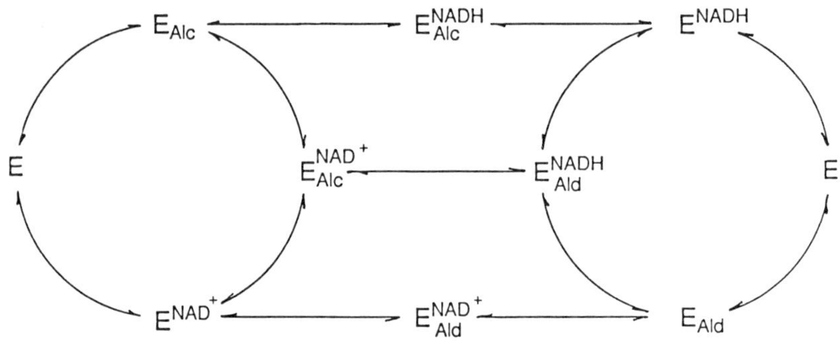

Scheme 3. Random kinetic scheme for the catalytic interconversion of primary alcohols and aldehydes.

Binding studies and steady-state kinetic measurements have failed to provide evidence for any nonequivalence of the two subunits in the LADH molecule [14]. Reports based on a transient state kinetic study of the reduction of certain aromatic aldehydes developed the conclusion that the reaction occurs in two distinct phases with exactly one half of the limiting reactant in each phase. This led to the proposal of "half-site reactivity" for the enzyme. However, analyses based on the assumption of equivalent and noninteracting sites indicate that the behavior of the enzyme is fully compatible with the ordered mechanism in Scheme 2 [61] with the biphasic time course attributable to the transient accumulation of an enzyme-NAD^+-alcohol complex [62]. More recently, isotope and substituent effects were found to be in disagreement with Scheme 2 and were interpreted to support some degree of subunit cooperativity [63]. However, it is not clear from the experimental evidence presently available that any kinetic nonequivalence of the two subunits of LADH is required [64].

6.2. Proton and Hydride Transfer

Steady-state kinetic evidence locates the hydride transfer process between substrate and coenzyme in the ternary complex conversion step of Scheme 2. Corroboration in the form of transient state kinetic studies clearly demonstrates that this step is subject to a primary isotope effect in reactions performed with either RCD_2OH or NADD [15]. Subsequent kinetic studies have focused on the mechanism of the net release of protons which are obligatory in alcohol oxidation by NAD^+. Such proton transfer occurs prior to, and is uncoupled from, the hydride transfer step. The process is initiated by NAD^+ attachment with the concomitant pK_a perturbation of the ionizing group that controls the catalytic site. The ionizing group has the dual function to assist the binding of substrates and to mediate proton transfer from bound alcohol to solution [12]. It has been suggested that this ionizing group is a water molecule ligated to zinc [14], whose conjugate base, a zinc-bound hydroxide, may participate in proton abstraction thus converting alcohol to alcoholate in a substrate-activating step [14]. Subsequent binding to zinc juxtaposes the alcoholate with the nicotinamide ring for efficient hydride transfer. Zinc-bound aldehyde is then released, followed by reduced coenzyme to complete the catalytic cycle.

Hydride transfer from ethanol to NAD^+ is dependent on the basic form of an ionizing group with a pK_a of 6.4 in the productive ternary complex [65]. Thus, deprotonation of the enzyme-NAD^+-alcohol complex represents an obligatory step preceding hydride transfer [66] (Scheme 4). The pH-dependent reactivity of the enzyme-NAD^+-alcohol complex may be attributed to the ionization of the enzyme-bound alcohol; the latter being directly coordinated to catalytic zinc ion [66]. Evidence in favor of Scheme 4 comes from the observation that the corresponding pK_a is dependent on the intrinsic acidity of the alcohol ligand. The idea that hydride transfer to aldehydes results in the formation of a zinc-coordinated alkoxide ion prior to proton transfer and product desorption is favored by studies on the effect of pH on the LADH-catalyzed reduction of trans-4(N,N-dimethylamino)cinnamaldehyde [67].

Transient state kinetic studies of substrate inhibition in horse LADH reactions may reflect synergistic effects on coenzyme desorption rates [68]. Zinc-chelating reagents such as 2,2'-bipyridine and 1,10-phenanthroline exhibit a limiting rate of ~250 sec^{-1} in their chromo-

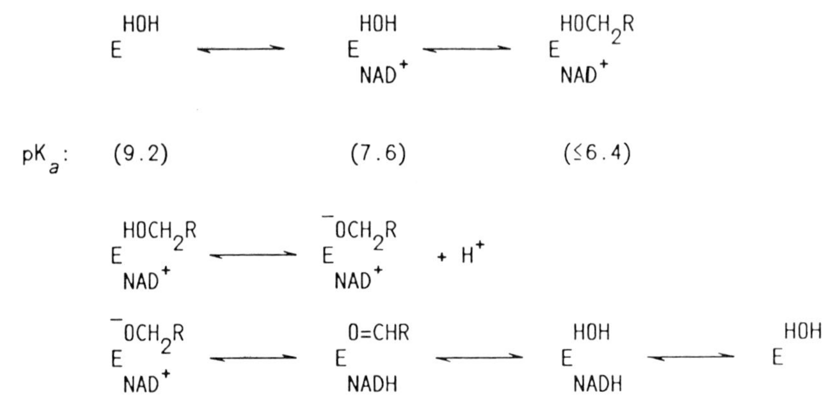

Scheme 4. Proton release/uptake associated with the formation of the ternary enzyme-NAD^+-alcohol complex.

phoric association with the enzyme. This is attributed to rapid outer sphere complex formation followed by a slow inner sphere coordination which is limited by the rate of dissociation of water bound to zinc [52,69,70]. Others argue that a limiting rate of 250 sec^{-1} is far too slow to represent the dissociation of a zinc-bound water and suggest instead structural rearrangements associated with an increase from 4 to 5 of the metal coordination number [53]. This suggestion is supported by the observation that bipyridine binding to Co(II) enzyme causes a substantial decrease in intensity of the 650-nm absorption band [71].

7. CATALYTIC MECHANISMS

Coordination of substrate to catalytic zinc ion facilitates hydride transfer from NADH to aldehyde [14,15]. By the principle of microscopic reversibility, coordination to the metal ion is also essential for the enzymic activation of alcohol substrates. A facile deprotonation step generates the alkoxide thus lowering the energy barrier for hydride transfer. This transfer takes place from the alkoxide ion to NAD^+ in what appears to be a completely solvent-free environment. The absence of solvent may be important in eliminating side reactions, as, for example, between the hydride donor and water [16]. Dehydration of the active site also favors inner sphere zinc coordination of substrate and enhances the electrostatic stabilization of zinc-bound alkoxide ion [16,72]. The latter effect could explain why the NAD^+-induced pK_a perturbation is greater with zinc-bound alcohols than with zinc-bound water. The ligation of alcohols and aldehydes to catalytic zinc ion brings the substrate into the correct juxtaposition for rapid hydride transfer and proton exchange. Alcohol binding to catalytic zinc ion in the presence of NAD^+ results in its immediate deprotonation, followed by direct hydride transfer. Conversely, hydride transfer from NADH is facilitated by the polarizing effect of catalytic zinc ion on the aldehyde substrate. Protonation of the resulting alkoxide ion triggers desorption of the alcohol product.

The open-to-closed conformational transition appears to be of crucial importance for correct positioning of reactants and for the creation of a favorable anhydrous environment for hydride transfer. Data for binary enzyme-alcohol complexes indicate that ligation of RCH_2OH to the catalytic zinc ion provides a major contribution to the net pK_a perturbation, $\Delta pK = 5\text{-}6$. The additional pK_a shifts contributed by NAD^+ amounts to a ΔpK of ~3. In many respects, the ionization properties of zinc-bound water appear to be analogous to those of zinc-bound alcohol, except that the additional pK_a shift, ΔpK, induced by NAD^+ amounts to only 1.6 with water. This latter behavior indicates that water bound to catalytic zinc ion remains unionized at physiological pH, thereby ensuring its facile displacement by alcohols. The differential pH effect induced by either NAD^+ or NADH binding to liver alcohol dehydrogenase derives from the ionization of zinc-bound water. Kinetic results, including reported solvent and reactant isotope effects [13,15,73,74], provide clear evidence that proton releases to solution during the oxidation of primary alcohols occurs at the ternary complex level. Crystallographic evidence for an almost completely dehydrated active site in productive ternary complexes means that the proton transfer must occur via a system of hydrogen bonds which link the hydroxyl group of zinc-bound alcohol to the imidazole group of His-51. The latter group is accessible to solvent (as well as to buffers).

Productive ternary complexes may contain a pentacoordinated zinc complex with water as a fifth ligand [50]. This conclusion is based on observations made with Co(II) enzyme in frozen solutions at 4°K. The physiological relevance of spectral results obtained at 4°K is not clear, but the data are compatible with the catalytic participation of a water molecule in a pentacoordinated zinc complex. Water thus held acts as an important constituent of the proton relay system.

8. SITE-DIRECTED MUTAGENIC STUDIES

The alcohol dehydrogenase isozymes of *Saccharomyces cerevisiae* are responsible for the interconversion of acetaldehyde and ethanol. Alcohol dehydrogenase I (ADH I) is the classical anaerobic fermentative enzyme, first isolated and characterized by Racker [75]. Alcohol dehydrogenase II (ADH II), another cytoplasmic form, is repressed by glucose and has a 10-fold lower Michaelis constant for ethanol than that of other yeast isozymes [76]. Both enzymes are composed of four identical 37,000 mol wt subunits. Although none of the yeast isozymes have yet been crystallized, insight into their predicted tertiary structures has been gained by comparison with the horse liver enzyme [15,17]. Both ADH I and II have different substrate and cofactor parameters. Dehydrogenases from mutant species have been isolated and sequenced [77]. A third yeast isozyme, alcohol dehydrogenase III, found in mitochondria, is also being characterized [78].

Extensive genetic analysis has resulted in the identification of yeast genes which encode structural information for both ADH I and II. These genes have been cloned [79,80] and a number of laboratories are using protein engineering involving systematic alteration of structure to gather direct information on the nature of enzyme catalysis.

Kinetic analysis of mutant enzymes combined with high-resolution structural data from protein x-ray crystallography should facilitate a penetrating assessment of the relationship between structure and function. In particular, the strength and nature of enzyme-substrate interactions, as well as the efficiency of the proton shuttle system and its detailed role in catalysis, can now be scrutinized.

Among the many alcohol dehydrogenase mutations that were carried out, Met-294 to Leu, His-47 to Arg, and His-51 to Gln, were confirmed by gene sequencing. The modified enzymes thus obtained were purified from host strains of yeast which did not produce wild-type enzymes [81]. The kinetic constants of the wild dehydrogenases from *S. cerevisiae* and these three mutant enzymes are being actively studied. Although many essential features of the active sites of the liver and yeast enzymes are similar, there are some fundamental differences. The adenine-binding pocket in the yeast enzyme is more polar than in the liver enzyme because His-47 replaces Arg-47 in the latter. The yeast enzyme has a 100-fold larger turnover number compared to the liver enzyme and the question arises as to exactly how amino acid substitution can account for the difference in k_{cat}.

In LADH, the hydrogen-bonded network connecting the buried substrate bromobenzyl alcohol to Ser-48 and to His-51 could function as a proton relay system with His-51 acting as a proton transfer group [24]. This system is essentially conserved in the yeast enzyme, although Ser-48 is replaced by Thr-48. Both enzymes show that a group with a pK_a value of 7 ± 1 should be unprotonated for maximal rate of alcohol oxidation [82-84]. Recent studies have systematically explored the exceptionally broad specificity displayed by LADH [85-91], while confirming the narrower specificities of the yeast isozymes I and II. Although alcohol dehydrogenases I, I-Leu, and III (which also has Leu-294) exhibit very similar Michaelis constants for ethanol, ADH II is decidedly different. Clearly, the substitution of Leu-294 for Met-294 cannot account for the difference between ADH I and ADH II.

ADH I has two histidine residues in the active site. His-47 and His-51; hence detailed interpretations of the pH dependency are complicated [82,83]. In the liver enzyme, with Arg-47 and His-51, a variety of studies have suggested that His-51 functions as a proton shuttle group. Indeed, the liver enzyme is said to be inactivated by diethylpyrocarbonate, with a dependence on pH that implicates a group of pK_a 9.6 [35]; in contrast, inactivation of the yeast enzyme suggests a group of pK_a 7.1 [92]. Changing His-47 to Arg-47 in ADH I reduces the turnover number by a factor of 6 but does not alter the pH-rate profile.

On the other hand, replacement of His-51 by Gln-51 results in a less active mutant enzyme and one that no longer exhibits a pK_a value of 7 [81]. This histidine is not obligatorily required for catalysis, although its replacement compromises catalytic activity and markedly alters the pH-rate profile [81,93]. Kinetic data from the mutant Gln-51 enzyme in particular support the hypothesis that His-51 helps to significantly increase the rate of proton shuttle that precedes hydride transfer to NAD^+ in alcohol oxidation and which follows hydride transfer from NADH in aldehyde reduction. The observation of a 30-fold decreased activity at pH 7 [81,93] provides a means for estimating the extent to which His-51 contributes to the proton shuttle. Thus, the major mode of proton transfer (~97%) at pH 7 would appear to occur between the proton shuttle group, His-51, and buffer. Since most buffers can partially penetrate into the active site cavity they thus provide a secondary, albeit less efficient (~3% at pH 7) form of proton shuttle [93]. Further studies are required to unravel the multiple participants in the proton relay mechanism and to explain the pH and buffer dependencies of this imperfect mutant. To learn more about the nature of alcohol dehydrogenase catalysis and the relationship between protein structure and function through the use of site-directed mutagenesis, we need to study enzymic systems that in the wild type are well characterized in structural [94,95], mechanistic [96-100], inhibitory [101], and energetic [89,91,102] terms. Many engineered proteins are far from inactive and the origins of their reduced activity are still being evaluated. Clearly, the existence of these less efficient and "imperfect" mutant enzymes provides an opportunity to obtain a better understanding of the anatomy of alcohol dehydrogenase catalysis.

ABBREVIATIONS AND DEFINITIONS

ADH	alcohol dehydrogenase (EC 1.1.1.1)
LADH	horse liver alcohol dehydrogenase (EC 1.1.1.1)
NAD^+	nicotinamide adenine dinucleotide (oxidized form)
NADH	nicotinamide adenine dinucleotide (reduced form)
YADH	yeast alcohol dehydrogenase

ACKNOWLEDGMENTS

Acknowledgment is made to the donors of the Petroleum Research Fund, administered by the American Chemical Society, for support of research on electrostatic interactions. The author's research activities on mechanisms of enzyme catalysis have received generous support from the National Science Foundation, National Institutes of Health of the U.S. Public Health Service, and the Muscular Dystrophy Association. The author thanks Drs. Anna Pocker and Bruce P. Ronald for many helpful discussions.

REFERENCES

1. H. Jörnvall, *Eur. J. Biochem.*, *16*, 25 (1970).
2. H. Jörnvall, *Eur. J. Biochem.*, *72*, 425 (1977).
3. H. Jörnvall, *Eur. J. Biochem.*, *72*, 433 (1977).
4. H. Eklund, B. Nordström, E. Zeppezauer, G. Söderlund, I. Ohlsson, T. Boiwe, B.-O. Söderberg, O. Tapia, C.-I. Brändén and Å. Åkeson, *J. Mol. Biol.*, *102*, 27 (1976).
5. T. Boiwe and C.-I. Brändén, *Eur. J. Biochem.*, *77*, 173 (1977).
6. H. Eklund, J.-P. Samama, and L. Wallén, *Biochemistry*, *21*, 4858 (1982).
7. H. Eklund, J.-P. Samama, and T. A. Jones, *Biochemistry*, *23*, 5982 (1984).
8. H. Dietrich and M. Zeppezauer, *J. Inorg. Biochem.*, *17*, 227 (1982).
9. C. Sartorius, M. Gerber, M. Zeppezauer, and M. F. Dunn, *Biochemistry*, *26*, 871 (1987).

10. W. Maret, I. Andersson, H. Dietrich, H. Schneider-Bernlöhr, R. Einarsson, and M. Zeppezauer, *Eur. J. Biochem., 98,* 501 (1979).

11. W. Maret, H. Dietrich, H.-H. Ruf, and M. Zeppezauer, *J. Inorg. Biochem., 12,* 241 (1980).

12. J. D. Shore, H. Gutfreund, R. L. Brooks, D. Santiago, and P. Santiago, *Biochemistry, 13,* 4185 (1974).

13. P. F. Cook, N. J. Oppenheimer, and W. W. Cleland, *Biochemistry, 20,* 1817 (1981).

14. C.-I. Brändén, H. Jörnvall, H. Eklund, and B. Furugren, in "The Enzymes", Vol. 11A, 3rd ed. (P. D. Boyer, ed.), Academic Press, New York, 1975, p. 103 ff.

15. J. P. Klinman, *Crit. Rev. Biochem., 10,* 39 (1981).

16. H. Eklund and C.-I. Brändén, in "Zinc Enzymes" (T. G. Spiro, ed.), John Wiley and Sons, New York, 1983, p. 124 ff.

17. H. Eklund and C.-I. Brändén, in "Biological Macromolecules and Assemblies", Vol. 3 (F. Jurnak and A. McPherson, eds.), John Wiley and Sons, New York, 1987, pp. 73-142.

18. G. Pettersson, *Crit. Rev. Biochem., 21,* 349 (1987).

19. T. G. Flynn and H. Weiner (eds.), "Enzymology of Carbonyl Metabolism", Alan R. Liss, New York, 1985.

20. H. Weiner and T. G. Flynn (eds.), "Enzymology and Molecular Biology of Carbonyl Metabolism", Alan R. Liss, New York, 1987.

21. E. Cedergren-Zeppezauer, J.-P. Samama, and H. Eklund, *Biochemistry, 21,* 4895 (1982).

22. E. Cedergren-Zeppezauer, *Biochemistry, 22,* 5761 (1983).

23. H. Eklund, J.-P. Samama, L. Wallén, C.-I. Brändén, Å. Åkeson, and T. A. Jones, *J. Mol. Biol., 146,* 561 (1981).

24. H. Eklund, B. V. Plapp, J.-P. Samama, and C.-I. Brändén, *J. Biol. Chem., 257,* 14349 (1982).

25. E. S. Cedergren-Zeppezauer, I. Andersson, S. Ottonello, and E. Bignetti, *Biochemistry, 24,* 4000 (1985).

26. E. Bignetti, G. L. Rossi, and E. Zeppezauer, *FEBS Lett., 1,* 17 (1979).

27. G. Schneider, E. Cedergren-Zeppezauer, S. Knight, H. Eklund, and M. Zeppezauer, *Biochemistry, 24,* 7503 (1985).

28. A. J. Sytkowski and B. L. Vallee, *Biochemistry, 17,* 2850 (1978).

29. A. J. Sytkowski and B. L. Vallee, *Proc. Natl. Acad. Sci. USA, 73,* 344 (1976).

30. R. Rudolph, J. Gerschitz, and R. Jaenicke, *Eur. J. Biochem., 87,* 601 (1978).

31. G. Schneider, H. Eklund, E. Cedergren-Zeppezauer, and M. Zeppezauer, *Proc. Natl. Acad. Sci. USA, 80,* 5289 (1983).

32. K. Dalziel, *Acta Chem. Scand., 11,* 397 (1957).

33. M. Zeppezauer, in "The Coordination Chemistry of Metalloenzymes" (I. Bertini, R. Drago, and C. Luchinat, eds.), Reidel, Boston, 1983, p. 99 ff.

34. W. Maret, M. Zeppezauer, A. Desideri, L. Morpurgo, and G. Rotilio, *FEBS Lett., 136,* 72 (1981).

35. M. Hennecke and B. V. Plapp, *Biochemistry, 22,* 3721 (1983).

36. B. V. Plapp, *J. Biol. Chem., 245,* 1727 (1970).

37. E. Zeppezauer, H. Jörnvall, and I. Ohlsson, *Eur. J. Biochem., 58,* 95 (1975).

38. K. H. Dahl and J. S. McKinley-McKee, *Eur. J. Biochem., 120,* 451 (1981).

39. K. H. Dahl, J. S. McKinley-McKee, and H. Jörnvall, *FEBS Lett., 71,* 287 (1976).

40. K. H. Dahl, J. S. McKinley-McKee, H. C. Beyerman, and A. Noordam, *FEBS Lett., 99,* 308 (1979).

41. K. H. Dahl and J. S. McKinley-McKee, *J. Inorg. Biochem., 15,* 79 (1981).

42. K. H. Dahl and J. S. McKinley-McKee, *Eur. J. Biochem., 118,* 507 (1981).

43. K. Chadka and B. V. Plapp, *Biochemistry, 23,* 216 (1984).

44. L. G. Lange, J. F. Riordan, and B. L. Vallee, *Biochemistry, 13,* 4361 (1974).

45. H. Jörnvall, L. G. Lange, J. F. Riordan, and B. L. Vallee, *Biochem. Biophys. Res. Commun., 77,* 73 (1977).

46. H. Jörnvall, J. Hempel, B. L. Vallee, W. F. Bosron, and T.-K. Li, *Proc. Natl. Acad. Sci. USA, 81,* 3024 (1984).

47. W. Maret, M. Zeppezauer, J. Sanders-Loehr, and T. M. Loehr, *Biochemistry, 22,* 3202 (1983).

48. I. Andersson, R. Bauer, and I. Demeter, *Inorg. Chim. Acta, 67,* 53 (1982).

49. I. Andersson, W. Maret, M. Zeppezauer, R. D. Brown III, and S. H. Koenig, *Biochemistry, 20,* 3424 (1981).

50. M. W. Makinen and M. B. Yim, *Proc. Natl. Acad. Sci. USA, 78,* 6221 (1981).

51. P. Andersson, J. Kvassman, A. Lindström, B. Oldén, and G. Pettersson, *Eur. J. Biochem., 113,* 425 (1981).

52. S. A. Evans and J. D. Shore, *J. Biol. Chem., 255,* 1509 (1980).

53. P. Andersson, J. Kvassman, A. Lindström, and B. Oldén, *Eur. J. Biochem., 108,* 303 (1980).

54. P. Andersson, J. Kvassman, A. Lindström, B. Oldén, and G. Pettersson, *Eur. J. Biochem., 114,* 549 (1981).

55. P. Andersson, J. Kvassman, B. Oldén, and G. Pettersson, *Eur. J. Biochem., 118,* 119 (1981).

56. G. Pettersson and H. Eklund, *Eur. J. Biochem., 165,* 157 (1987).

57. I. Bertini, G. Lanini, C. Luchinat, C. Haas, W. Maret, and M. Zeppezauer, *Rev. Port. Quin., 27,* 257 (1985).

58. W. Maret and M. Zeppezauer, *Biochemistry, 25,* 1584 (1986).

59. M. F. Dunn, *Struct. Bond., 23,* 61 (1976).

60. P. Andersson, J. Kvassman, B. Oldén, and G. Pettersson, *Eur. J. Biochem., 139,* 519 (1984).

61. G. Pettersson, *Eur. J. Biochem., 69,* 273 (1976).

62. J. Kvassman and G. Pettersson, *Eur. J. Biochem., 69,* 279 (1976).

63. M. F. Dunn, S. A. Bernhard, D. Anderson, A. Copeland, R. G. Morris, and J. P. Rogue, *Biochemistry, 18,* 2346 (1979).

64. P. Andersson and G. Pettersson, *Eur. J. Biochem., 122,* 559 (1982).

65. R. L. Brooks, J. D. Shore, and H. Gutfreund, *J. Biol. Chem., 244,* 2382 (1972).

66. J. Kvassman and G. Pettersson, *Eur. J. Biochem., 87,* 417 (1978).

67. R. G. Morris, G. Saliman, and M. F. Dunn, *Biochemistry, 19,* 725 (1980).

68. M. T. Kamlay and J. D. Shore, *Arch. Biochem. Biophys., 222,* 59 (1983).

69. M. J. Gilleland and J. D. Shore, *Biochem. Biophys. Res. Commun., 40,* 230 (1970).

70. M. Frolich, D. J. Creighton, and D. S. Sigman, *Arch. Biochem. Biophys., 189,* 471 (1978).

71. C. Syvertsen and J. S. McKinley-McKee, *J. Inorg. Biochem., 22,* 43 (1984).

72. P. Andersson, J. Kvassman, B. Oldén, and G. Pettersson, *Eur. J. Biochem., 133,* 651 (1983).

73. J. Schmidt, J. Chen, M. DeTraglia, D. Minkel, and J. T. McFarland, *J. Am. Chem. Soc., 101,* 3634 (1979).

74. K. B. Taylor, *Biochemistry, 22,* 1040 (1983).

75. T. Young, V. Williamson, A. Taguchi, M. Smith, A. Sledziewski, D. Russell, J. Osterman, C. Denis, D. Cox, and D. Beier, in "Genetic Engineering of Microorganisms for Chemicals" (A. Hollaender, R. D. DeMoss, S. Kaplan, J. Konisky, D. Savage, and R. W. Wolfe, eds.), Plenum Press, New York, 1982, p. 335 ff.

76. C. Wills, *Nature, 261,* 26 (1976).

77. C. Wills and H. Jörnvall, *Nature, 279,* 734 (1979).

78. E. T. Young and D. Pilgrim, *Mol. Cell. Biol., 5,* 3024 (1985).

79. J. L. Bennetzen and B. D. Hall, *J. Biol. Chem., 257,* 3018 (1982).

80. D. W. Russell, M. Smith, V. M. Williamson, and E. T. Young, *J. Biol. Chem., 258,* 2674 (1983).

81. B. V. Plapp, A. J. Ganzhorn, R. M. Gould, D. W. Green, and A. D. Hershey, in "Enzymology and Molecular Biology of Carbonyl Metabolism", Vol. 3 (H. Weiner and T. G. Flynn, eds.), Alan R. Liss, New York, 1987, pp. 227-236.

82. J. P. Klinman, *J. Biol. Chem., 250,* 2569 (1975).

83. C. J. Dickenson and F. M. Dickinson, *Biochem. J., 147,* 303 (1975).

84. J. Kvassman and G. Pettersson, *Eur. J. Biochem., 103,* 565 (1980).

85. Y. Pocker and K. W. Raymond, in "Alcohol and Aldehyde Metabolizing Systems", Vol. 4 (R. G. Thurman, ed.), Plenum Press, New York, 1980, pp. 137-150.

86. Y. Pocker, K. W. Raymond, and W. H. Thompson III, in "Zinc Enzymes" (I. Bertini, C. Luchinat, W. Maret, and M. Zeppezauer, eds.), Birkhäuser, Boston, 1986, pp. 435-449.

87. H. Dutler, A. Ambar, and J. Donatsch, in "Zinc Enzymes" (I. Bertini, C. Luchinat, W. Maret, and M. Zeppezauer, eds.), Birkhäuser, Boston, 1986, pp. 471-483.

88. M. F. Dunn, A. K. H. MacGibbon, and K. Pease, in "Zinc Enzymes" (I. Bertini, C. Luchinat, W. Maret, and M. Zeppezauer, eds.), Birkhäuser, Boston, 1986, pp. 485-505.

89. Y. Pocker, H. Li, and J. D. Page, in "Advances in Biomedical Alcohol Research" (K. D. Lindros, R. Ylikahri, and K. Kiianmaa, eds.), Pergamon Press, Oxford, 1987, pp. 181-185.

90. Y. Pocker and S. C. DeRoy, in "Enzymology and Molecular Biology of Carbonyl Metabolism", Vol. 3 (H. Weiner and T. G. Flynn, eds.), Alan R. Liss, New York, 1987, pp. 179-187.

91. Y. Pocker, H. Li, and J. D. Page, in "Enzymology and Molecular Biology of Carbonyl Metabolism", Vol. 3 (H. Weiner and T. G. Flynn, eds.), Alan R. Liss, New York, 1987, pp. 217-225.

92. C. J. Dickenson and F. M. Dickinson, *Biochem. J., 161,* 73 (1977).

93. Y. Pocker, S. C. DeRoy, H. Li, and J. D. Page, *Genetic Physico-Chemical Approaches for Analysis of Biological Catalysts, 1,* 99 (1986).

94. H. Eklund, A. Jones, and G. Schneider, in "Zinc Enzymes" (I. Bertini, C. Luchinat, W. Maret, and M. Zeppezauer, eds.), Birkhäuser, Boston, 1986, pp. 377-392.

95. E. S. Cedergren-Zeppezauer, in "Zinc Enzymes" (I. Bertini, C. Luchinat, W. Maret, and M. Zeppezauer, eds.), Birkhäuser, Boston, 1986, pp. 393-415.

96. M. Zeppezauer, in "Zinc Enzymes" (I. Bertini, C. Luchinat, W. Maret, and M. Zeppezauer, eds.), Birkhäuser, Boston, 1986, pp. 417-434.

97. Y. Pocker, K. W. Raymond, and W. H. Thompson III, in "Zinc Enzymes" (I. Bertini, C. Luchinat, W. Maret, and M. Zeppezauer, eds.), Birkhäuser, Boston, 1986, pp. 435-449.

98. G. Pettersson, in "Zinc Enzymes" (I. Bertini, C. Luchinat, W. Maret, and M. Zeppezauer, eds.), Birkhäuser, Boston, 1986, pp. 451-464.

99. M. W. Makinen and W. Maret, in "Zinc Enzymes" (I. Bertini, C. Luchinat, W. Maret, and M. Zeppezauer, eds.), Birkhäuser, Boston, 1986, pp. 465-470.

100. A. K. H. MacGibbon, S. C. Koerber, K. Pease, and M. F. Dunn, *Biochemistry, 26,* 3058 (1987).

101. Y. Pocker and K. W. Raymond, *Alcohol, 2,* 3 (1985).

102. Y. Pocker, S. C. DeRoy, H. Li, and J. D. Page, *Proc. Symp. Protein Soc., 1,* 115 (1987).

10

Probing the Mechanism of Action of Carboxypeptidase A by Inorganic, Organic, and Mutagenic Modifications

David S. Auld, James F. Riordan, and Bert L. Vallee
Center for Biochemical and Biophysical Sciences and Medicine
Harvard Medical School
Seeley G. Mudd Building, 250 Longwood Avenue
Boston, MA 02115

1. INTRODUCTION

Chemical modifications have been employed for decades as a means to study enzyme mechanisms. Metalloenzymes are particularly suitable in this regard, since they can be modified both inorganically, i.e., by metal substitution, and organically, i.e., at amino acid side chains. In fact, substitution of different metals for that present in a native metalloenzyme could be viewed as the inorganic equivalent of the substitution of a specific, amino acid residue by site-directed mutagenesis. Organic modification benefits from an extensive repertoire of residue-specific reagents and since "active site" residues are often hyperreactive, it is possible at times to achieve modification with high selectivity. This can be augmented further by high-resolution chromatography, including affinity chromatography, to obtain a uniquely labeled product. This, incidentally, is by far the most direct and widely employed means to identify such active site residues.

Many organic modifications alter the chemical characteristics of an amino acid residue so drastically that it is difficult to interpret their effects on enzymatic activity unequivocally. However, the chemistry of some amino acids is exceptional in this regard. In the case of tyrosine, the properties of the phenolic hydroxyl group can be either perturbed by a wide range of ortho substitutions or abolished altogether by directly blocking the OH. Similar considerations pertain to the active site substitution of a metal with coordination or ligand exchange properties markedly different from those of the native metal ion. Importantly, chemical labeling can introduce a unique probe into the protein which can serve to signal critical details of both conformation and catalysis. Metal substitution can provide similar opportunities, and when the two are performed jointly the information potential is enhanced synergistically. Clearly, the entire range of site-directed mutagenic modifications now feasible enlarges the scope of such studies enormously.

In the following we will summarize and reexamine the effects of inorganic, organic, and mutagenic modifications on the specificity and mechanism of action of carboxypeptidase A (CPD-A). Highlights of past and present CPD-A research are presented here not only to draw mechanistic conclusions but to indicate progress, the approaches that have made it possible, and the current state. It is hoped that this assessment will facilitate extensions of past efforts and encourage further inquiry.

2. INORGANIC MODIFICATIONS

2.1. Effect of Metal Substitution on Activity

The metal atom is indispensible to the catalytic activity of CPD-A, which normally contains 1 g-at of zinc per molecular weight of 34,600 [1]. When zinc is removed, the enzyme is completely inactive, whereas when it or of a number of other divalent metal ions are readded the dual activities of the enzyme toward peptides and esters are restored to varying degrees [2]. The cobalt-substituted enzyme can be up to 10 times as active as the native zinc enzyme toward a particular peptide substrate, but the activities of these two enzymes toward esters are virtually the same. On the other hand, the activity of the manganese, cadmium, and mercury enzymes toward ester substrates is much greater than that toward the corresponding peptide substrates [2-4].

A comparison of the kinetic parameters for the hydrolysis of several tri- and tetrapeptides and their depsipeptide analogs by the Zn(II) and Cd(II) enzymes clearly establishes the origin of this

TABLE 1

Kinetic Parameters for Matched Ester and Peptide Pairs by Zinc and Cadmium CPD-A[a]

Substrate	Zinc enzyme		Cadmium enzyme	
	k_{cat}	$10^4 K_m$	k_{cat}	$10^4 K_m$
Bz-Gly-Gly-L-OPhe	30,000	3.3	34,000	79.0
Bz-Gly-Gly-Gly-L-OPhe	31,000	3.4	45,000	29.0
Dns-Gly-Gly-Gly-L-OPhe	11,000	0.25	28,000	2.2
Bz-Gly-Gly-L-Phe	1,200	10.0	41	8.0
Bz-Gly-Gly-Gly-L-Phe	2,600	37.0	86	41.0
Dns-Gly-Gly-Gly-L-Phe	4,200	8.0	400	8.1

[a]Assay conditions: 25°C, pH 7.5, 1.0 M NaCl [4]. Units of k_{cat} and K_m are min^{-1} and M, respectively.

differential response (Table 1). The increased esterase activity of the Cd enzyme relative to the Zn enzyme is reflected in slightly higher k_{cat} values [4]. However, the corresponding K_m values are 10- to 20-fold higher than those for the Zn enzyme. Exactly the opposite relationship is found for the peptide substrates. The decreased peptidase activity of the Cd enzyme is due to drastically reduced k_{cat} values, i.e., 10- to 30-fold, while the K_m values are unchanged.

Since the K_m values obtained for the Zn and Cd enzymes acting on peptide substrates are nearly identical, this implies that metal substitution does not affect peptide binding. Stopped-flow fluorescence assays of the N-dansylated peptide substrate, Dns-$(Gly)_3$-L-Phe, verify this conclusion [4]. This technique allows direct examination of the enzyme-substrate (ES) complex and the results demonstrate that the reduced activity of the Cd enzyme and the inactivity of the apoenzyme toward peptides is clearly due to an effect on catalysis, not to a lack of peptide binding.

Extending earlier studies we recently found Cu(II)CPD-A to exhibit from 0.5 to 2.5% (vs. oligopeptides) and 1.8 to 24% (vs. depsipeptides) of the activity (k_{cat}/K_m) of the zinc enzyme [5]. The apoenzyme substituted with alkaline earth metals and the remaining transition metal ions are completely inactive toward both peptides and esters [2]. Electron density difference maps show that—with the exception of Hg(II)—in all catalytically active metal derivatives of CPD-A the metal ion occupies the same site as Zn(II) [6].

2.2. Cobalt as a Spectral Probe of Metal Coordination

The spectroscopic properties of some substituent metal ions, particularly Co(II) [7,8], render them excellent probes of the active site of CPD-A. They can be used to study the interaction of the enzyme with substrates and inhibitors, and to monitor concomitant conformational changes in the course of catalysis.

Cobalt CPD-A is active both as a peptidase and an esterase. At 20°C its visible absorption spectrum has a shoulder near 500 nm and maxima at 555 and 572 nm, both with molar absorptivities of about 150 (Fig. 1). The extrema of the magnetic circular dichroism (MCD) spectrum at room temperature correspond to four absorption bands, at 478, 503, 537, and 577 nm (Fig. 1), that are resolved in the absorption spectrum at liquid helium temperature [9].

Comparison of spectral data of cobalt CPD-A and of other cobalt enzymes with those of cobalt complex ions give first indications of the features of their metal-protein interaction, revealing similarities and specifying characteristic differences [7]. The intensity of the cobalt CPD-A absorption spectrum indicates a low-symmetry metal environment [9,10]. Compared with Co(II) complexes of known geometry, the band intensities are greater than those normally associated with octahedral complexes [11], close to those reported for certain trigonal bipyramidal complexes [12], and within the range observed for tetrahedral complexes [13,14]. The band energies are high, consistent with strong metal-ligand interactions.

The overall band pattern of the cobalt CPD-A spectrum, especially as seen in MCD, is most similar to that of tetrahedral Co(II)

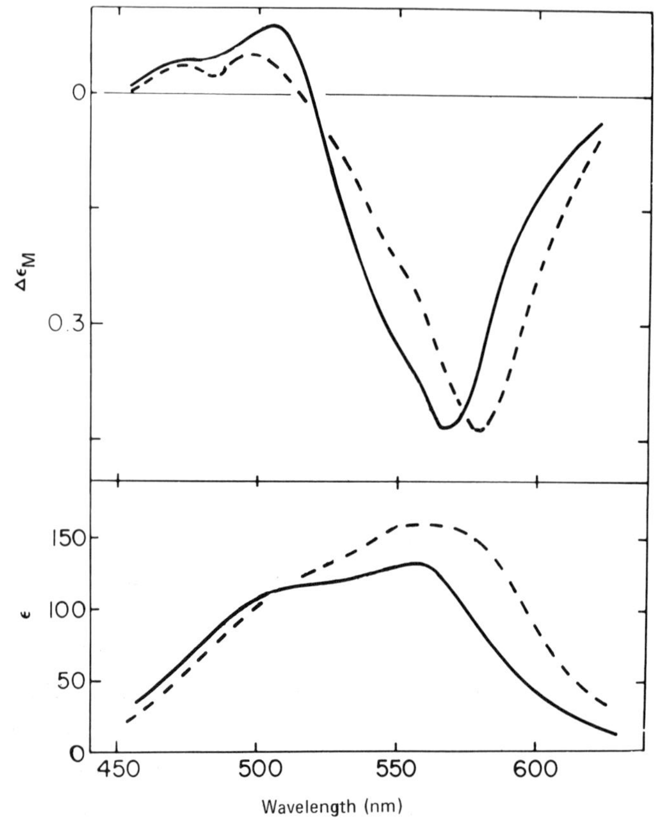

FIG. 1. Magnetic circular dichroic (above) and electronic absorption (below) spectra of cobalt(II) CPD-A in the absence (---) or presence (—) of Gly-L-Tyr [9].

complexes [14,15], though the energy span of the enzyme's visible spectrum is a few thousand cm^{-1} greater, conceivably due to asymmetrical coordination. In cobalt complexes that are basically tetrahedral, the near-infrared spectrum can be especially diagnostic of coordination geometry [16]. The two-banded pattern, 940 and 1510 nm ($\varepsilon \sim 20$) [9], split by about 4000 cm^{-1}, is consistent with a distorted coordination, possibly axial in nature.

The paramagnetism of the d^7 Co(II) ion allows characterization of its complexes by electron paramagnetic resonance (EPR) spectroscopy, and such spectra are particularly sensitive to distortions in the symmetry and geometry of the cobalt coordination sphere, which are reflected in the g values and degree of hyperfine splitting [17-19]. Jointly, the absorbance, MCD, and EPR spectra of cobalt CPD-A [9,20] have been thought to be most consistent with an asymmetrically coordinated cobalt atom, bound firmly in a distorted tetrahedrallike environment. This is in agreement with the x-ray structure of the zinc enzyme [21,22], which also indicates that the metal coordination is 4 or 5 depending on whether one or both oxygens of the γ-carboxyl group of Glu-72 are coordinated to the zinc [6,23]. It has been proposed that the active site of the enzyme is "poised for catalysis", its intended biological function, a situation referred to as the entatic state [24].

2.3. Cobalt as a Probe of Active Site Ligand Interactions

A variety of substances can bind to, and thereby perturb, the spectral properties of cobalt CPD-A, thus providing information about the nature of metal coordination in the enzyme. Enzyme inhibitors change the absorption and circular dichroic (CD) spectra characteristically. Acetate, butyrate, phenylacetate, β-phenylpropionate, and indole-3-acetate each generate characteristic spectral shifts depending on inhibitor concentration. In particular, β-phenylpropionate and phenylacetate, whose kinetic effects are complex, induce multiple spectral changes, apparently corresponding to different modes of inhibitor binding. At low concentrations these agents act noncompetitively toward peptide hydrolysis and induce only small spectral changes. At higher concentrations they are competitive inhibitors and they cause different, more extensive spectral shifts [9,25-27]. Moreover, under these latter conditions the absorption band at 590 nm becomes optically active. These effects occur at β-phenylpropionate concentrations comparable to those at which the agent is thought to displace metal-bound water from manganese carboxypeptidase, as judged by nuclear magnetic resonance (NMR) data [28]. The formation of an

EI_2 complex is consistent with the observation that β-phenylpropionate can protect two active site tyrosyl residues of carboxypeptidase A from acetylation [29].

Addition of Gly-Tyr to Co(II)CPD-A causes the two absorption maxima of the enzyme to coalesce into a single broad peak at 560 nm, with a prominent shoulder at 520 nm (Fig. 1). The 940-nm maximum splits into two bands, with peaks at 850 and 1150 nm, and the maximum at 1570 nm shifts to 1420 nm [9]. Slight changes in band shape are also observed in the MCD spectrum (Fig. 1). Similar but less extensive changes are induced by L-Lys-Tyr-NH_2 [9]; however, it is not hydrolyzed and unlike Gly-L-Tyr does not induce conformational changes in the crystalline enzyme [30].

The electronic absorption spectrum of cobalt CPD-A is optically active. The CD spectrum has a negative ellipticity band at 538 nm and a shoulder at about 500 nm (Fig. 2). Inhibitors and substrates

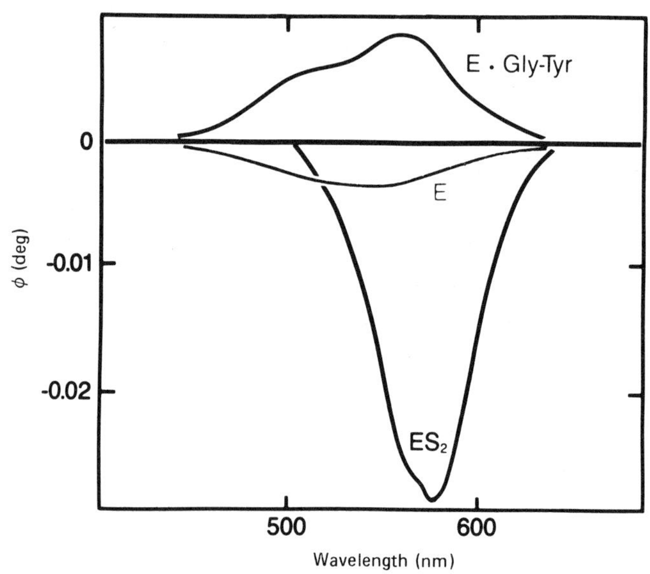

FIG. 2. Circular dichroic spectra of cobalt CPD-A (E), its Gly-Tyr complex [9], and the ES_2 intermediate of Z-Sarc-Ala-Phe, formed during equilibrium-trapping experiments [31].

also change this spectrum characteristically. Thus, Gly-Tyr induces *positive* CD bands at 505 and 550 nm [9] while the ES_2 intermediate for peptide substrates such as Dns-Ala-Ala-Phe and Z-Sarc-Ala-Phe induce a remarkably strong *negative* ellipticity band centered at 578 nm (Fig. 2) [31]. The corresponding MCD spectrum for the peptide ES_2 intermediate [31] suggests that in this state the metal coordination geometry is tetrahedrallike: its shape corresponds to that of many tetrahedral cobalt complex ions [32]. Given that the spectral perturbations for inhibitors and pseudosubstrates differ markedly from those observed with rapidly turned over peptide substrates, there is clearly a need for structural studies on ES complexes in order to correlate structure and function.

2.4. Cryospectrokinetic Studies

One approach to the structural and kinetic characterization of catalytic intermediates is to lengthen their lifetimes by examining enzymatic rates at subzero temperatures. Thus at -17°C, changes in the spectra of the cobalt enzyme can be followed during catalysis allowing direct visualization of the formation and breakdown of the intermediates of oligopeptides and their depsipeptide (ester) analogs [20,33].

The cobalt spectra of peptide and ester intermediates identify catalysis-related changes in the active site metal coordination sphere characteristic of each type of substrate. The spectra of the ES_2 intermediates generated with all peptides examined so far are clearly distinct from those found with esters (Fig. 3). Thus, the spectral features of ES_2 intermediates formed with peptides which differ in length, internal residues, sequences, N-terminal blocking groups, and C-terminal amino acids all have an absorption maximum at 570 ± 3 nm and a band of low intensity at 473 ± 6 nm. In contrast, the spectra of the ES_2 intermediates for the corresponding ester analogs have two maxima of unequal intensity at 598 ± 3 and 551 ± 2 nm as well as a prominent shoulder at 523 ± 3 nm [33].

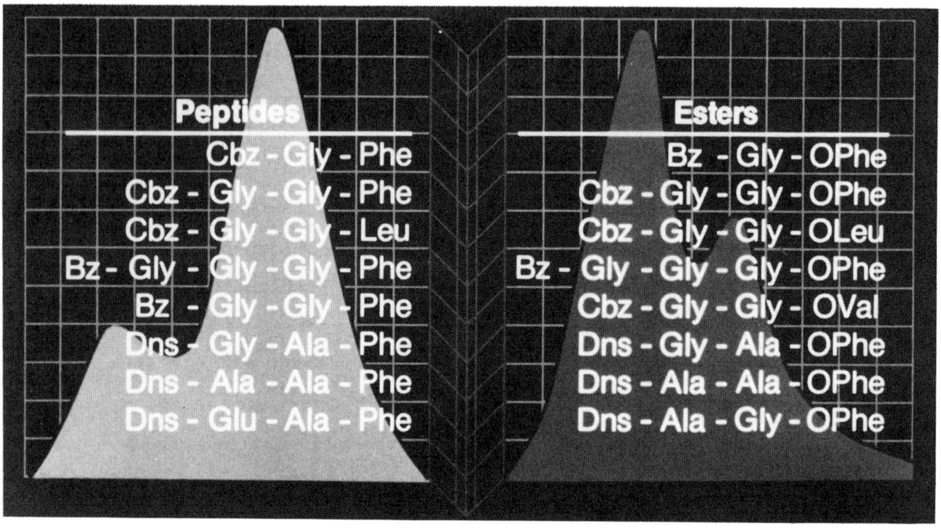

FIG. 3. Schematic representation of the cobalt absorption spectra of the intermediates generated with peptides (left) and esters (right).

The combined results of equilibrium trapping [31], cryoquench [34], and cryospectrokinetic [20,33,35] studies at zero to subzero temperatures indicate that in peptide ES_2 intermediates the scissile bond is not broken. Thus amide bond hydrolysis occurs during or after the rate-limiting step and is preceded by coordination of the substrate to the metal. On the other hand, in depsipeptide hydrolysis the susceptible ester bond is already cleaved in the ES_2 ester intermediate which is more properly denoted EP_1P_2 where P_1 and P_2 are the products of the forward reaction (or the substrates for the reverse reaction). The rate-limiting step for depsipeptides thus involves release of products which may occur directly or following a mandatory conformational change [20,31,33-35].

2.5. Metal-Anion Interactions

While a number of studies of the cobalt enzyme have demonstrated that binding of substrates, pseudosubstrates, and specific inhibitors alters the symmetry of metal binding to the enzyme, simple anions such as halides and others of the spectrochemical series either fail to perturb the cobalt spectrum at pH 7 or affect it only when added at relatively high concentrations [9]. The accessibility of the metal atom to large, specific substrate and inhibitor ligands together with its inaccessibility to small anions indicates an unusual, i.e., entatic, metal-binding site that is poised for ligand exchange or addition [24], yet also includes some elements of protein structure that protect the metal coordination sphere of CPD-A from disruption by simple anions.

Recent studies have shown that binding of the C-terminal products of peptide and ester hydrolysis, e.g., L-Phe and L-phenyl lactate (L-OPhe) and aromatic carboxylic acids, opens up the active site, allowing access of anions to the metal [36]. Binding of an anion and a carboxylate inhibitor to the enzyme is strongly synergistic. Thus, in the presence of L-Phe the K_D for azide is lowered 350-fold [36]. The absorption spectra of tight-binding ternary complexes are characteristic and indicate that the inorganic anion binds to the metal (Fig. 4). Azide/carboxylate inhibitor complexes display a typical charge transfer band near 310 nm with an ϵ near 1,200 M^{-1} cm^{-1} [36]. Substitution of a -H by -OH or $-NH_2$ (converting β-phenylpropionate to L-OPhe or L-Phe, respectively) promotes a striking increase in the molar absorptivities of the ternary azide complex d-d transitions but changes the band positions very little (Fig. 4). The increase in spectral intensity presumably reflects a minor steric rearrangement of the metal ligands induced by the somewhat bulkier hydroxyl and amino groups. Aside from their mechanistic implications, these findings are also of importance to anyone employing high concentrations of commercial preparations of the enzyme that have not been purified by affinity chromatography. These preparations of CPD-A can contain proteases that can hydrolyze the apoenzyme to generate free amino acids which then interfere with spectral studies subsequent to reconstitution of the apoenzyme with cobalt [37].

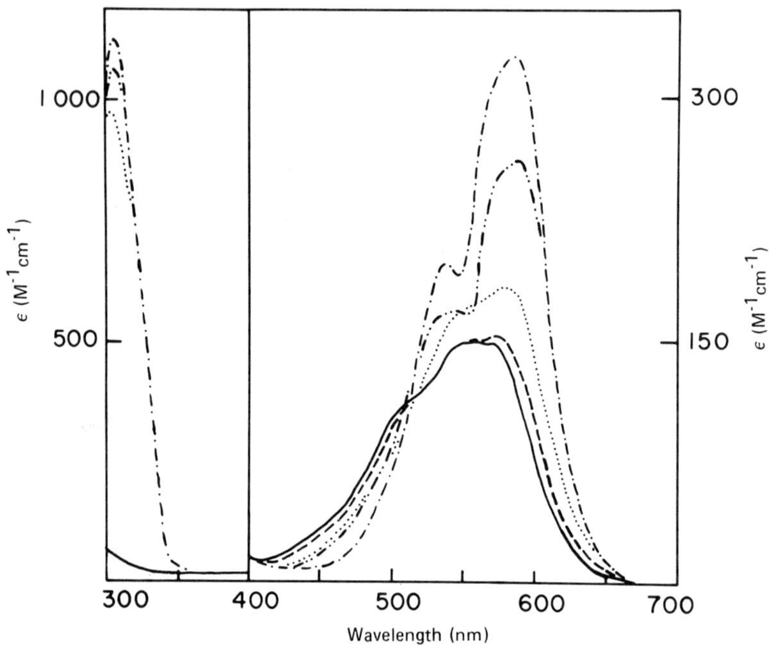

FIG. 4. Electronic absorption spectrum of cobalt(II) CPD-A alone (——), in the presence of azide (---), and their ternary complexes with L-phenylalanine (-·-·), L-β-phenyllactate (-···-), and β-phenylpropionate (····). (Reproduced from Ref. 36 by permission of the American Chemical Society.)

2.6. Cobalt(II)-^{13}C Interactions

The paramagnetic properties of cobalt have served to gain information on the carboxylate-binding site of L-Phe and other carboxylate inhibitors in the absence and presence of azide [38,39]. Thus, the interaction of the ^{13}C-enriched carboxylate of L-Phe with the unpaired electron of the cobalt enzyme generates a NMR spectrum which exhibits significant line broadening when compared to that of the zinc enzyme. Such an interaction can be dipolar and is discernible up to a carbon-cobalt(II) distance of 0.8 nm [40]. Upon binding of ^{13}C-labeled L-Phe to the cobalt enzyme the longitudinal and transverse relaxation rates, T_{1P}^{-1} and T_{2P}^{-1}, are enhanced significantly compared to the zinc

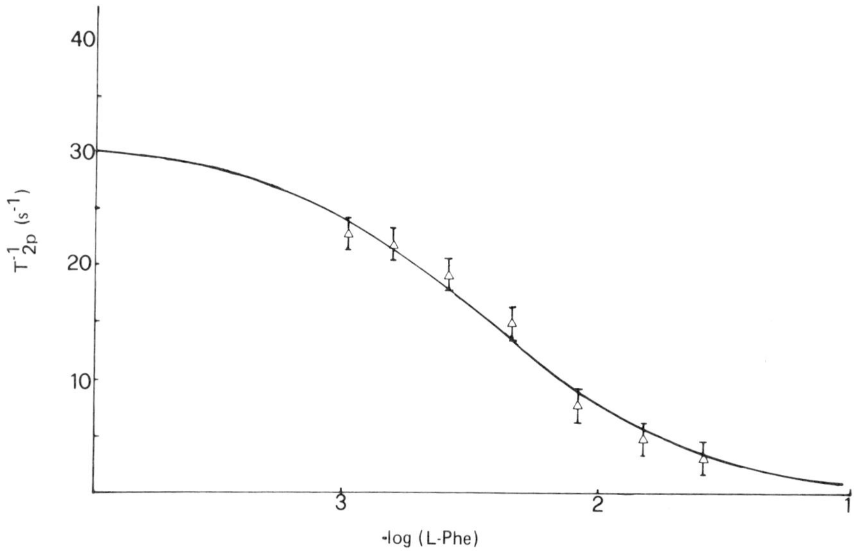

FIG. 5. T_{2P}^{-1} values of the carboxylic ^{13}C nucleus of L-Phe in the presence of M Co(II)CPD-A as a function of L-Phe concentration [38]. (Reproduced from Ref. 38 by permission of Elsevier Science Publishing Co. Inc.)

enzyme, allowing both determination of an affinity constant for inhibitor binding K and calculation of the Co(II)-$^{13}COO^-$ distance. The L-Phe concentration dependence of T_{2P}^{-1} yields an affinity constant of 290 ± 60 M^{-1} (Fig. 5). The calculated distances of Co-^{13}C for the enzyme-Phe and the enzyme-Phe-N_3^- complexes are 0.39 ± 0.04 and 0.42 ± 0.04 nm [38]. Both values are too great for direct coordination of the carboxylate group to the metal atom. These results in conjunction with those of electronic absorption studies [36] on these complexes lead to the conclusion that N_3^-, but not the amino acid carboxylate, is bound to the metal. In addition, ^{13}C NMR studies of ^{13}C-enriched carboxylates of acetate and phenylacetate have shown that the paramagnetic cobalt atom in the enzyme strongly broadens the ^{13}C signal in the enzyme-di-inhibitor (EI_2) ternary complex [39]. The degree of line broadening in the fully formed complex is typical of the coordination of a carboxylate group to the metal. On the other hand, the Co-$^{13}COO^-$ line broadening in the (EI) binary complex indicates that the carboxylate group binds 0.5-0.7 nm from the metal, a distance consistent with its interaction with one or more arginyl residues, e.g., Arg-145, Arg-127, or Arg-71 [39].

Recent rapid-scanning cryospectrokinetic studies of cobalt CPD-A have identified and characterized inhibitor-enzyme-substrate (IES) ternary complexes formed by the interaction between the enzyme, a peptide substrate, and a noncompetitive inhibitor [41]. Generation of a cobalt absorption spectrum distinct from that induced by the peptide or inhibitor alone signals the formation of the IES complex. Remarkably, the cobalt spectra of the IES complexes match those previously found for the steady-state intermediates of the cobalt enzyme with ester (depsipeptide) substrates alone [20,33]. As indicated above, chemical quenching studies have demonstrated that in the ester intermediates the scissile bond is broken [34]. These findings imply that a peptide and a noncompetitive inhibitor of its hydrolysis occupy the same binding loci as the products of depsipeptide hydrolysis. NMR studies with suitably ^{13}C-labeled inhibitor and/or peptide carboxyl groups should help to determine whether in these complexes it is the carboxyl group of the inhibitor or the substrate that is bound to the metal.

2.7. Cobalt as a Probe of pH-Dependent Processes

The absorption spectrum of Co(II)CPD-A is altered by lowering pH. Since high concentrations of salt (typically ≥1 M NaCl) are required to solubilize sufficient enzyme for these studies, it seemed likely that the spectral perturbation might reflect chloride binding. Indeed, the spectral differences at mildly acid pH values (e.g., 5.2) in the presence of Cl^-, Br^-, and NO_3^- have been shown to be due to an acquired anion sensitivity effect [42]. Chloride displaces water from the active site zinc atom at low pH due to protonation of Glu-270 as measured by ^{35}Cl NMR [43]. These data have been inferred to show that water remains coordinated to zinc throughout the pH region from 5 to 10, and that its stability with respect to chloride displacement is related to the hydrogen-bonding capacity generated by ionization of Glu-270. An equivalent interpretation would be that a protonated Glu-270 is hydrogen-bonded to metal-bound hydroxide, thus accounting for the stability [3].

The alkaline pH dependence of the cobalt enzyme's peptidase activity and its spectral changes at 625 nm are superimposable [9], indicating that a single group with a pK_a of 8.8 may be responsible for both effects. In view of the results of a number of studies on Tyr-248-modified enzymes (see Secs. 3.4 and 4.3), this alkaline pH dependence seems to reflect the state of ionization of this residue and its proximity to the active site metal. If the ionization $EH \rightleftharpoons E + H^+$ were indeed to reflect the ionization of $M(L_n)(H_2O) \rightleftharpoons M(L_n)(OH^-) + H^+$, the acidity of such a coordinated water molecule, judging from complex metal ions, might be a function of the metal to which it is coordinated. However, the values of pK_{EH} for zinc, cobalt, and manganese CPD-A are essentially identical and, hence, inconsistent with such a postulate [44].

2.8. Exchange Inert Co(III) as a Probe of Catalysis

The earliest mechanistic proposals for CPD-A, and indeed almost all succeeding proposals, envision that the active site zinc atom serves as a Lewis acid polarizing the carbonyl group of the susceptible peptide bond of the substrate, thereby making it more labile to hydrolysis [45]. The metal has also been thought to provide a coordinated OH^- that could serve as a nucleophile attacking the substrate carbonyl carbon in the catalytic process, a so-called metal-hydroxide mechanism. The existence of such mechanisms have been demonstrated to occur in the hydrolysis of amide or ester bonds in model compounds [46-48] and, indeed, to be far more effective than the zinc-carbonyl mechanism [46]. The metal-hydroxide acyl intermediate that forms in these reactions has been referred to as a metal-carboxyl anhydride and would be indistinguishable kinetically from an anhydride involving Glu-270 [47].

Since the Co(III) enzyme is inactive, even for a single turnover, the metal-hydroxide mechanism, as generally defined, cannot pertain in the case of CPD-A [49]. Further, ligand exchange in the inner coordination sphere of the metal must be an essential step in the hydrolysis of substrates. It might be argued that the change in geometry—in this case to octahedral on oxidation of Co(II) to Co(III)—would reorient the metal-bound hydroxyl group relative to the substrate, thereby preventing hydrolysis. However, metal substitutions resulting in a range of coordination geometries all generate active carboxypeptidases [26]. These include Zn(II), which is either distorted four-coordinate- [21,22] or five coordinate-like [23]; Co(II), distorted tetrahedral- or five-coordinate-like [32,50-52]; Ni(II), octahedral [51]; and VO(IV), square-pyramidal [53]. In addition, recent crystallographic studies have placed zinc and nickel at opposite ends of a series of increasingly square-pyramidal coordination geometries in CPD-A derivatives in the order Zn, Co, Mn, Cd, and Ni [6]. On the other hand, the results of Co(II) oxidation to Co(III) do not rule out a metal-hydroxide nucleophilic reaction as part of a two-step catalytic mechanism involving initial inner sphere coordination of the substrate to the metal.

3. ORGANIC MODIFICATIONS

3.1. Primary Sequence

Of the 306 amino acid residues of bovine CPD-A [54] only 86 residues are totally conserved when compared to bovine [55] and crayfish CPD-B [56], rat CPD-A1 [57], and *Streptomyces griseus* CPD [58] (Table 2). The zinc-binding ligands, His-69, Glu-72 and His-196 (bovine CPD-A numbering) are among the invariant amino acid residues which include several that have been assigned binding and catalytic roles, i.e., Glu-270, Tyr-248, Tyr-198, Arg-127, Arg-145, and Phe-279. A number of other tyrosines (12, 165, 204, and 265) and arginines (71, 124, and 130) are also fully conserved. Many of these residues have not as yet been assigned any specific role in the function of the enzyme.

TABLE 2

Comparisons of Primary Structures of Carboxypeptidases

	10			
1.	MDWT	SYHDYDEINA	WLDSLATDYP	ELASVEDVGL
2.	---TTGHSYE	KYNNWETIEA	WTEQVASENP	DLISRSAIGT
3.	ARSTNTFNYA	TYHTLDEIYD	FMDLLVAEHP	QLVSKLQIGR
4.	ALSTDSFNYA	TYHTLDEIYE	FMDLLVAEHP	QLVSKIQIGN
5.	DFPPADS	RYHNYAEMNA	AIDARIAANP	SIMSKRVIGK
	50			
	SYEGRTMKLL	KLGKGGADKP	IIFIDGGIHA	REWIAPSTVT
	TFLGNTIYLL	KVGKPGSNKP	AVFMDCGFHA	REWISPAFCQ
	SYEGRPIYVL	KFSTGGSNRP	AIWIDLGIHS	REWITQATGV
	TFEGRPIHVL	KFSTGGTNRP	AIWIDTGIHS	REWVTQASGV
	TYQGRDVIAV	K		
		100		
	YIVNEFVSN-	--SATYDDIL	SNVNFYVMPT	INPDGYAYTF
	WFVREAVRTY	GREIHMTEFL	DKLDFYVLPV	VNIDGYIYTW
	WFAKKFTENY	GQNPSFTAIL	DSMDIFLEIV	TNPNGFAFTH
	WFAKKVTKDY	GQDPTFTAVL	DNMDIFLEIV	TNPDGFAYTH
			150	
	TDDRLWRKTR	SETGSVLGCKG	ADPNRNWSFH	WDEVGASDSP
	TTNRMWRKTR	STRAGSS-CTG	TDLNRNFDAG	WCSIGASNNP
	SENRLWRKTR	SVTSSSL-CVG	VDANRNWDAG	FGKAGASSSP
	KTNRMWRKTR	SHTQGSL-CVG	VDPNRNWDAG	LGKAGASSNP
				200
	CSDIYAGPEP	FSEVEMRNVR	DQILEYAANIK	VYLTFHSYSQ
	CSETYCGSAA	ESEKESKAVA	DFIRNHLSSIK	AYLTIHSYSQ
	CSETYHGKYA	NSEVEVKSIV	DFVKNH-GNFK	AFLSIHSYSQ
	CSETYRGKFP	NSEVEVKSIV	DFVTSH-GNIK	AFISIHSYSQ
	LWMYPWGFTS	DLPDDWQDLD	TLATNAVDAL	TAVHGTRYEI
	MMLYPYSYDY	KLPKNNVELN	TLAKGAVKKL	ASLHGTTYSY
	LLLYPYGYTT	QSIPDKTELN	QVAKSAVAAL	KSLYGTSYKY
	LLLYPYGYTS	EPAPDQAELD	QLAKSAVTAL	TSLHGTEFKY
	250			
	GSSTNTIYAA	AGGSDDWAKG	EGGVKYAYTIE	LRDTGNYGFL
	GPGATTIYPA	SGGSDDWAYD	QG-IKYSFTFE	LRDKGRYGFV
	GSIITTIYQA	SGGSIDWSYN	QG-IKYSFTFE	LRDTGRYGFL
	GSIIDTIYQA	SGSTIDWTYS	QG-IKYSFTFE	LRDTGLRGFL
		300		
	LPENQIIPTG	EETFEGV-KVV	ANFVKDTYS	
	LPESQIQPTC	EETMLAI-KYV	TSYVLEHL-	
	LPASQIIPTA	QETWLGVLTIM	EHTVNN---	
	LPASQIIPTA	EETWLALLTIM	DHTVKHPY-	

1, Crayfish B [56]; 2, bovine B [55]; 3, bovine A [54]; 4, rat A1 [57]; 5, *S. griseus* [58]. Common residues are underlined.

3.2. Glutamic Acid Modification

Modification of Glu-270 either by carbodiimide or alkylating reagents completely abolishes both peptidase and esterase activities [59,60] (Table 2). Glu-270 has been viewed variously as a general base promoting attack of water at the scissile carbonyl or as a nucleophile forming a covalent anhydride with the acyl segment of a substrate. Neither role has been established satisfactorily.

Reaction of Glu-270 with the water-soluble carbodiimide 1-cyclohexyl-3-(2-morpholinoethyl)carbodiimide metho-p-toluenesulfonate confers anion sensitivity on the cobalt enzyme's absorption spectrum at neutral pH leading to the reasonable inference that Glu-270 contributes importantly to the anion insensitivity of the native enzyme [42,43]. It could do this either by serving as an electrostatic barrier to approaching anions or by interacting directly with the metal-bound water through a hydrogen bond. The latter possibility is particularly attractive in terms of a plausible catalytic mechanism and has been offered as a rationalization for other observations. It is consistent with the 4.5-Å distance between the carboxylate of Glu-270 and the metal [61], and with calculations of coordination geometry from perturbed angular correlation spectra on the Cd(II) enzyme and angular overlap theory [62]. The nuclear quadrupole interaction parameters for the various molecular species detected by this technique indicate that two of these species differ by the equivalent of a half unit of charge on the solvent molecule that functions as a metal ligand. This has been accounted for in terms of a proton bridge shared between this metal-coordinated water molecule and an ionized carboxyl group [62].

Kinetic studies are also consistent with this assignment. The acid limb of the pH vs. k_{cat}/K_m profiles is perturbed by metal substitution and shifts from 6.1 for the zinc enzyme to 5.3 and 6.4 for cobalt and manganese carboxypeptidases, respectively [63]. This suggests that the acid pK_a is due to the metal-bound H_2O molecule. Its kinetically determined heat of ionization, 3.1-3.4 kcal/mol, is not that expected for a carboxyl group [27,44]. The pH-k_{cat} profiles for the zinc and cobalt enzyme catalyzed hydrolysis of Bz-Gly-Gly-Phe are characterized by pK_a values of 6.3 and 5.6, respectively, and a ΔH_i value of 6.8 kcal/mol [44]. Similar activation energies have been observed for this pK_a in kinetic studies of ester hydrolysis at low temperature [64]. Values of ΔH_i in the range of 6-8 kcal/mol have been reported for the ionization of the water ligand in five-coordinate metal complexes of models [65]; a further decrease in the coordination number would be expected to strengthen the metal-ligand bonding interactions and facilitate proton ionization [66]. Alternatively, since the metal-bound water and the carboxylate of Glu-270 are within hydrogen-bonding distance of one another, the pK_a at 6.0 could be regarded as a cooperative ionization of these two groups.

3.3. Tyrosine Modification

CPD-A was the first enzyme shown to have a functional tyrosyl residue at its active center. O-Acetylation of the enzyme with acetic anhydride or N-acetylimidazole [67,68] substantially alters the steady-state kinetic behavior of the enzyme acting on the "classical" peptide and ester substrates Z-Gly-Phe and Bz-Gly-OPhe, respectively. Peptidase activity is markedly reduced while esterase activity is increased more than sixfold, the latter due to an effect on ester binding. Both of these reagents acetylate approximately five tyrosyl residues. However, the inhibitor β-phenylpropionate prevents the functional effects of acetylation and also protects two of the tyrosyl residues from being modified. Treatment of the acetylated enzyme with a nucleophile such as hydroxylamine restores native activity concomitant with the deacylation of two O-acetyltyrosyl residues.

Iodination, succinylation, photooxidation, and several other reactions also alter tyrosyl residues in CPD-A with analogous catalytic consequences [29]. In most instances, however, the number of residues modified was not examined. Low concentrations of diazotized arsanilic acid (DAA), 5-diazo-^{1}H-tetrazole (DHT), or tetranitromethane (TNM) modify mainly a single tyrosine of the enzyme, Tyr-248 [69-72], but the functional responses to these modifications differ.

Remarkably, all of these ortho-substituted Tyr-248 derivatives exhibit either elevated or unaltered esterase activity, while peptidase activity is altered to different extents (Table 3). DHT has a minimal effect on peptidase activity [69], DAA lowers it to approximately 60% [70] and nitration with TNM to about 15% [71]. Prior acetylation completely prevents the functional effects either of azo coupling or nitration [69]. Hence, the tyrosines modified by DAA, DHT, and TNM are among those that are acetylated. However, since the former reagents modify but one tyrosine, these data do not establish whether one or more tyrosines participate in activity.

Nitration of the enzyme that has been coupled previously with DHT does not affect the esterase activity of this azoenzyme but it decreases its peptidase activity to that characteristic of nitrocarboxypeptidase, 15% (Table 3). The product contains one azo- and one nitrotyrosyl residue per molecule [69]. Clearly, modification of either tyrosine can affect the activity of CPD-A.

A variety of means have been employed to identify the specific tyrosines modified by these different reagents. Immunoaffinity chromatography has been used to identify Tyr-248 as the residue specifically modified by DAA and TNM, respectively [70,71]. This same technique was used to show that DHT also modifies Tyr-248 but that subsequent nitration affects Tyr-198 [69]. A double-labeling procedure with ^{125}I and ^{131}I was helpful in identifying Tyr-248 as one of the residues susceptible to iodination [72].

3.4. Selective Modification of Tyr-248

The selective modification of Tyr-248 by all of these reagents appears to be related to its location in the three-dimensional structure of the enzyme. The proximity of the phenolic hydroxyl group to a positive charge—contributed either by the metal atom or the guanidino group of Arg-145 in solution or possibly His-120

TABLE 3

Chemical Modification of Tyrosine and Glutamic Acid Residues of CPD-A

Residue-modified Bz-Gly-OPhe	Chemical agent	Activity (% native enzyme) Peptidase Z-Gly-Phe	Activity (% native enzyme) Esterase
Tyrosine			
Tyr (probably includes 248)	N-Acetylimidazole	<2	600
Tyr (probably includes 248)	Acetic anhydride	<2	600
Tyr-248 + other Tyr	Iodine	<2	500-600
Tyr-248	5-Diazo-^{1}H-tetrazole	100	200
Tyr-248	Tetranitromethane (TNM)	15	180
Tyr-248	Diazotized arsanilic acid	60	100
Tyr-248 and Tyr-198	5-Diazo-^{1}H-tetrazole + TNM	15	200
Tyr-248	Diazotized arsanilic acid + [Co(III)(EDDA)]	0	200
Glutamic Acid			
Glu-270	Cyclohexylmorpholinoethyl carbodiimide (CMC)	0	0
Glu-270	N-Bromoacetyl-N-methyl-L-Phe	0	0
Glu-270	Woodward's Reagent K	0	0

of a neighboring molecule in the crystal lattice—would lower its pK_a and thereby enhance electrophilic substitution in the ortho position. In a like manner, nitration of Tyr-198 may specifically occur in DHT-modified CPD-A but not in the native enzyme, perhaps because a conformational change induced by azo coupling brings Arg-71 [69] or Arg-127 close to Tyr-198.

Azo-Tyr-248 CPD-A has the following desirable features essential for mechanistic studies [73,74]: (1) it is *highly active* toward esters and peptides, (2) it contains a site specific probe whose spectra can reflect azo-Tyr-248 itself as well as an intramolecular metal complex between Zn and azo-Tyr-248, and (3) the rate constant for dissociation of the azo-Tyr-248-Zn complex, 64,000 sec^{-1}, is orders of magnitude faster than the catalytic step itself (0.01-500 sec^{-1}). (The failure of previous studies to detect this rapid process led to erroneous conclusions [75].) On stopped-flow mixing of the enzyme with products or rapidly turned over substrates at pH 8.5, the azo-Tyr-248-Zn complex is disrupted to form azotyrosine within the 3-msec flow time, i.e., long before significant hydrolysis can occur. The results of these dynamic studies specify a course of catalytic events in solution different from that inferred from x-ray structure analysis of crystals. In solution, upon binding of substrate Tyr-248 is displaced outward, away from the zinc [73], not inward, as postulated on the basis of crystallography [21,22,76].

The position of Tyr-248 in the crystal structure of the CPD-A• Gly-Tyr complex has led to the suggestion that this residue serves as a proton donor in the catalytic mechanism [21,22,61,76]. The DAA modified Tyr-248 enzyme can be converted into a Co(III) (ethylenediamine-N,N'-diacetato) derivative [77]. This modification blocks the phenolic oxygen of Tyr-248 eliminating it as a proton donor. The modified enzyme is completely devoid of peptidase activity, but still binds peptides. In addition it has either markedly reduced k_{cat} values or higher K_m values toward ester substrates. It was suggested [77] that the loss of peptidase activity is due to the inability of the tyrosine to act as a proton donor in peptidase catalysis as originally inferred from structural analysis of the crystalline enzyme [21,22,61,76], but this has since been shown not to be so (see below).

The kinetically determined heat of ionization for the group with a pK_a of 9, 5.5 kcal/mol, is essentially that expected for a phenol [44]. The results of nitration provide strong evidence that this group is a tyrosyl residue. Thus, using TNM, Tyr-248 can be nitrated under mild conditions [71] and reduced subsequently to the aminophenol with hydrosulfite. The chromophoric properties of ortho-substituted tyrosine derivatives allow spectral determination of the phenolic pK_a. The effect of pH on the kinetic parameters can therefore be related directly to the state of ionization of this group. Thus, the alkaline side of the pH profile of a CPD-A derivative containing 1.67 nitrotyrosines is shifted to 6.6 [78].

The role of Tyr-248 in catalysis has been examined at the hands of the nitrated enzyme. A purified, homogeneous mononitro-Tyr-248 derivative has been obtained by affinity chromatography [27]. Its pH-rate profiles indicate that the limiting value of k_{cat} is fivefold less than that of the native enzyme but the limiting value of K_m is essentially unchanged [27], (Table 4). The pH k_{cat}/K_m profile for Dns-Ala-Ala-Phe hydrolysis by NO_2-Tyr-248 CPD-A, markedly different

TABLE 4

Kinetic and Spectral Properties of Tyr-248-CPD Derivatives[a]

	Kinetic					Spectral
Enzyme	k_{cat}	K_m	k_{cat}/K_m	pK_{EH_2}	pK_{EH}	pK_a
Native	95	19	5.0	6.2	8.9	—
NO_2-Tyr-248-CPD	17	14	1.2	6.2	7.1	7.0
NH_2-Tyr-248-CPD	35	30	1.2	6.2	9.2	9.4

[a]Dns-Ala-Ala-Phe, 1 M NaCl, 20°C. Units for pH independent values of k_{cat}, K_m, and k_{cat}/K_m are, respectively, sec^{-1}, μM, and $\mu M^{-1}sec^{-1}$ [27].

from that of the native enzyme, is characterized by an alkaline pK_a of 7.1 which corresponds to the spectrally determined NO_2-Tyr pK_a of 7.0 (Table 4). Hydrosulfite reduction of NO_2-Tyr-248 to NH_2-Tyr-248 restores the kinetic alkaline pK_a value to 9.2, and shifts the spectral pK_a to 9.4 (Table 4). The near identity of the alkaline pK_a values for the nitro- and amino enzyme's pH-activity profile, 7.1 and 9.2, respectively, to the corresponding spectrally determined pK_a values of 7.0 and 9.4, respectively, indicates that the alkaline pK_a reflects the ionization state of Tyr-248 in the modified enzymes and, by inference, links the native enzyme's alkaline pK_a (8.9) to Tyr-248. These studies further indicate that the consequences of ionization of Tyr-248 result in loss of peptide binding to the enzyme.

The k_{cat}-pH profile for Dns-Ala-Ala-Phe does not reflect ionization of the phenolic group. Since the rate-determining step for peptide hydrolysis involves cleavage of the scissile bond [34], the pH-independent behavior of k_{cat} indicates that Tyr-248 acts neither as a general acid nor as a general base catalyst in this step. Thus, mechanistic proposals featuring Tyr-248 acting as a general acid catalyst [21,22,61,76,77] are of dubious significance (see below).

The precise participation in the mechanism of Tyr-198, the only other tyrosyl residue near the active site, is also questionable. X-ray crystallographic studies have assigned it a binding function: involvement in secondary substrate-binding modes and responsibility for substrate activation and inhibition. Thus far it has not been possible to chemically modify this tyrosine specifically without prior or simultaneous modification of Tyr-248 [69]. Hence, this hypothesis has not been tested experimentally.

4. SITE-DIRECTED MUTAGENIC MODIFICATIONS

Site-directed mutagenesis is a powerful means to study the role of specific amino acid residues in the mechanism of enzyme action. This approach has recently been used specifically for rat CPD-A1 [57,79] to investigate the role of the phenolic hydroxyl groups of Tyr-248 and Tyr-198 in catalysis.

4.1. Comparison of Primary Structure and Kinetic Properties of Rat CPD-A1 and Bovine CPD-A

Rat CPD-A cDNA clones have been isolated from a cDNA library prepared from rat pancreatic mRNA. Thus far two pancreatic carboxypeptidases have been detected by this means [80] but the sequence information of only one of them, A1, has been reported [57]; it displays 78% homology with that of the bovine enzyme. The amino acids considered important to substrate binding and catalysis as well as the metal-binding ligands are all totally conserved (Table 2). However, even with such high homology the two enzymes exhibit kinetic differences which are readily apparent. The rat A1 enzyme catalyzes the hydrolysis of Bz-Gly-Phe, Z-Gly-Gly-Phe, and o-(trans-p-chlorocinnamoyl)-β-phenyl lactate (ClCin-OPhe) with 5- to 20-fold lower K_m and k_{cat} values compared to the bovine enzyme [79]. For the rat A1 enzyme substrate inhibition occurs at a Z-Gly-Gly-Phe concentration of approximately 10 times K_m compared to 100 times its K_m for the bovine enzyme [26,79].

4.2. Catalytic Properties of Rat (Phe-198) and (Phe-248) CPD-A1

Site-directed mutagenesis of the rat CPD-A1 cDNA followed by expression of the mutagenically altered template has led to the individual and selective or concurrent replacements of Tyr-248 and Tyr-198 by Phe, providing direct means to test the essentiality of their phenolic hydroxyl groups in catalysis [79]. The Tyr codon (TAT) was converted to Phe (TTT) within the rat CPD-A cDNA pCO1260 by oligonucleotide-directed mutagenesis [81,82]. For the three substrates examined, Z-Gly-Gly-Phe, BzGly-OPhe, and ClCin-OPhe, the kinetic

TABLE 5

Kinetic Parameters for Wild-Type Rat-CPD-A1 and Its Mutant (Phe-198) and (Phe-248) A1 Forms[a]

Substrate	Wild	Phe-198-		Phe-248-		Phe-248 Phe-198-	
				k_{cat} (sec^{-1})			
Z-Gly-Gly-Phe	50	34	(68)	18	(36)	13	(26)
Bz-Gly-OPhe	1040	1050	(100)	1090	(110)	1250	(120)
ClCin-OPhe	14	2.7	(19)	3.3	(24)	0.68	(4.9)
				K_m (μM)			
Z-Gly-Gly-Phe	32	35	(110)	160	(500)	220	(690)
Bz-Gly-OPhe	99	120	(120)	105	(110)	140	(140)
ClCin-OPhe	26	17	(66)	230	(880)	200	(770)
				$k_{cat}/K_m \times 10^{-5}$ (M^{-1} sec^{-1})			
Z-Gly-Gly-Phe	16	9.9	(62)	1.1	(6.9)	0.57	(3.6)
Bz-Gly-OPhe	105	90	(86)	104	(99)	87	(83)
ClCin-OPhe	5.2	1.6	(31)	0.15	(2.9)	0.034	(0.65)

[a]Assay conditions: 50 mM Tris, 0.5 M NaCl, 25°C. The values in parentheses are the percent of the kinetic parameter ratio of mutant CPD-A1 to wild type CPD-A1 [83,84].

parameters of the Phe-198 mutant enzyme are changed only slightly [84]. The greatest change occurs for ClCin-OPhe where k_{cat} decreases to 19% of that of the native enzyme (Table 5). For Bz-Gly-OPhe there is essentially no change in activity while the k_{cat} value for Z-Gly-Gly-Phe decreases to 68% of the native enzyme. These results are very similar to those obtained with the Tyr-248-DHT, Tyr-198-nitrated modified enzyme [69] (Table 3). The nitration of Tyr-198 does not change the activity of the enzyme toward Bz-Gly-OPhe but decreases that toward Z-Gly-Phe to 15% of the native enzyme (Table 3). The dependence of k_{cat} on pH over the range 5-9 for the rat (Phe-198) (pK_a of 6.0) enzyme-catalyzed hydrolysis of Z-Gly-Gly-Phe is very similar to those of the wild-type rat (pK_a of 5.7) and bovine (pK_a of 5.9) enzymes (Fig. 6) [63,84].

The changes in kinetic parameters for the mutant Phe-248 enzyme are slightly more marked [83] (Table 5). For Z-Gly-Gly-Phe

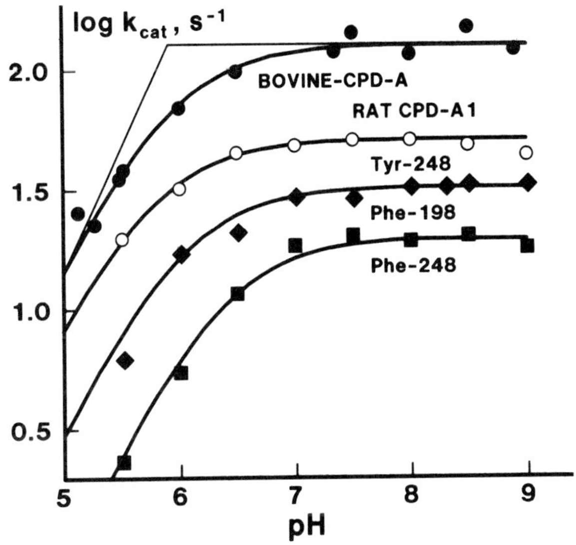

FIG. 6. The pH dependence of k_{cat} for the bovine CPD-A (●), wild-type rat A1 (○), rat (Phe-198) A1 (♦), and rat (Phe-248) A1 (■) enzymes catalyzed hydrolysis of Z-Gly-Gly-Phe [63,83,84].

and ClCin-OPhe the k_{cat} values of the Tyr-248 and Phe-248 rat enzymes are decreased three- and fourfold, respectively, but those for Bz-Gly-OPhe are virtually identical (Table 5). For Bz-Gly-OPhe the k_{cat}/K_m values of the Phe-248 mutant remain essentially unchanged, but for Z-Gly-Gly-Phe and ClCin-OPhe hydrolysis they are decreased to 6.9 and 2.9% of the native enzyme, respectively (Table 5). The overall kinetic results of this mutagenic modification are very similar to those for the chemical modification of Tyr-248 with an ortho nitro or a diazoarsanilic acid group [70,71]. In the latter cases activity toward Bz-Gly-OPhe increases twofold while peptidase activity decreases from 15 to 60% of the native enzyme (Table 3). The dependence of k_{cat} values on pH for the Phe-248 mutant-catalyzed hydrolysis of Z-Gly-Gly-Phe hydrolysis are again very similar to the wild-type rat and bovine enzymes [63,83] (Fig. 6). In fact, the maximal k_{cat} values for the rat wild-type and Phe-248 mutants differ by the same factor, 2.5-fold, as do those for the bovine CPD-A and rat wild-type enzymes (Fig. 6).

These results reiterate that the phenolic hydroxyl moiety is not obligatory for the hydrolysis of peptide substrates and hence Tyr-248 clearly does not act as a general acid [79,83] or general base in catalysis.

The percent change in the kinetic parameters of the Phe-248, Phe-198 mutant enzyme are essentially the product of the changes observed for the singly modified enzymes (Table 5). The change in catalytic parameters is greatest for ClCin-OPhe where k_{cat} decreases 20- and the K_m value increases 7.7-fold (Table 6) [84]. The k_{cat}/K_m value for Bz-Gly-OPhe is virtually unchanged while those for Z-Gly-Gly-Phe and ClCin-OPhe are decreased to 3.6 and 0.65% of the native enzyme, respectively. These results again are reminiscent of the acetylated CPD-A derivatives where activity toward Z-Gly-Gly-Phe decreases to 7% [26] and Z-Gly-Phe to less than 2% [68] (Table 3). On the other hand, the acetylated enzyme displays sixfold increased activity toward Bz-Gly-OPhe, largely due to a displaced substrate inhibition profile [68] (Table 3).

4.3. Influence of Tyr-248 on Peptide Binding

The mutagenesis and chemical modification studies of CPD-A have proved that the state of ionization of Tyr-248 cannot be crucial to the catalytic step of peptide hydrolysis [29,83,84]. Cryokinetic studies of the bovine enzyme have shown that at this step the peptide substrate is coordinated to the active site metal, and that bond cleavage occurs during or after the rate-determining step [31,33-35]. It is therefore justified to ask whether or not the state of ionization of Tyr-248 has any relevance to the substrate-binding steps which lead to the formation of the peptide ES_2 intermediate, i.e., within the K_m time domain [86].

Rat (Phe-248) CPD-A1 exhibits five- and ninefold increased K_m values for Z-Gly-Gly-Phe and ClCin-OPhe, and a 70-fold increased K_i value for the potato inhibitor, suggesting that Tyr-248 plays some role in ligand binding [79].

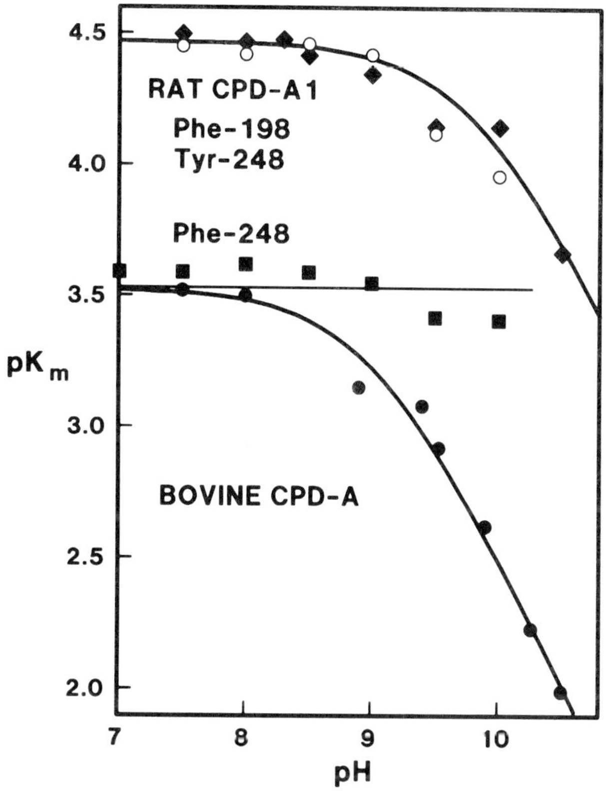

FIG. 7. The pH dependence of pK_m of Z-Gly-Gly-Phe for bovine CPD-A and rat CPD-A1 wild-type and mutant enzymes [63,83,84]. The line for the Phe-248 mutant represents the average value of K_m for the pH range 7-10.

The pH dependence of pK_m for the mutant and wild-type enzymes reveals a possibly deleterious role of the ionization of Tyr-248 (Fig. 7). Over the pH range from 7 to 10.5 the pH profiles of pK_m for the bovine CPD-A, wild-type rat A1, and Phe-198 mutant enzymes all exhibit the same general features for the peptide Z-Gly-Gly-Phe (Fig. 7). Peptide-binding steps as reflected in pK_m values is strongest at pH 7-8 and decreases above pH 9 with a pK_a of 9.0 for the bovine enzyme and 9.8 for the rat A1 and Phe-198 enzymes (Fig. 7). In marked contrast for the Phe-248 mutant, pK_m is not pH-dependent over the pH range 7-10. These results are consistent with the results of chemical modification and kinetic studies of modified Tyr-248 derivatives [27]. Thus, ionization of the NO_2-Tyr-248 and NH_2-Tyr-248 enzymes, and, by inference, the native Tyr-248 enzyme leads to loss in peptide binding to the enzyme. This may be due to an electrostatic interaction between the phenolate anion and some positive charge on the enzyme which could prevent peptide binding, e.g., an arginine or the metal. In the Phe-248 mutant no such ionization can occur and thus a change in pK_m as a function of pH cannot be expected and, indeed, none is seen (Fig. 7).

Clearly, the results of these mutagenesis experiments are critical to mechanistic interpretations. The results, while quite analogous to those obtained by chemical modifications of Tyr-248, add significantly to these by eliminating the OH group of tyrosine from consideration as crucial to any mechanistic scheme.

5. CONCLUSIONS

Work on the function-structure relationships of bovine CPD-A has now spanned five decades in which there has been decisive progress in methodology, technology, and theory pertinent to the understanding of the structural basis of the action of enzymes and their mechanisms. CPD-A was the first zinc metalloenzyme whose mechanism was examined by inorganic and organic modifications [45] and later by site-directed mutagenic techniques [79]. The substitution of different transition metals for the native zinc ion resulted in a series of metalloenzymes whose characteristic activities toward ester and peptide substrates helped to define the role of the metal in catalysis [2-4]. In particular, both exchange-labile Co(II) and exchange-inert Co(III) have served to examine inner and outer sphere mechanisms in CPD-A action [3,49], and the special probe properties of cobalt and cadmium enzymes have been especially helpful in elucidating details of substrate binding [7,9,19,62]. Cryospectrokinetic studies of the cobalt-substituted enzyme have, in fact, enabled metallointermediates in the mechanism of action of CPD-A to be characterized [20,31,33-35].

Selective chemical modifications have pointed to Glu-270, Arg-145, and Tyr-248 as having roles in the catalytic process. However, the precise manner in which each participates has never been defined. All tyrosine modifications, for instance, decrease peptidase activity to varying degrees while increasing esterase activity or leaving it unaltered [3,67-71]. Modifications directed at the phenolic OH group have had the greater effect on activity, although in no case do they abolish it. O-acylation, for example, merely weakens the binding of both peptide and ester substrates to the enzyme. Hence it would seem that tyrosine is not indispensible to activity. Nevertheless, ortho nitration of Tyr-248 does provide a sensitive monitor of the acid-base properties of its phenolic OH group which directly affects catalysis [27].

Recently, site-directed mutagenesis has provided a novel and powerful means to study the role of specific amino acid residues in enzyme mechanisms, and this has given new impetus to efforts to solve such problems with CPD-A. Guided by the results of both conventional chemical approaches and x-ray crystallography, substitution of Tyr-248 or Tyr-198 by Phe have been singled out [83,84]. Rat (Phe-248) CPD-A1 is enzymatically active and, thus, mechanisms that assign a "crucial" role to Tyr-248 in catalysis cannot be valid [79,83]. Mutagenic and chemical modifications have also raised questions about the inferred importance of the 12-Å movement of Tyr-248 on Gly-Tyr binding to the crystalline enzyme [21,22,66,76]. As Tyr-248 is clearly not essential to the mechanism of CPD-A action, this movement can no longer be thought of as a verification of the induced fit theory of enzyme catalysis [85].

Finally, the results of these studies pertain also to carboxypeptidase B, thermolysin, angiotensin-converting enzyme, and aminopeptidase M, among other zinc metalloproteins whose mechanisms have been proposed to be analogous to that for CPD-A [86].

ABBREVIATIONS

CD	circular dichroism
CPD-A	carboxypeptidase A
DAA	diazotized arsanilic acid
DHT	5-diazo-^{1}H-tetrazole
EI	enzyme-inhibitor complex
EPR	electron paramagnetic resonance
ES	enzyme-substrate complex
MCD	magnetic circular dichroism
NMR	nuclear magnetic resonance
TNM	tetranitromethane

REFERENCES

1. B. L. Vallee and H. Neurath, *J. Am. Chem. Soc., 76,* 5006 (1954).
2. J. E. Coleman and B. L. Vallee, *J. Biol. Chem., 235,* 390 (1960).
3. B. L. Vallee, A. Galdes, D. S. Auld, and J. F. Riordan, in "Metal Ions in Biology: Zinc Enzymes", Vol. 5 (T. G. Spiro, ed.), John Wiley and Sons, New York, 1984, p. 25 ff.
4. D. S. Auld and B. Holmquist, *Biochemistry, 13,* 4355 (1974).
5. A. Schäffer and D. S. Auld, *Biochemistry, 25,* 2476 (1986).
6. D. C. Rees, J. B. Howard, P. Chakrabarti, T. Yeats, B. T. Hsu, K. D. Hardman, and W. N. Lipscomb, in "Zinc Enzymes" (I. Bertini, C. Luchinat, W. Maret, and M. Zeppezauer, eds.), Birkhäuser, Basel, 1986, p. 155 ff.
7. B. L. Vallee and B. Holmquist, in "Methods for Determining Metal Ion Environments in Proteins: Structure and Function of Metalloproteins" (D. W. Darnall and R. G. Wilkins, eds.), Elsevier, New York, 1980, p. 27 ff.
8. I. Bertini and C. Luchinat, in "Advances in Inorganic Biochemistry", Vol. 6 (G. L. Eichhorn and L. G. Marzilli, eds.), Elsevier, New York, 1984, p. 71 ff.
9. S. A. Latt and B. L. Vallee, *Biochemistry, 10,* 4263 (1971).
10. C. J. Ballhausen, "Introduction to Ligand Field Theory", McGraw-Hill, New York, 1962.
11. R. L. Carlin, in "Transition Metal Chemistry", Vol. 1 (R. L. Carlin, ed.), Marcel Dekker, New York, 1965, p. 1 ff.

12. A. Dori and H. B. Gray, *Inorg. Chem., 7,* 889 (1968).

13. F. A. Cotton and R. H. Soderberg, *J. Am. Chem. Soc., 84,* 872 (1962).

14. W. DeW. Horrocks, Jr., J. N. Ishley, B. Holmquist, and J. S. Thompson, *J. Inorg. Biochem., 12,* 131 (1980).

15. R. G. Denning and J. A. Spencer, *Symp. Faraday Soc., 3,* 84 (1969).

16. D. M. L. Goodgame and M. Goodgame, *Inorg. Chem., 4,* 139 (1965).

17. B. R. McGarvey, in "Transition Metal Chemistry", Vol. 3 (R. L. Carlin, ed.), Marcel Dekker, New York, 1966, p. 89 ff.

18. J. R. Pilbrow, *J. Magn. Reson., 31,* 479 (1978).

19. I. Bertini, C. Luchinat, and M. S. Viezzoli, in "Zinc Enzymes" (I. Bertini, C. Luchinat, W. Maret, and M. Zeppezauer, eds.), Birkhäuser, Basel, 1986, p. 27 ff.

20. K. F. Geoghegan, A. Galdes, R. A. Martinelli, B. Holmquist, D. S. Auld, and B. L. Vallee, *Biochemistry, 22,* 2255 (1983).

21. J. A. Hartsuck and W. N. Lipscomb, in "The Enzymes", Vol. 3, 3rd ed. (P. D. Boyer, ed.), Academic Press, New York, 1971, p. 1 ff.

22. F. A. Quiocho and W. N. Lipscomb, *Adv. Prot. Chem., 25,* 1 (1971).

23. D. C. Rees and W. N. Lipscomb, *J. Mol. Biol., 160,* 475 (1982).

24. B. L. Vallee and R. J. P. Williams, *Proc. Natl. Acad. Sci. USA, 59,* 498 (1968).

25. D. S. Auld, S. A. Latt, and B. L. Vallee, *Biochemistry, 11,* 4994 (1972).

26. D. S. Auld and B. L. Vallee, *Biochemistry, 9,* 602 (1970).

27. D. S. Auld, K. J. Larsen, and B. L. Vallee, in "Zinc Enzymes" (I. Bertini, C. Luchinat, W. Maret, and M. Zeppezauer, eds.), Birkhäuser, Basel, 1986, p. 133 ff.

28. G. Navon, R. G. Shulman, B. J. Wylua, and T. Yamane, *J. Mol. Biol., 57,* 15 (1970).

29. B. L. Vallee and J. F. Riordan, *Brookhaven Symp. Biol., 21,* 91 (1968).

30. T. A. Stoitz, M. L. Ludwig, F. A. Quiocho, and W. N. Lipscomb, *J. Biol. Chem., 242,* 4662 (1967).

31. K. Geoghegan, A. Galdes, G. Hanson, B. Holmquist, D. S. Auld, and B. L. Vallee, *Biochemistry, 25,* 4669 (1986).

32. B. Holmquist, T. A. Kaden, and B. L. Vallee, *Biochemistry, 14,* 1454 (1975).

33. D. S. Auld, A. Galdes, K. F. Geoghegan, B. Holmquist, R. A. Martinelli, and B. L. Vallee, *Proc. Natl. Acad. Sci. USA, 82,* 5041 (1984).

34. A. Galdes, D. S. Auld, and B. L. Vallee, *Biochemistry, 25,* 646 (1986).

35. A. Galdes, D. S. Auld, and B. L. Vallee, *Biochemistry, 22,* 1888 (1983).

36. R. Bicknell, A. Schäffer, I. Bertini, C. Luchinat, B. L. Vallee, and D. S. Auld, *Biochemistry, 27,* 1050 (1988).

37. R. Bicknell, A. Schäffer, D. S. Auld, J. F. Riordan, R. Monnani, and I. Bertini, *Biochem. Biophys. Res. Commun., 133,* 787 (1985).

38. C. Luchinat, R. P. Monnanni, S. Roelens, B. L. Vallee, and D. S. Auld, *J. Inorg. Biochem., 32,* 1 (1988).

39. I. Bertini, R. P. Monnanni, G. C. Pellacani, M. Sola, B. L. Vallee, and D. S. Auld, *J. Inorg. Biochem., 32,* 13 (1988).

40. I. Bertini and C. Luchinat, "NMR of Paramagnetic Molecules in Biological Systems", Benjamin Cummings, Boston, 1986.

41. D. S. Auld, K. Geoghegan, A. Galdes, and B. L. Vallee, *Biochemistry, 25,* 5156 (1986).

42. K. F. Geoghegan, B. Holmquist, C. A. Spilburg, and B. L. Vallee, *Biochemistry, 22,* 1847 (1983).

43. R. S. Stephens, J. E. Jentoft, and R. G. Bryant, *J. Am. Chem. Soc., 96,* 8041 (1974).

44. D. S. Auld and B. L. Vallee, *Biochemistry, 10,* 2892 (1971).

45. B. L. Vallee, J. F. Riordan, and J. E. Coleman, *Proc. Natl. Acad. Sci. USA, 49,* 109 (1963).

46. D. A. Buckingham, D. M. Foster, and A. M. Sargeson, *J. Am. Chem. Soc., 92,* 6151 (1970).

47. M. A. Wells and T. C. Bruice, *J. Am. Chem. Soc., 99,* 5341 (1977).

48. R. Breslow, D. E. McClure, R. S. Brown, and J. Eisenach, *J. Am. Chem. Soc., 97,* 194 (1975).

49. H. E. Van Wart and B. L. Vallee, *Biochemistry, 17,* 3385 (1978).

50. L. C. Kuo and M. W. Makinen, *J. Am. Chem. Soc., 257,* 24 (1982).

51. R. C. Rosenberg, C. A. Root, and H. C. Gray, *J. Am. Chem. Soc., 97,* 21 (1975).

52. I. Bertini, G. Canti, and C. Luchinat, *J. Am. Chem. Soc., 104,* 4943 (1982).

53. R. J. DeKoch, D. J. West, J. C. Cannon, and N. D. Chasteen, *Biochemistry, 13,* 4347 (1974).

54. R. A. Bradshaw, L. H. Ericsson, K. A. Walsh, and H. Neurath, *Proc. Natl. Acad. Sci. USA, 63,* 1369 (1969).

55. K. Titani, L. H. Ericsson, K. A. Walsh, and H. Neurath, *Proc. Natl. Acad. Sci. USA, 72,* 1666 (1975).

56. K. Titani, L. H. Ericsson, S. Kumar, F. Jakob, H. Neurath, and R. Zwilling, *Biochemistry, 23,* 1245 (1984).

57. C. Quinto, M. Quiroga, W. F. Swain, W. C. Nikovits, Jr., D. N. Standring, R. L. Pictet, P. Valenzuela, and W. J. Rutter, *Proc. Natl. Acad. Sci. USA, 79,* 31 (1982).

58. D. J. Strydom and T. J. Bazzone, *J. Prot. Chem., 4,* 141 (1985).

59. G. M. Hass and H. Neurath, *Biochemistry, 10,* 3535 (1971).

60. H. Nau and J. F. Riordan, *Biochemistry, 14,* 5285 (1975).

61. W. N. Lipscomb, *Tetrahedron, 30,* 1725 (1974).

62. R. Bauer, G. Christensen, J. T. Johansen, J. L. Bethune, and B. L. Vallee, *Biochem. Biophys. Res. Commun., 90,* 679 (1979).

63. D. S. Auld and B. L. Vallee, *Biochemistry, 9,* 4352 (1970).

64. M. W. Makinen, L. C. Kuo, J. J. Dymowski, and S. Jaffer, *J. Biol. Chem., 254,* 356 (1979).

65. P. Wooley, *Nature, 258,* 677 (1975).

66. G. Zundel, "Hydration and Intermolecular Interaction", Academic Press, New York, 1969, p. 69 ff.

67. J. F. Riordan and B. L. Vallee, *Biochemistry, 12,* 1460 (1973).

68. R. T. Simpson, J. F. Riordan, and B. L. Vallee, *Biochemistry, 2,* 616 (1963).

69. L. Cueni and J. F. Riordan, *Biochemistry, 17,* 1834 (1978).

70. J. T. Johansen, D. M. Livingston, and B. L. Vallee, *Biochemistry, 15,* 2584 (1972).

71. G. Muszynska and J. F. Riordan, *Biochemistry, 15,* 46 (1976).

72. O. A. Roholt and D. Pressman, *Proc. Natl. Acad. Sci. USA, 58,* 280 (1967).

73. L. W. Harrison, D. S. Auld, and B. L. Vallee, *Proc. Natl. Acad. Sci. USA, 72,* 3930 (1975).

74. L. W. Harrison and B. L. Vallee, *Biochemistry, 17,* 4359 (1978).

75. F. A. Quiocho, C. H. McMurray, and W. N. Lipscomb, *Proc. Natl. Acad. Sci. USA, 69,* 2850 (1972).

76. W. N. Lipscomb, G. N. Reeke, J. P. Hartsuck, F. A. Quiocho, and P. H. Bethge, *Phil. Trans. Roy. Soc. London Ser. B, 257,* 177 (1970).

77. M. S. Urdea and J. I. Legg, *J. Biol. Chem., 254,* 11868 (1979).

78. W. L. Mock and J.-T. Chen, *Arch. Biochem. Biophys. 203,* 542 (1980).

79. S. J. Gardell, C. S. Craig, D. Hilvert, M. S. Urdea, and W. J. Rutter, *Nature, 317,* 551 (1985).

80. A. M. Boulet, C. R. Erwin, and W. J. Rutter, *Proc. Natl. Acad. Sci. USA, 83,* 3599 (1986).

81. M. J. Zoller and M. Smith, *DNA, 3,* 479 (1984).

82. C. S. Craig, C. Largman, T. Fletcher, S. Roczniah, P. J. Barr, R. Fletterick, and W. J. Rutter, *Science, 228,* 291 (1985).

83. D. Hilvert, S. J. Gardell, W. J. Rutter, and E. T. Kaiser, *J. Am. Chem. Soc., 108,* 5298 (1986).

84. S. J. Gardell, D. Hilvert, J. Barnett, E. T. Kaiser, and W. J. Rutter, *J. Biol. Chem., 262,* 576 (1987).

85. D. E. Koshland, *Proc. Natl. Acad. Sci. USA, 44,* 98 (1958).

86. D. S. Auld, in "Enzyme Mechanisms" (M. I. Page and A. Williams, eds.), Roy. Soc. Chem., 1987, p. 256 ff.

11

Site-Directed Mutagenesis of *E. coli* Alkaline Phosphatase: Probing the Active Site Mechanism and the Signal Sequence-Mediated Transport of the Enzyme

John E. Butler-Ransohoff,* Debra A. Kendall,
and Emil Thomas Kaiser
Laboratory of Bioorganic Chemistry and Biochemistry
The Rockefeller University
Box 113, 1230 York Avenue
New York, NY 10021-6399

1. INTRODUCTION

Bacterial alkaline phosphatase is present in the periplasmic space of *Escherichia coli* where it functions in response to phosphate

*Present affiliation: Bayer AG, Pharmaforschungszentrum, 5600-Wuppertal, Federal Republic of Germany.

deprivation. It is synthesized as a precursor with an amino terminal hydrophobic signal peptide that mediates the export of the protein to the periplasm. Cleavage of the signal peptide yields the mature form of this dimeric zinc(II) metalloenzyme. The mature enzyme has a dimer molecular weight of 94,058 and its complete amino acid sequence has been determined.

The catalytic mechanism of alkaline phosphatases has been investigated in numerous kinetic [1] and structural [2] studies. The widely accepted kinetic scheme for the action of *E. coli* alkaline phosphatase has four steps encompassing an essential phosphoryl enzyme intermediate [Eq. (1)] [3].

$$E + ROP_i \underset{k_{-1}}{\overset{k_1}{\rightleftharpoons}} E \cdot ROP_i \underset{k_{-2}[ROH]}{\overset{k_2}{\rightleftharpoons}} E - P_i \underset{k_{-3}}{\overset{k_3}{\rightleftharpoons}} E \cdot HOP_i \underset{k_{-4}}{\overset{k_4}{\rightleftharpoons}} E + HOP_i \tag{1}$$

During the course of the catalytic reaction the phosphoryl group ($-P_i = -PO_3^-$) is attached covalently to serine-102* in the active site. Subsequently, water or an alternative nucleophilic acceptor such as an alcohol dephosphorylates the phosphoryl enzyme.

The crystal structure has been determined to 2.8 Å resolution [2]. It shows the enzyme as a dimer with two reactive centers 32 Å apart. The topology of the protein fold of each subunit is of the α/β class of proteins. Each subunit has an active site of stringent steric constraints that contains a cluster of three functional metal sites. Two metal sites, M1 and M2, occupied by zinc are 3.9 Å apart. The third one, 5 and 7 Å apart from M2 and M1, respectively, is usually occupied by magnesium. The structural disposition of those metals suggests that, despite the lack of protein acid and protein base functions, the zinc atoms can activate both nucleophiles, serine-102 and water, necessary for double in-line nucleophilic displacement

*Here we use the numbering scheme for isozyme 3 which has as its amino terminal residue Thr. In publications describing signal sequence mutations, however, the numbering scheme for isozyme 1 is typically used in which the amino terminal Thr (position 2) is preceded by Arg (position 1).

on phosphorus. Furthermore, along with arginine-166 they stabilize the transition state during bond formation and bond cleavage [2]. Evidence for the formation of a phosphoryl enzyme intermediate has been provided by several studies of the pre-steady-state kinetics [4] and by the isolation of the ^{32}P-labeled intermediate at pH = 5 where it is stable [5]. At neutral and alkaline pH, dissociation of the product phosphate from the noncovalent complex, E•P, is rate limiting [6], precluding the direct detection of this intermediate. A study of the stereochemical course of phosphoryl group transfer catalyzed by alkaline phosphatase at pH 9.5 showed it to occur with retention of configuration at phosphorus [7]. This was interpreted to be consistent with a double-displacement mechanism involving the transient formation of a phosphoryl enzyme intermediate under these conditions.

Since the enzyme phosphorylation step is not rate limiting at neutral and alkaline pH, there is little change in the enzymatic rate as the pK_a of the leaving group is varied over six orders of magnitude, e.g., from 4-nitrophenyl phosphate to methyl phosphate [3]. A host of thorough investigations of the mechanism os alkaline phosphatase based on metal ion substitution (i.e., the behavior of metalloalkaline phosphatases) and on the use of ^{31}P, ^{113}Cd/^{111}Cd, ^{13}C, ^{19}F, and ^{35}Cl NMR has been described [8]. In this chapter we report how we have used site-directed and cassette mutagenesis to study three diverse aspects of this enzyme, i.e., to explore the consequences of varying the crucial nucleophile in the active site, to probe the role of the potential electrophile arginine-166 in the catalytic mechanism, and to investigate the relationships of primary sequence and secondary structure to the function of the signal peptide of alkaline phosphatase.

2. CONSTRUCTION AND PREPARATION OF ALKALINE PHOSPHATASE MUTANTS

E. coli alkaline phosphatase is an ideal system with which amino acid substitutions can be made using recombinant DNA technology. The expression of the alkaline phosphatase gene is regulated with respect to the phosphate level of the extracellular medium. Consequently, mutants with potentially low activity can be constructed, and not selected against, by growing bacteria which harbor the *phoA*-containing plasmid in a phosphate-rich medium. Conversely, enzyme production can be maximized by shifting the bacteria to a phosphate-depleted medium. Since the mature enzyme is transported to the periplasm, enzyme isolation requires the release of the contents of the periplasmic space by cold-osmotic shock treatment [9] and purification from few other proteins. This is readily accomplished by Affi-Gel-Blue- and DEAE- umn column chromatography [10].

2.1. Mutagenesis Methodology

Standard mutagenesis techniques have been employed to make amino acid substitutions in the active site of alkaline phosphatase [11]. Synthetic oligonucleotide primers containing the DNA sequence for the desired amino acid substitution were annealed to an M13mp19 vector containing *phoA*. Following elongation and ligation of the primer, the double-stranded plasmid was used to transform an *E. coli* host. Mutant plasmids were detected by colony screening with the same ^{32}P-labeled oligomer. For most studies, the mutant *phoA* genes were then subcloned into pBR322 between the BamHI and HindIII sites.

Construction of the mutant 2IL02 signal sequence involved nine amino acid substitutions. The rapid development of this mutant was accomplished by creating unique restriction sites flanking the region of the desired mutations for use in cassette mutagenesis [12]. Digestion at these sites permitted the removal of the wild-type encoding segment and the mutant was generated by subsequent ligation with synthetic oligonucleotides coding for the new sequence.

Whenever possible, mutations were made that would involve the creation or removal of a restriction site for initial screening purposes. All mutations were verified by direct DNA sequencing according to the method of Sanger et al. [13].

3. MUTATIONS IN THE ACTIVE SITE RESIDUE 102

3.1. Replacement of Serine-102 with Cysteine

Chemical modification and recombinant DNA methods have been successfully employed in converting catalytically important nucleophilic residues in an enzyme into other potentially reactive nucleophilic groups. So far, the genetic engineering of nucleophiles in enzyme active sites has focused mainly on catalytic species involved in acyl transfer reactions such as the enzyme β-lactamase (serine → cysteine) [14], (serine → threonine) [15].

The examples studied to date showed low activity toward all but the most highly activated substrates. In most acyl transfer reactions the catalytic pathway involves formation and decomposition of tetrahedral intermediates and requires rapid proton transfer. The latter requirement may be responsible for the low catalytic activity of acyltransferase serine mutants, since changes in the pK for the ionization of the nucleophilic group on small alterations in geometry at the active site may disrupt one of the proton transfer steps in such a way that it might become rate limiting. With mutation of the alkaline phosphatase active site residue Ser-102 to the potential nucleophile Cys we hoped to retain substantial activity. If transient formation of a phospho-Cys-102 species occurred, the pentacoordinate intermediates that might lie along the pathway to and from this species would be quite unstable. Rapid proton transfer may not be a requirement for the efficient breakdown of the pentacoordinate intermediates, if they are formed.

Five Ser-102 mutants of *E. coli* alkaline phosphatase (BAP) were generated by site-directed mutagenesis: Ser → Cys (S102C), Ser → His (S102H), Ser → Thr (S102T), Ser → Leu (S102L), Ser → Ala (S102A). The amino acid substitutions, their abbreviations, and their Michaelis-Menten parameters are shown in Table 2. We shall first discuss the cysteine mutant S102C [16] since this one mutant has been thoroughly characterized and its mechanism elucidated [17]. The other mutants will be discussed in the following parts of this section.

TABLE 1

Kinetic Parameters for the Action of Wild-Type and Cysteine-102 Alkaline Phosphatase on 4-Nitrophenyl Phosphate and 2,4-Dinitrophenyl Phosphate (50 mM Tris, 50 mM NaCl, pH 7.5, 25°C)

	4-Nitrophenyl phosphate			2,4-Dinitrophenyl phosphate		
Enzyme	k_{cat} (sec^{-1})	K_m (μM)	k_{cat}/K_m (M^{-1} sec^{-1})	k_{cat} (sec^{-1})	K_m (μM)	k_{cat}/K_m (M^{-1} sec^{-1})
Wild-type	19	5.5	$5.5 \cdot 10^6$	18	5.6	$3.2 \cdot 10^6$
S102C	4.6	41	$1.1 \cdot 10^5$	15	35	$4.3 \cdot 10^5$

Source: Data from Ref. 16.

S102C is an effective catalyst in the hydrolysis of phosphate monoesters [16]. Contrary to the natural serine-containing enzyme, the cysteine mutant is selective toward various substrates. 4-Nitrophenyl phosphate is a better substrate than 4-methylumbelliferyl phosphate and both are superior to β-naphthyl phosphate. This order of reactivity corresponds to the leaving group tendency of the product alcohols. At 10 mM substrate, 1 M α,α,α-tris(hydroxymethyl)methylamine (Tris) (pH = 8) the relative rate ratios with the cysteine mutant are 1:0.37:0.06, respectively, whereas the wild-type enzyme hydrolyzes these substrates at nearly equal rates [3]. The comparison presented in Table 1 clearly indicates that selectivity reflects a k_{cat} effect: the k_{cat} and K_m values of the wild-type enzyme for both 2,4-dinitrophenyl phosphate and 4-nitrophenyl phosphate are nearly identical, whereas the mutant enzyme has a k_{cat} value for 2,4-dinitrophenyl phosphate that is roughly three times greater than the corresponding value for the hydrolysis of 4-nitrophenyl phosphate, while the K_m values for both substrates are very similar. This observation suggests that, unlike the behavior of the wild type enzyme, the rate-determining step of the reaction catalyzed by the cysteine mutant is the phosphorylation of the active site thiol group, rather than either dephosphorylation or product release. Further evidence was provided by the fact that the V_{max} values of the cysteine mutant are independent of the concentration of Tris in the reaction buffer. Tris acts as an alternate phospho group acceptor from the wild-type phosphoenzyme, and higher concentrations of Tris increase the rate of the wild-type catalyzed reaction as expected if dephosphorylation is rate limiting. With the mutant enzyme if phosphorylation is rate controlling, no rate acceleration at higher Tris concentrations would be expected.

Most of the characteristics of the mutant S102C alkaline phosphatase are consistent with the view that the number and sequence of catalytic steps have been conserved. However, the possibility that the mutant enzyme does not follow the same catalytic mechanism as the wild type had to be considered, particularly since the active site thiol failed to react with a series of thiol-modifying reagents such as 5,5'-dithiobis(2-nitrobenzoic acid), 6,6'-dithiodinicotinic acid, 4,4'-dipyridyldisulfide, 4-chloro- and 4-hydroxymercuribenzoate, Hg^{2+} salts, iodoacetate, and iodoacetamide. Considering the stringent steric requirements of the active site of alkaline phosphatase and based on a computer graphics analysis using the published crystal structure data [2], it is possible to rationalize the lack of reactivity of the first five reagents: none of these seems to fit properly in the active site. This argument does not apply, however, to the last three of the series. Their lack of inhibitory power can best be explained by assuming a tight coordinative interaction between the thiol group and one or both of the active site zinc atoms. Two further experiments raised questions about the role of a phosphoenzyme intermediate in the mechanism of the cysteine mutant. In contrast to observations on the wild-type enzyme, no phosphoenzyme could be detected by precipitating S102C alkaline phosphatase in the presence of $^{32}P_i$ at low pH [16]. No "burst" was detected over a wide range of pH values when mixing S102C alkaline phosphatase and high concentrations of its best substrate, 2,4-dinitrophenyl phosphate, in a Durrum-Gibson stopped-flow apparatus. With the wild-type enzyme the appearance of this stoichiometric burst at low pH values is strong evidence for the formation of a phosphoenzyme during the catalytic cycle [4].

If an intermediate is formed, evidence for its existence may be obtained by the observation of similar partitioning for attack by a given set of nucleophiles on several different substrates. Thus, experiments with the wild-type phosphoenzyme involving a competition between attack by water or Tris give a constant ratio of the products arising from these reactions that is independent of the substrate leaving group, as is required by a mechanism that proceeds via a common reaction intermediate [18]. In contrast, the S102C alkaline phosphatase as well as several others of the residue 102 mutants show a variable product ratio (Table 3). Such variability does not disprove the role of a common phosphoenzyme intermediate since the substrate leaving group could still be enzyme-bound during the phosphoryl transfer step to the trapping nucleophile. However, the above enumeration of results suggests that the path of catalysis followed by the mutant enzyme could be different from that obeyed by the wild-type enzyme; the thiol group may not participate directly in the catalytic act and the S102C alkaline phosphatase might use a carbonic anhydrase-type mechanism in which water (or Tris) ligated to zinc(II) is the primary nucleophile.

At this point only the analysis of the stereochemical course of the phospho group transfer could give unambiguous information as to which of the two possible mechanisms is followed by the mutant enzyme. On the basis that enzyme-catalyzed substitution steps at phosphorus are "in-line" displacements [19], the two mechanisms have different stereochemical outcomes. If the reaction indeed requires the formation of a covalent intermediate and the phosphoryl group is subsequently transferred to water or to an alternate nucleophilic acceptor, then the configuration at phosphorus would be retained as the consequence of two consecutive inversions. On the contrary, the configuration at phosphorus would be inverted if the mutant-catalyzed solvolysis occurred in a single substitution step. The transphosphorylation from 4-nitrophenyl-(*R*p)-$^{16}O,^{17}O,^{18}O$)phosphate to (*S*)-propane-1,2-diol catalyzed by the S102C mutant proceeds with retention at phosphorus: the product was found to be 81 ± 5% *R*p (Fig. 1) [17].

Peak number	1	2	3	4	5	6	7	8
Observed	15.5	29.4	39.7	15.5	14.3	41.0	27.9	16.8
Predicted for *R*	13.6	25.8	45.0	15.6	13.6	45.0	25.8	15.6

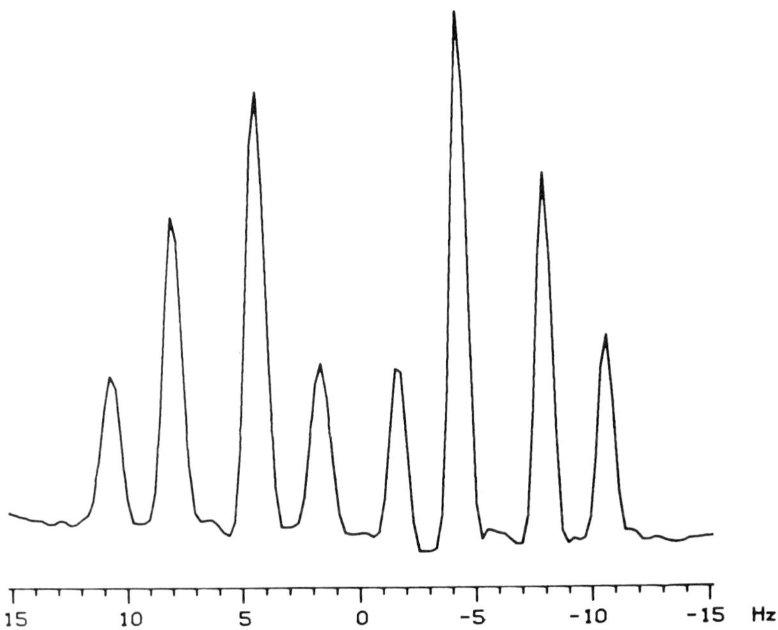

FIG. 1. ^{31}P NMR spectrum of the cyclic phosphate triesters derived from the phospho-(S)-propane-1,2-diols. These were obtained by transphosphorylation catalyzed by mutant S102C alkaline phosphatase using 4-nitrophenyl(R_p)-[^{16}O,^{17}O,^{18}O]-phosphate as substrate. The observed and predicted peak integrals for the ^{31}P NMR spectrum are also shown.

This result shows that the transferred phosphorus atom undergoes an even number of displacements: the most probable pathway involves the two inversions and a single phosphoenzyme intermediate.

The observation that the catalytic efficiency of the S102C enzyme is not affected by the Tris concentration but is sensitive to the nature of the phenolic group is consistent with the hypothesis that the Ser to Cys mutation has changed the rate-determining step, from release of product phosphate to phosphorylation of the enzyme. This explains why no kinetic burst could be observed. The unexpected dependence of the partition ratios on the chemical nature of the substrate leaving group very likely reflects a situation in which the leaving group remains in the active site long enough to influence the selective binding of the alternate nucleophilic acceptor. Finally, the stereochemical course of the reaction on phosphorus provides strong evidence that the chemical path of catalysis has been conserved in this mutant. This result shows that for an enzyme catalyzing a phosphoryl group transfer it is possible to change the active site nucleophile and still observe high k_{cat} values.

3.2. Replacement of Ser-102 with Other Potential Nucleophiles (His, Thr)

Although histidine is known to be the active nucleophile in several transphosphorylases, the catalytic efficiency of the His-102 mutant (S102H) is the lowest of all mutants listed in Table 2. The S102H is also the only mutant where no dependence of the [ROH]/P_i product ratio on the substrate leaving group could be detected [20]. The mean value of the product ratio (in 1 M Tris as defined in Table 3) of three different substrates is 2.38 ± 0.10. This is similar to the measured mean value for the wild-type enzyme 2.49 ± 0.12. The constancy of the product ratio suggests the formation of a phosphoenzyme intermediate for the His mutant; on the other hand, the similarity of these values for the wild-type and His-AP could be interpreted as an indication that His-AP could well be an inactive enzyme with a small contamination of wild-type enzyme. However, the K_m values are significantly different and this is good evidence that we are looking at a real (if strongly diminished) activity, presumably originating from a phosphohistidine enzyme intermediate. Another indication of a conserved catalytic mechanism is that the V_{max} of His-AP is enhanced by the addition of Tris as an alternate nucleophilic acceptor to the

TABLE 2

Michaelis-Menten Parameters of the Mutant Alkaline Phosphatases as Determined Using Lineweaver-Burk's Double Reciprocal Plot (4-Nitrophenyl Phosphate, 1 M Tris, pH 8.05, 25°C)

Enzyme	10^2 k_{cat} (sec^{-1})	K_m (μM)	10^{-3} k_{cat}/K_m (M^{-1} sec^{-1})
BAP	3100 ± 50	19.5 ± 0.2	1588
S102C	417 ± 20	895 ± 50	4.7
S102L	7.00 ± 0.07	17.6 ± 1.0	4.0
S102T	1.75 ± 0.02	14.6 ± 1.1	1.2
S102H	3.00 ± 0.06	28.8 ± 2.8	1
S102A	3.55 ± 0.70	27.6 ± 2.8	1.3

Source: Data from Refs. 16, 20, and 21.

reaction media. As with the wild-type-catalyzed reaction, one can easily show that the increase in reaction rate is not reflected as an increase in the rate of inorganic phosphate appearance, but exclusively as an increase in the rate of formation of phosphorylated Tris.

A phosphoryl transfer experiment under conditions similar to those used for the Cys mutant with 1,2-propanediol showed <<1% transphosphorylation and thus precluded the study of the stereochemistry of this mutant-catalyzed reaction.

Not enough data are available on the threonine-102 mutant (S102T) to make any mechanistic arguments. Its very low k_{cat} could reflect a severe steric impediment to an efficient phosphorylation of the enzyme.

3.3. Replacement of Ser-102 with Amino Acids Lacking a Nucleophilic Side Chain (Ala, Leu)

Most strikingly, preliminary studies of both residue 102 mutants [20, 21] S102A and S102L show that these species have an unexpectedly high activity in catalyzing the hydrolysis of several phosphate monesters.

TABLE 3

$[ROH]/[P_i]$ Ratios in 1 M Tris pH = 8.0 (25°C)

Enzyme	4-Nitrophenyl phosphate	Methylumbelliferyl phosphate	β-Naphthol phosphate
BAP	2.57 ± 0.05	2.59 ± 0.12	2.31 ± 0.05
S102C	6.39 ± 0.35	4.06 ± 0.39	3.33 ± 0.05
S102L	1.84 ± 0.09	2.75 ± 0.25	3.71 ± 0.24
S102H	2.48 ± 0.11	2.23 ± 0.08	2.44 ± 0.06

Their specific activities are both enhanced by high Tris concentration in the assay solutions, and in both cases this is due solely to an increase in the rate of Tris phosphorylation and not to an increase in the rate of inorganic phosphate formation. The partition ratio for the competition between water and Tris for attack at the phosphoryl group was measured at 1 M Tris with three substrates for S102L and it varies significantly (Table 3). This could be interpreted as an argument against a common phosphoenzyme intermediate and the activity might be explained in terms of a water(Tris)-on-zinc mechanism. However, as we have shown in the case of S102C, this variability of the product ratios is a necessary but not sufficient condition for establishing such an alternate catalytic path. At present the very low transphosphorylation ratio using butane-1,3- or propane-1,2-diol as acceptors has precluded the analysis of the stereochemical course of transphosphorylation catalyzed by S102L. These experiments were done using 4-nitrophenyl phosphate as substrate. We can expect better results using β-naphthyl phosphate as substrate, in which case this stereochemical experiment shall shed light on the intriguing catalytic activity of these "phosphatases without nucleophiles". It might be conceivable that another, as yet unrecognized nucleophile in the active site is recruited by these enzymes in a "normal" phosphorylation-dephosphorylation mechanism. However, a careful examination of the crystal structure of alkaline phosphatase from *E. coli* [2] makes this latter possibility seem unlikely.

4. MUTATIONS IN THE ACTIVE SITE RESIDUE 166

While the catalytic mechanism of alkaline phosphatase has been thoroughly investigated, the role of specific residues other than serine-102 is not fully understood. One of the characteristic features of alkaline phosphatase is the lack of sensitivity of the k_{cat}/K_m values for monophosphate ester hydrolysis toward the leaving group of the substrate. A very small change in effective charge on the leaving group's oxygen in the course of the enzyme phosphorylation step has been observed. This is consistent with substantial electrophilic interaction of groups in the active site with the leaving group [3]. The role of this electrophile could be filled by either a zinc atom or by the side chain of arginine-166. Alternatively, the role of the guanidinium side chain of Arg-166 could be to stabilize the pentacoordinate transition state. We have used site-directed mutagenesis to substitute the wild type's arginine-166 for lysine (R166K) and for glutamine (R166Q) [22]. The lysine substitution was chosen as a conservative change which would maintain the charge and the hydrogen-bonding ability of the wild-type arginine. The glutamine substitution was aimed at conserving the hydrogen-bonding ability while deleting the charge and thus any electrophilic functionality at residue 166. A hydrogen bond between residue 166 and aspartate-101 could be important in keeping serine-102 in the right orientation for nucleophilic attack on the substrate's phosphorus [2]. We found that when arginine-166 is replaced by lysine and glutamine, the mutant enzymes formed are active phosphatases. This indicates that the positive charge at residue 166 is not required for catalysis. Table 4 compares the values of k_{cat} and K_m for both mutants.

It is remarkable that the k_{cat} value at pH = 8 and 50 mM Tris for the mutant lacking the positive charge in the active site (R166Q) is smaller by just a factor of 3 than the corresponding value for the mutant that conserves the charge (R166K). However, the effect of the charge at residue 166 on the K_m value is more pronounced and the K_m for the R166Q mutant increases by an order of magnitude.

TABLE 4

Michaelis-Menten Parameters of the Lys-166 and Gln-166 Alkaline Phosphatases (2,4-Nitrophenyl Phosphate, 25°C, pH = 8.05)

	1 M Tris		0.05 M Tris	
Enzyme	k_{cat} (sec^{-1})	K_m (μM)	k_{cat} (sec^{-1})	K_m (μM)
BAP	31.0	19.5	—	—
R166K	11.7	74	4.4	5.2
R166Q	15.9	2690	1.4	65.9

Roughly the same behavior is observed at 1 M Tris (Table 4) with fairly similar k_{cat} values and widely differing K_m values. Whether or not the increased K_m values for the R166Q mutant reflect a diminishing binding constant for the Michaelis complex will be clarified by the dissection of the microscopic rate constants by pre-steady-state measurements. Binding of the substrate to the enzyme seems to involve primarily the phosphoryl group and electrostatic interactions appear to be likely to play a dominant role in the binding process. Besides the metal ions, residue 166 is the only potential electrophile in or near the active site of alkaline phosphatase.

Other important features of the wild-type enzyme are conserved in both mutants. For each mutant the k_{cat} values for two different substrates are very similar, showing the same lack of dependence on the substrate leaving group that has been so thoroughly documented for the wild-type enzyme [3]. Based on the Michaelis-Menten parameters shown in Table 5 for two substrates of widely different pK values, a Bronsted plot of log (k_{cat}/K_m) vs. the pK of the leaving group would give a slope (β_{LG}) of -0.36 for both mutants. Although only two substrates were examined, it is clear that the change in the effective charge on the oxygen of the leaving group during enzyme phosphorylation is small and only slightly larger than the β_{LG} = -0.2 measured for the wild-type enzyme with 18 aryl monophosphate esters [3].

The pH profiles of both k_{cat}/K_m and K_m closely resemble those measured for the wild-type enzyme; the only difference is a slightly

TABLE 5

The Leaving Group Dependence on the Activity of R166K and R166Q Alkaline Phosphatases (pH = 8.0, 50 mM Tris, Ionic Strength Kept Constant with NaCl at 200 mM)

	4-Nitrophenyl phosphate (pKa = 7.16)			Phenyl phosphate (pKa = 9.99)		
Enzyme	k_{cat} (sec^{-1})	K_m (μM)	k_{cat}/K_m (M^{-1} sec^{-1})	k_{cat} (sec^{-1})	K_m (μM)	k_{cat}/K_m (M^{-1} sec^{-1})
R166K	4.4	5.2	$8.4 \cdot 10^5$	6.8	81.1	$8.4 \cdot 10^4$
R166Q	1.4	65.9	$2.1 \cdot 10^4$	1.1	575.0	$1.9 \cdot 10^3$

narrowed optimal pH range. That the activity observed for the mutants cannot be due to contaminating wild-type enzyme was substantiated by sequencing the pertinent region in the alkaline phosphatase gene using the DNA isolated from the same cells used for enzyme production.

Preliminary studies of the pre-steady-state kinetic behavior of both mutants suggest that the mechanism followed by these mutants is the same as in the wild-type enzyme. We have observed biphasic liberation of dinitrophenolate upon mixing a 10-fold excess of 2,4-dinitrophenyl phosphate and each of the mutant enzymes (5.3 μM) at pH = 5.5 in a Durrum-Gibson stopped-flow spectrophotometer. The initial rapid liberation of dinitrophenolate lasted about 100 msec with the lysine-166 mutant and about 120 msec with the glutamine-166 mutant. Both were followed by slow, zero order hydrolysis of the excess substrate. This is consistent with the formation of a phosphorylated intermediate, as reported for the wild type [4]. The formation of a phosphoenzyme intermediate is also evident by the effect of Tris concentration on the catalytic turnover (Table 4): a significant increase in the k_{cat} value with increasing Tris concentration for both mutants would be expected if the dephosphorylation of the phosphorylated intermediate is rate limiting.

Our results with lysine and glutamine alkaline phosphatases indicate that the catalytic efficiency of alkaline phosphatase does not require the presence of a positive charge at residue 166. Furthermore, arginine-166 is not the electrophile required to explain the observed small change in effective charge on the leaving group oxygen of the phosphate monoester during the course of catalysis. This role is most likely fulfilled by one of the active site zinc atoms.

5. STRUCTURAL OPTIMIZATION OF THE SIGNAL PEPTIDE: POTENTIAL FOR TRANSPORT ENHANCEMENT

The examples described above illustrate the use of single amino acid substitutions to explore the role of specific functional groups in catalytic activity. In contrast, "global mutations" are often more useful in examining the conformation of an entire segment of a peptide or protein. Here we use the alkaline phosphatase signal peptide as an example to show how multiple amino acid substitutions can be used to distinguish between the importance of primary sequence and secondary structure, and how optimization of structural features can be used to enhance function.

E. coli alkaline phosphatase is synthesized as a precursor containing an amino terminal signal sequence. This signal peptide functions in export of alkaline phosphatase from the cytoplasm. It is then cleaved and the mature form of the enzyme resides in the *E. coli* periplasm. The alkaline phosphatase signal sequence is representative of signal peptides in general; signal peptides share little primary sequence homology but are typically characterized by a positively charged amino terminus, a central region rich in hydrophobic residues, and a cleavage site recognized by a signal peptidase. Physical studies using chemically synthesized signal peptides suggest that the residues in the core region are important for interaction with the *E. coli* inner membrane and may do so through the formation of an α-helix [22].

To explore the potential role of a hydrophobic α-helix in signal peptide function, we have produced a new signal sequence containing nine consecutive leucines in the core region [23]. Construction of this mutant sequence involved using mutagenesis to make four amino

Wild Type: M K Q S T I A L A L L P L L F T P V T K A

Mutant 13A: M K Q S T I $\underline{L}$ L $\underline{L}$ L L $\underline{L}$ L L $\underline{L}$ T P V T K A

Mutant 2I102: M K Q S T I $\underline{I}$ $\underline{I}$ $\underline{I}$ $\underline{I}$ $\underline{I}$ $\underline{I}$ $\underline{I}$ $\underline{I}$ $\underline{I}$ T P V T K A

FIG. 2. The sequences of mutant and wild-type alkaline phosphatase signal peptides. The amino acid substitutions made to produce the mutant sequences are underlined.

acid substitutions: two alanines, one phenylalanine, and one proline were all replaced with leucine (mutant 13A, see Fig. 2). Leucine was chosen for this approach as it is hydrophobic, has a substantial propensity to adopt an α-helical conformation in polymers [24], and, fortuitously, already occurs five times in the alkaline phosphatase signal sequence. The length of the mutant signal sequence as well as the composition of the amino and carboxy terminal segments remained identical to the wild-type peptide. The function of this mutant signal peptide was evaluated using an *E. coli* host harboring the mutant *phoA* gene. Treatment of whole cells with lysozyme in a hypertonic medium permitted the release of the periplasmic contents. Analysis of this fraction revealed that the enzyme was correctly transported to this compartment when synthesized as a precursor containing the mutant 13A sequence. Pulse-chase studies further demonstrated that the mutant precursor was processed to the mature form of the enzyme more rapidly than the wild type. These results indicate that optimization of a hydrophobic α-helix is sufficient for transport and processing, and that specific residues are not required in the hydrophobic domain of the alkaline phosphatase signal peptide.

To explore further the relative contribution of hydrophobicity and α-helix potential in signal peptide function, we constructed a second mutant in which the core region is composed exclusively of isoleucines (mutant 2IL02, see Fig. 2) [12]. Although isoleucine is an isomer of leucine and should contribute a comparable degree of hydrophobicity, synthetic polymers of isoleucine adopt β-sheet structures in nonpolar environments [24]. We were initially surprised, therefore, to find that this mutant signal sequence also functioned exceptionally well in transport of the enzyme. The mutant signal peptide was cleaved very rapidly and the mature form of alkaline phosphatase was correctly localized to the *E. coli* periplasm. Since it is unlikely that a signal peptide functions as an aggregate in vivo, we think it is probable that the signal sequence containing polyisoleucine adopts an α-helix. For a monomeric signal peptide, α-helix formation provides for hydrogen bonding of the polar amides along the peptide backbone. This will be particularly important if there is an overall driving force for formation of a hydrophobic surface [25].

The signal sequence of both mutants 13A and 2IL02 involve the repetition of one amino acid in the core segment and consequently these polymeric units should readily form regularly packed structures. The natural sequence probably does this less well; it is composed of a variety of several different amino acids including a proline, which tends to disrupt structural uniformity. Optimization of this structural unit, however, resulted in enhanced transport function. This approach could also be applied to the mature enzyme. Using the x-ray structure of the native enzyme [2], segments could be chosen for replacement with new structural units. The new sequences could be designed to form more regular α-helical or β-sheet structures, and to modify the length and location of these units with the goal of redesigning alkaline phosphatase to perform new functions.

6. CONCLUSION AND OUTLOOK

We have shown through our work on alkaline phosphatase that for an enzyme catalyzing a phosphoryl group transfer reaction, it is possible to change the active site nucleophile and still observe high k_{cat} values. The question as to what is the actual mechanism followed by the mutant enzymes has been answered for the cysteine mutant (S102C).

The stereochemical course of the reaction catalyzed by this mutant shows a conserved chemical path of catalysis, while a change in the energy of the transition states is evident from its substrate selectivity and kinetic behavior with alternate acceptors.

The other mutants at residue 102, in particular those with no nucleophilic side chain, force us to consider the possibility that some of these mutants use a carbonic anhydrase-type mechanism in which water (or Tris) ligated to zinc(II) is the primary nucleophile. Further evaluation of the effect of these amino acid substitutions awaits the study of the stereochemical course of the phospho group transfer catalyzed by its best exponent: leucine-102 (S102L).

The generation and kinetic characterization of the lysine- and glutamine-166 mutants allows the conclusion that the charge at residue 166 is neither essential for efficient catalysis nor is it required to explain the small change in effective charge on the leaving oxygen of the substrate observed for catalysis by both mutants as well as by wild-type alkaline phosphatase.

It appears as if the coordinative environment of the active site zinc atoms may be the crucial feature in nature's design of a phosphohydrolase. It is tempting to speculate that by changing the ligand field of these cations, one could fine-tune the activity of this enzyme and recreate the activity of other zinc enzymes such as carbonic anhydrase (hydratase), angiotensin-converting enzyme (peptidase), and even—by combining site-directed mutagenesis and chemical modification—alcohol dehydrogenase (oxidoreductase).

ACKNOWLEDGMENTS

The authors wish to acknowledge the participation of Drs. S. S. Ghosh, S. E. Rokita, and S. C. Bock in many of the studies described.

REFERENCES

1. J. E. Coleman and J. F. Chlebowski, *Adv. Inorg. Biochem., 1,* 1 (1979).
2. J. M. Sowadski, M. D. Handschumaker, H. M. Krishna-Murty, B. A. Foster, and H. W. Wyckhoff, *J. Mol. Biol., 186,* 417 (1985).
3. A. D. Hall and A. Williams, *Biochemistry, 25,* 4784 (1986).
4. S. H. D. Ko and F. J. Kezdy, *J. Am. Chem. Soc., 89,* 7139 (1967).
5. C. Milstein, *Biochem. J., 92,* 410 (1964).
6. M. Cocivera, J. McManaman, and I. B. Wilson, *Biochemistry, 19,* 2901 (1980).
7. R. S. Jones, A. L. Kindman, and J. R. Knowles, *Nature, 275,* 564 (1978).
8. J. E. Coleman and P. Gettins, in "Metal Ions in Biology", Vol. 5 (T. G. Spiro, ed.), John Wiley and Sons, New York, 1983.
9. H. C. Neu and L. A. Heppel, *J. Biol. Chem., 240,* 3685 (1965).
10. J. E. Butler-Ransohoff, D. A. Kendall, and E. T. Kaiser, *Proc. Natl. Acad. Sci. USA, 85,* 4276 (1988).
11. M. J. Zoller and M. Smith, *DNA, 3,* 479 (1984).
12. D. A. Kendall and E. T. Kaiser, *J. Biol. Chem., 263,* 7261 (1988).
13. F. Sanger, S. Nicklen, and A. R. Coulson, *Proc. Natl. Acad. Sci. USA, 74,* 5463 (1977).
14. I. S. Sigal, B. G. Harwood, and R. Arentzen, *Proc. Natl. Acad. Sci. USA, 74,* 7157 (1982).
15. G. Dalbadie-Macfarland, L. W. Cohen, A. D. Riggs, E. Marin, K. Itakura, and J. H. Richards, *Proc. Natl. Acad. Sci. USA, 79,* 6409 (1982).
16. S. S. Ghosh, S. C. Bock, S. E. Rokita, and E. T. Kaiser, *Science, 231,* 145 (1986).
17. J. E. Butler-Ransohoff, D. A. Kendall, S. Freeman, J. R. Knowles, and E. T. Kaiser, *Biochemistry, 27,* 4777 (1988).
18. H. Barrett, R. Butler, and I. B. Wilson, *Biochemistry, 8,* 1042 (1969).
19. J. R. Knowles, *Ann. Rev. Biochem., 49,* 877 (1980).
20. J. E. Butler-Ransohoff, D. A. Kendall, and E. T. Kaiser, unpublished results.
21. S. E. Rokita, D. A. Kendall, and E. T. Kaiser, unpublished results.
22. M. S. Briggs, L. M. Gierasch, A. Zlotnick, J. D. Lear, and W. F. DeGrado, *Science, 228,* 1096 (1985).

23. D. A. Kendall, S. C. Bock, and E. T. Kaiser, *Nature, 321,* 706 (1986).

24. H.-A. Arfmann, R. Labitzko, and K. G. Wagner, *Biopolymers, 16,* 1815 (1977).

25. D. M. Engelman, T. A. Steitz, and A. Goldman, *Ann. Rev. Biophys. Chem., 15,* 321 (1986).

12

Site-Directed Mutagenesis of Heme Proteins

Patrick S. Stayton, William M. Atkins, Barry A. Springer, and Stephen G. Sligar
Departments of Biochemistry and Chemistry
University of Illinois
415 Roger Adams Laboratory
1209 West California Street
Urbana, IL 61801

1. INTRODUCTION

The experimental inquiry into the molecular basis for heme-containing metalloprotein function has produced a rich and distinguished scientific history. A plethora of detail has emerged, firmly grounded in well-characterized molecular structures, and extending to rigorous kinetic and chemical descriptions of enzymatic activity. In addition to studies directly related to physiological function, the heme proteins have proved to be valuable model systems, particularly in the important fields of electron transfer and oxygen activation chemistry. The existence of a vast library of structural and functional characterizations, coupled with the importance and relevance of their activities, makes the heme proteins an extremely attractive target for dissection using site-directed mutagenesis. Indeed, the state of progress with these proteins is so advanced that specific genetic manipulation probably represents the next frontier of study and breakthrough in our understanding of their molecular function.

The information content for protein function is inherent in the primary amino acid sequence. This information is sufficient to direct the attainment of a three-dimensional lattice displaying physiochemical specificity in the context of a living cell. In its broadest sense the much bandied term "structure-function relationship" refers to the multitude of information contained in the primary protein structure, and the goal of a "structure-function study" is to crack the code so that the relationship of the part—namely, the amino acid—may be determined relative to the whole.

In very general terms, this mapping from sequence to function may be divided into two broad categories: first, the attainment of tertiary structure, and second, the activity resulting from that tertiary structure. Into the first category would fall the process of protein folding and, in the case of the heme proteins, the incorporation of the iron protoporphyrin IX prosthetic group. The second category, for which most of the mutagenesis work has been reported, is concerned with delineating amino acid function with respect to the context of the given three-dimensional structure. It is useful to think of the primary sequence in terms of coded information, where some amino acids may function primarily in the storage of folding information, some may function in storing information for secondary structures, some may function in storing the three-dimensional structural information, some may store the information for the actual chemical catalysis, some the information for binding other proteins, and some for binding cofactors. Although the experimental structure-function study of heme proteins in the aforementioned sense must involve divisions such as these, it is clear that they are ultimately artificial unless these parts are related to a more global description, since a single amino acid may be, and probably is, involved in several structural and functional features.

Site-directed mutagenesis provides an ideal method for determining the information content of an amino acid within the context of the protein since it may be selectively altered while, ideally, the rest of the protein is kept constant. Information in all the previously mentioned forms is in principle obtainable. For example, even with a protein of unknown tertiary structure, the mutagenesis of an amino acid can be used to determine whether it is important in binding a cofactor, etc. This type of information is important in delineating smaller, more manageable functional portions of a long primary sequence. However, without detailed structural information it is difficult to assess the specific feature responsible for the monitored function. For the given example, it would be difficult to determine whether the mutation resulted in loss of direct binding of the cofactor or simply resulted in a structural change removing the direct binding contact at another site. The full potential of site-directed mutagenesis technology is only realized in systems where precise structural data are available.

Although it has been only 3 years since the first publication of a structure-function study utilizing site-directed mutagenesis (utilizing the heme protein-cytochrome c), the proliferation of work makes it impossible to include even the entire body of heme protein investigations in this review. We have thus generally limited this chapter to the heme enzyme systems for which crystal structures are available. We acknowledge work on the other heme-linked systems, such as the membrane-associated cytochromes, photosynthetic reaction center, bacterial and mitochondrial oxidases, and rhodopsin and other ion pumps/channels, but space limits us to consideration of cytochromes c, c peroxidase, b_5, P-450, and the oxygen-binding systems myoglobin and hemoglobin.

2. EXPRESSION OF HEME PROTEINS

The realization of the potential represented by site-directed mutagenesis technology is dependent on the efficient production of protein by microorganisms. The expression of proteins, especially foreign proteins, is in general a very poorly understood process in *Escherichia coli* and yeast. There are several levels at which control of protein production may be effected, with a correspondingly large number of cellular factors which may contribute to protein stability or degradation. It is thus impossible at present to accurately predict a priori whether a given protein will be efficiently expressed in *E. coli*. Nevertheless, there are many expression systems available that offer rational features aimed at maximizing the potential for expression of heterologous proteins. In this section we will briefly describe the systems used to obtain production of the proteins covered in this chapter.

Two general strategies are available to the molecular biologist who desires genetic control of a given protein. Traditionally, the gene of interest has been isolated by any of a variety of molecular cloning techniques. A second means of obtaining the desired gene is by the method of total gene synthesis. This latter technique has become feasible with the advent of efficient and inexpensive synthesis of oligonucleotides. Both of these approaches have been used to obtain the genes for a variety of heme proteins.

Expression of foreign proteins in *E. coli* requires conserved features in the 5' untranslated DNA sequence. Chief among these are those representing the RNA polymerase entry site, or promoter region, the ribosome-binding (Shine-Dalgarno) sequence, and the spacer region between this feature and the initiating methionine codon. Several strong promoters are commercially available, including the lac, tac, and λ Pl promoters. The ribosome-binding sites and spacer regions are usually derived from sequences directing highly expressed *E. coli* proteins, with the former maximizing homology with the 16S RNA of the ribosomal subunit.

Human myoglobin and human hemoglobin genes, however, are not expressed in *E. coli*, even when placed behind strong promoters and conserved ribosome-binding sites [1,2]. A clever expression system has been developed by Nagai and coworkers for these proteins based on the production of a fusion protein [1]. Here, the globin sequence is placed downstream from the DNA coding for the 31 N-terminal amino acids of the highly expressed λ cII protein. In these constructions the DNA-coding sequence for a four amino acid protease recognition sequence separates the cII and globin sequences. The hybrid protein produced from this construction thus contains a protease cleavage site just prior to the globin sequence. The hybrid protein is expressed by *E. coli* and the globin can be liberated by subsequent cleavage using Factor Xa or trypsin. The apoglobins produced by this method can be reconstituted with heme and are identical by a variety of criteria with native human hemoglobin and myoglobin.

A yeast CcP expression system [3] required that mutations first be made in *E. coli*. Thus, the CcP gene was subcloned into the phage plasmid pEMBL so that single-stranded mutagenesis could be performed. The mutations were isolated from *E. coli* and cloned back into the yeast vector YEp13 for expression in *Saccharomyces cerevisiae*. The construction included the wild-type CcP promoter, terminator, and N-terminal signal sequence. This system thus contained information for transport of the precursor protein to the mitochondria, signal peptide cleavage, and heme incorporation. The mature holoprotein was purified from a yeast mutant containing an interruption in the chromosomal copy of CcP.

Cytochrome c peroxidase has also been expressed in *E. coli* [4]. The original yeast clone was manipulated to obtain the mature coding sequence minus the signal peptide. An initiator methionine codon was introduced and the subsequent construction was cloned behind the highly efficient dihydrofolate reductase promoter in pUC8. This construction led largely to the expression of apocytochrome c peroxidase in *E. coli*. The purified apoprotein was reconstituted using bovine hemin and the resulting holoprotein displayed the wild-type CcP characteristics.

Yeast iso-1-cytochrome c has been expressed in yeast using the shuttle vector YEp213 [5]. This vector is capable of replicating in both *E. coli* and *S. cerevisiae*. A yeast strain incapable of producing genomic iso-1-cytochrome c or iso-2-cytochrome c was utilized to ensure purification of the desired protein. This strain cannot be grown on glycerol or lactate until introduction of the plasmid-borne iso-1-cytochrome c. When the mutant-containing transformants were cured of plasmids, the transformed phenotype reverted to the parent phenotype, indicating that no genomic reversion had been selected.

Both cloning and total gene synthetic approaches to the manipulation of heme proteins have been utilized by Sligar and coworkers [6,8-10]. In the synthetic system a series of overlapping and complementary oligonucleotides are chemically synthesized and annealed in vitro (Fig. 1). The product is ligated enzymatically and cloned into the *E. coli* pUC vectors under the control of the lac promoter. The synthetic sequence includes the ribosome binding site from the *Pseudomonas putida* cytochrome P-450cam 5' region [6]. An advantage of this system is the ability to incorporate codons reflecting the available tRNA pools from *E. coli*. *E. coli* exhibits a bias in selection of codons for highly expressed gene products which has been shown to reflect intracellular tRNA concentration [7].

It is widely appreciated that codon content can indeed be crucial to efficient expression, and this is further supported by the observation that a human myoglobin cDNA clone does not appreciably express in

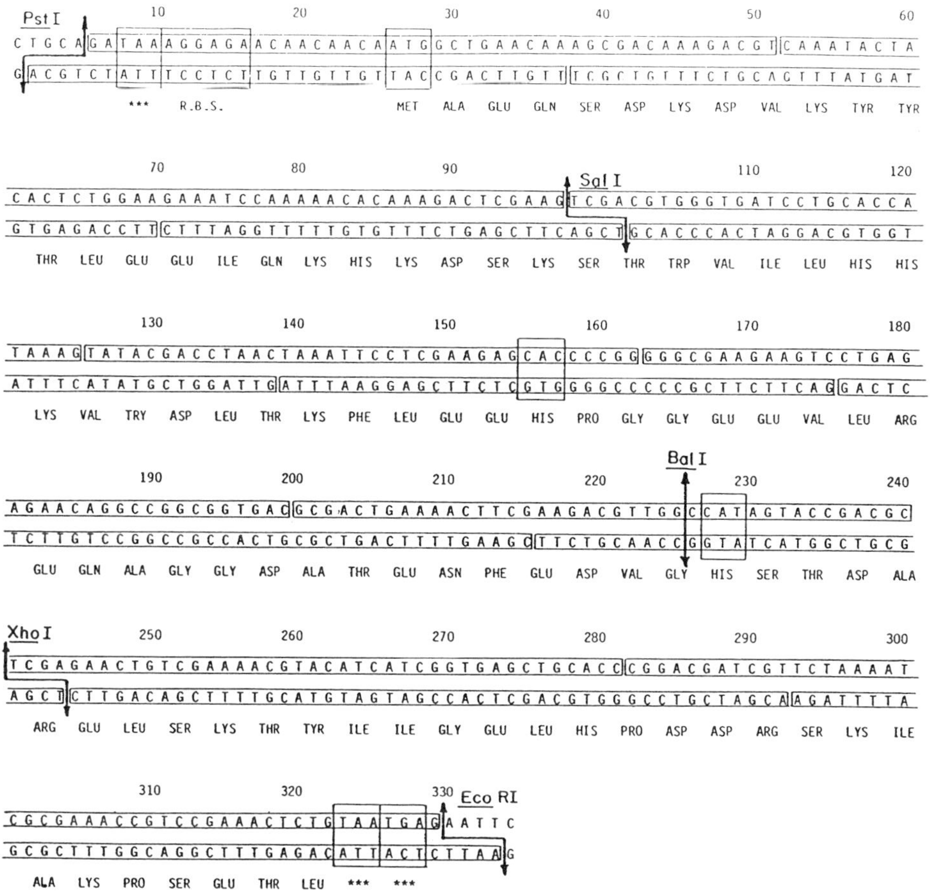

FIG. 1. The synthetic DNA sequence for rat liver cytochrome b_5 [9]. The individual oligonucleotides that were annealed and ligated are indicated. Also shown is the 5'-untranslated region containing the ribosome-binding site and stop codons to prevent read through translation, and the restriction sites available for cassette mutagenesis. The synthetic gene for sperm whale myoglobin is designed in a similar fashion [8].

E. coli, while the synthetic sperm whale myoglobin incorporating high codon bias expresses at high levels [8]. Analysis of codon usage frequency shows a high degree of clustered "poor" codons in the case of the natural human gene. Mammalian cytochrome b_5 has been expressed in an essentially identical system [9]. Both of these proteins are isolated in their respective native, holo forms, indicating that the heme biosynthetic pathway in *E. coli* is capable of high production levels. That optimized codon content is not sufficient to that ensure bacterial expression is shown by the failure of horse heart cytochrome c and poplar plastocyanin, designed with the identical philosophy, to express in *E. coli*.*

The cytochrome P-450cam gene was cloned by mutant complementation [10]. Its 5' nontranslated region is the basis for our synthetic clones and it is also expressed from the lac promoter in pUC. Though it is not expressed in *E. coli* at the very high levels of synthetic cytochrome b_5 and myoglobin, total yields of 1-2% of total cell protein are obtained. The codon bias displayed in the related gram negative *Pseudomonas* is close to that of *E. coli*.

3. CYTOCHROME c

Cytochrome c was the first protein to be genetically engineered in the attempt to better define protein structure-function relationships via site-specific mutagenesis [11]. This protein's rich history of amino acid sequence collection and structure determinations [12-20] made it a natural candidate for such studies. The key biological role of cytochrome c as a redox mediator has been explored in great detail [21-24] and it was this facet that was first targeted for study by site-specific mutagenesis methods. Cytochrome c has served a central role in the field of biological electron transfer [25-27] and it remains the object of much attention, particularly in conjunction with its diprotein complexes with cytochrome b_5 [28,29] and cytochrome c peroxidase (CcP) [30-34]. This section will focus on the role site-directed mutants of cytochrome c have played in eluci-

*Synthetic genes for horse heart cytochrome c (K. K. Rodgers and S. G. Sligar, unpublished results) and poplar plastocyanin (P. S. Stayton and S. G. Sligar, unpublished results) have been prepared and cloned in pUC with the same 5'-untranslated region in addition to the incorporation of high-bias codons. These proteins are not produced by *E. coli* with this expression system.

dating the mechanisms by which protein structure modulates the electron transfer process.

Biological electron transfer events may be roughly divided into two general processes. The first of these represents the set of protein-protein association/dissociation reactions whereby the donor and acceptor molecules interact to determine the geometry and reaction coordinates for electron transfer. This category would include the important arena of biological specificity, reflecting recognition patterns and the molecular nature of the dynamic diprotein electron transfer complexes. The second process is the electron transfer event itself. This area includes not only the kinetic description of the redox reaction, but also such important factors as the effects of driving potential, intervening medium, and distance. Both of these broad divisions have been addressed with site-directed cytochrome c mutants; we will first review work directly related to the latter process and then look at very recent work designed to probe the interrelationships between complex formation and electron transfer rates.

3.1. Phe-87 and Electron Transfer

Most of the cytochrome c site-directed mutagenesis has been focused on a single aromatic residue hypothesized to directly mediate electron transfer with cytochrome c peroxidase (Fig. 2). Phenylalanine 87 of yeast iso-1-cytochrome c is conserved throughout the phylogenetic series (Phe-82 in horse heart cytochrome c). This residue was thus targeted for mutagenesis by single-stranded oligonucleotide technology and the phenylalanine was initially changed to serine, tyrosine, and glycine [1]. In addition to the studies on the electron transfer properties of these mutants, work has been initiated to document phenylalanine-heme spectroscopic interactions [35].

The mutated cytochromes were capable of complementing a yeast strain lacking the iso-1 and iso-2 genes when grown on a nonfermentable carbon source, suggesting that cytochrome oxidase was reduced in vivo with these mutants. Activity determinations based on steady-

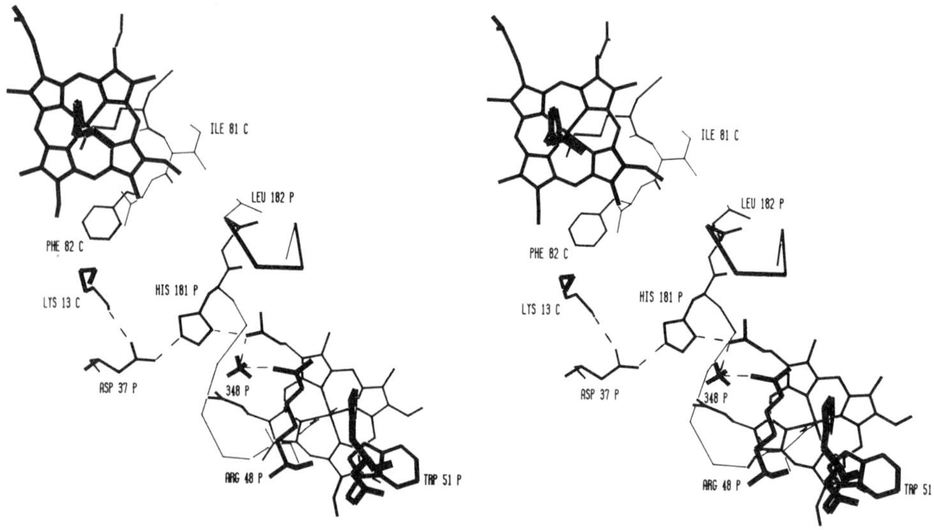

FIG. 2. A microscopic stereo view of the cytochrome c-CcP complex. Conserved Phe-82 (Phe-87 in yeast iso-1-cytochrome c) is shown in the hypothetical modeled structure. His-181 of CcP has also been a proposed participant in the electron transfer event. (Reproduced from Ref. 168 by permission of Marcel Dekker, Inc.)

state measurements indicated that the Ser, Tyr, and Gly mutant proteins were, respectively, 70, 30, and 20% as active as wild-type cytochrome c. The measured redox potentials for the Ser and Gly mutants were 50 mV lower than the wild-type protein and the Tyr mutant. This was attributed to greater solvent access at the heme prosthetic group with the smaller, aliphatic mutants, supporting a previously hypothesized role for Phe-87 as a modulator of solvent access to the heme. It was concluded from this work that Phe-87 is not essential for function and that its primary function is in regulation of the polarity of the heme environment.

These same mutant proteins were later studied in an elegant system developed by Hoffman and coworkers [36]. This characterization was much more revealing at a molecular level because a specific electron transfer event could be monitored. The system is based on the presence of a Zn protoporphyrin-substituted cytochrome c peroxidase.

A three-step cycle is initiated upon flash photoproduction of the Zn triplet state. The last step represents a reaction analogous to the biological system with thermal electron transfer from reduced cytochrome c to the Zn π-cation radical of substituted CcP. The rate constants derived from this system yielded a remarkable finding—cytochrome c with phenylalanine or tyrosine at position 87 transferred electrons to the Zn π-cation radical with a rate four orders of magnitude greater than the serine and glycine mutants [37]. Conversely, the rate constant for electron transfer between the triplet Zn protoporphyrin and ferric cytochrome c is insensitive to the amino acid at position 87.

The apparent sensitivity of the rate constant representing electron transfer between ferrous cytochrome c and the Zn π-cation radical to the presence of an aromatic amino acid at position 87 is suggestive of a superexchange [38] mechanism. The insensitivity of the forward rate constant to the identity of residue 87 could be due to the different chemical process represented by the nonphysiological photoexcited triplet Zn protoporphyrin electron transfer to ferric cytochrome c. More intriguing, however, is the possibility that protein conformational gating, for which some evidence has been given [39-41], is responsible for establishing a favorable electron transfer conduit.

This experimental approach has recently been expanded to include the additional yeast iso-1-cytochrome c mutants, Leu-87 and Ile-87, to more closely approximate the native structure [42]. In a series of meticulous and exquisitely controlled experiments, the rate of electron transfer between ferrous cytochrome c and the Zn π-cation radical was unequivocally confirmed to be 10,000 times faster when an aromatic residue was present at position 87, compared to every aliphatic amino acid variant mentioned. The chemical nature of the reactive intermediate responsible for the observed transient was clearly and quantitatively demonstrated to be the CcP Zn π-cation radical, Fe(II) cytochrome c complex. This was accomplished using the Ser-82 mutant protein for which the Zn protoporphyrin triplet decay was fast compared to the subsequent back electron transfer from Fe(II) to Zn π-cation radical. The kinetic difference spectrum unambiguously displayed the buildup of the aforementioned complex and returned to the initial Zn protoporphyrin ground/ferric cytochrome c resting state in strict first order fashion.

That the observed electron transfer rates were not dominated by a gross conformational change subsequent to intermediate formation, such as the heme ligand exchange observed at high pH with ferrous cytochrome c [43], was demonstrated using a preformed complex of ferrous cytochrome c and Zn protoporphyrin CcP. Photolysis was carried out in the presence of $[Co(III)(NH_3)_5Cl]^{2+}$, which irreversibly and rapidly oxidizes the triplet Zn to the π-cation radical. The subsequent rate of electron transfer from ferrous iron to the radical displays the same dependence on residue 87 identity previously noted.

These important results are clearly suggestive of an electron transfer mechanism based on coupled π-electron systems and superexchange theory, though the possibility of structural gating cannot be ruled out. As Hoffman et al. noted, the mutations could conceivably eliminate the subset of complex geometries responsible for the fast electron transfer rates. Hoffman also points out that the electronic and conformational models need not be exclusive, i.e., the superexchange model only requires that the aromatic residue be properly oriented and this could be coincident with the conformational selection of certain efficient electron transfer geometries involving this residue.

3.2. Surface Charge Mutations

A different slant to the study of electron transfer within the cytochrome c-cytochrome c peroxidase complex has recently been reported [44]. This set of experiments utilized a number of cytochrome c mutations at residues implicated in establishing complex formation (Fig. 3). The surprising results suggest that the electron transfer rates between ferrous cytochrome c and compound I of cytochrome c peroxidase are not optimized in the electrostatically stabilized complexes. Indeed,

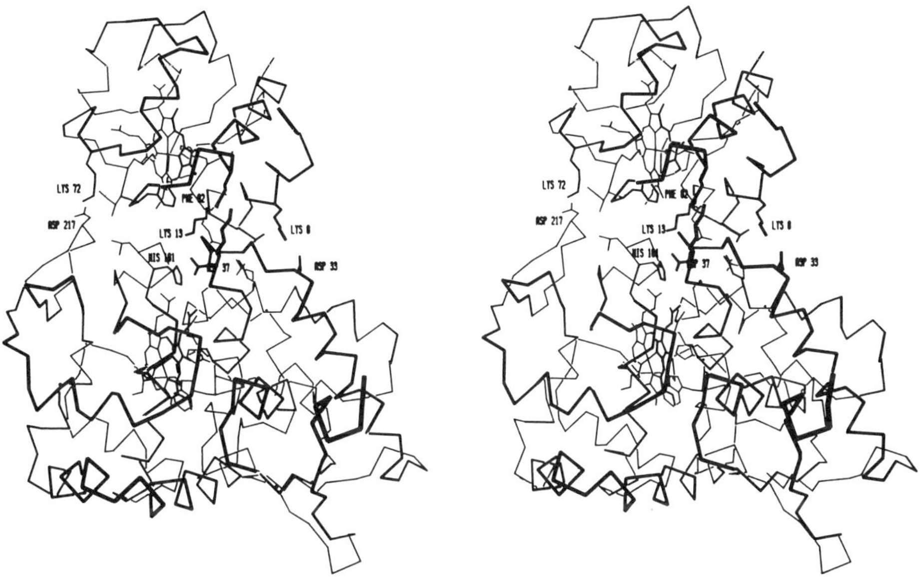

FIG. 3. Stereoscopic view of the cytochrome c-CcP complex. The proposed electrostatic contacts are indicated. (Reproduced from Ref. 168 by permission of Marcel Dekker, Inc.)

enhanced electron transfer rates were observed at higher ionic strengths with both wild-type and mutant proteins.

Electron transfer was mediated by the flash photolysis of various flavin analogs [45,46]. The effect of the individual mutations on complex orientation was probed kinetically by comparing the rate of ferric cytochrome c reduction within the preformed complex. While there were no changes in these rates with the uncomplexed proteins, significant differences were observed for some of the CcP(III)-complexed mutants. In particular, an increased reduction rate was measured for the R13I and K27Q mutants, suggesting a greater degree of heme exposure within these complexes. The Q16S, Q16K, and K72D mutants displayed rate constants similar to the wild-type cytochrome c-CcP(III) complex. When compound I of CcP was used in the preformed complex there were no significant cytochrome c reduction rate differences. This suggests that there are subtle conformational differences in the complexes derived from CcP(III) vs. CcP(compound I).

The intracomplex electron transfer rates between the ferrous cytochrome c proteins and compound I of CcP increase with increasing ionic strength. Moreover, at both low and high ionic strengths, the electron transfer rate of K13I is faster than the wild-type cytochrome c. Similarly, at low ionic strengths K72D exhibits a rate constant greater than that of the wild-type cytochrome c. These results suggest that the complex conformation most productive for fast electron transfer is electrostatically inhibited at low ionic strength. The Q16I and K27Q mutants display lower electron transfer rates than the wild-type cytochrome c indicating that these residues do make a favorable contribution to productive complex formation. On the other hand, the K13I cytochrome c displays an increased rate at both ionic strengths, suggesting that this wild-type residue may actually inhibit the attainment of the optimal functional complex.

It is clear from these results that models of diprotein complex formation incorporating the assumption that the electrostatically stabilized species represents the optimal alignment of heme centers for electron transfer must be treated tentatively. It is similarly clear from this work and from related reports that the structure of this complex varies in topological features dependent on cytochrome source, presence or absence of specific amino acid residues, and solution conditions, e.g., ionic strength. It should be noted that there is no direct structural evidence for specific interface geometries in this or any other diprotein complex. It is entirely possible, for example, that the complex or set of complexes present does not approximate the structure represented in computer-modeling studies. Conversely, the structure responsible for the faster electron transfer rates at high ionic strength may indeed be very similar to the computer-modeled wild type complex. Clearly, the arena of protein-protein association research burgeons with questions and the need for more precise structural determinations.

3.3. Structural Studies

A large number of site-specific mutants of yeast iso-1-cytochrome c have been isolated using traditional genetic methods. These mutant cytochrome c molecules have been generated and utilized to probe structural requirements—particularly along phylogenetic lines—and also to explore the nature of protein amino terminal processing. This work provides a good example of how the quality and number of conclusions derived from study of site-specific mutant proteins scales directly with the number of replacements at each position. The availability of several amino acid variants at the site of interest for each mutation makes interpretations much more meaningful. A review of this work has been presented recently [47], and the reader is directed to it for a more detailed account.

Sherman et al. [48] isolated missense mutations of yeast iso-1-cytochrome c displaying a loss of in vivo function. These proteins have been analyzed in relation to phylogenetic sequence data, using as a "road map" inferred tertiary structure from other species, and the recently determined three-dimensional crystal structure of iso-1-cytochrome c itself [49]. They can be divided into two groups of critical residues that interact either with the heme cofactor or are necessary for the correct folding of other portions of the molecule. It was found that with the exception of proline replacements, these missense mutants occur at invariant or highly conserved positions. As might be expected, replacement of groups in van der Waals contact with the heme prosthetic group led to a nonfunctional product. Interestingly, replacement of Leu-37, Trp-64, and Leu-73 resulted in loss of function, suggesting that packing of the protein surface around the heme is an extremely important structural feature.

Folding requirements were also elucidated with these mutant proteins. Several glycine residues were found to be essential due to their unique ability to pack into the protein interior, i.e., replacement with any larger side chains prevented proper folding. An important result from this work is that evolutionary invariance does not necessarily imply functional invariance, i.e., replacements at some phylogenetically invarient residues did not lead to loss of function. The isolation of mutants displaying a loss-of-function phenotype complements studies with functional second site revertants. Such a study with invariant proline-76 led to the formulation of a method to analyze the folding requirements of a specific protein subsite based primarily on the size of various side groups [50].

A slightly different aspect of structure-function studies focused on the invariant lysine-77 residue [51]. Single-stranded oligonucleotide-directed mutagenesis was used to alter this lysine to an arginine residue. Yeast iso-1-cytochrome c is normally trimethylated at this position and this modification has been hypothesized to function in the transport of the protein into the mitochondria and in the association of cytochrome c with the cytochrome c oxidase complex. The mutant protein was synthesized at normal rates, and in vivo and in vitro analyses demonstrated normal or slightly enhanced activities, suggesting that the trimethylation modification is not necessary for the normal function of this protein. This is yet another example of an evolutionary invariant residue being functionally replaced.

The availability of site-specific amino terminal mutants has enabled the determination of the specificity of methionine aminopeptidase [52]. These results, coupled with an extensive literature search, have led to the suggestion that this peptidase has a universal specificity based on the identity of the penultimate amino acid residue. Methionine was found to be cleaved when the penultimate site was occupied by alanine, glycine, and threonine, and partially cleaved when valine was present. It was not removed when arginine, asparagine, glutamine, isoleucine, leucine, lysine, or methionine was found at the adjacent position. It was further suggested that the three amino terminal residues are at least partially responsible for the specificity of acetyltransferase. Later work has suggested that the penultimate amino acid feature responsible for determining whether methionine is processed may be its size as reflected in the radius of gyration. Such a conclusion is supported by recent work [53,54]. These findings are of tremendous importance to the study of protein turnover and the expression of heterologous genes in *Escherichia coli*.

4. CYTOCHROME c PEROXIDASE

One of the first heme proteins subjected to structure-function studies using recombinant DNA technology was yeast cytochrome c peroxidase (CcP). A wealth of structural and spectroscopic data concerning CcP has accumulated in the last two decades. Several mechanistic questions remain unanswered, however, and CcP is ripe for site-directed mutagenesis to shed light on these issues [55,56]. As will be seen in the following discussion, much remains to be explained concerning CcP catalysis even with the data provided by these mutants.

CcP catalyzes the one-electron oxidation of two equivalents of ferricytochrome c by hydrogen peroxide and may play a physiological role in maintaining the redox state of cytochrome c. In addition, the reduction of H_2O_2 by the CcP-cytochrome c system may function in scavenging cellular H_2O_2 [57]. CcP was initially cloned and expressed in yeast [58], where the expression of cloned and mutant CcP's in genetically altered strains of yeast has simplified subsequent work. These strains contain an integrated interruption of the CcP genomic sequence, allowing for the uncontaminated expression from yeast shuttle vectors containing cloned CcP's and the natural promoter [59]. The protein has also been expressed in *E. coli* by engineering the 5' terminus of *S. cerevisiae* CcP such that the final product contained no leader sequence, with translation initiated from an artificially incorporated methionine codon [60]. In yeast, mature CcP is obtained only after removal of the leader sequence by intramitochondrial proteolytic cleavage. Thus, in addition to the mutations aimed at mechanistic understanding, CcP demonstrates the utility of designing the regulated expression of heme proteins in host organisms. Removal of the leader sequence at the genetic level and incorporation of the 5'-methionine codon allows for expression of approximately 15 mg/liter culture. However, the authors point out that 90% of the translated product is apoprotein [60].

The x-ray crystal structure of CcP is known to high resolution and several features of the active site need to be mentioned before a summary reaction mechanism can be fully understood [61,62] (Fig. 4). The prosthetic group in both the plant and yeast peroxidase is noncovalently bound protoporphyrin IX, with a fifth axial ligand of histidine, and with a sixth coordination position occupied by an aquo ligand. In this sense these peroxidases closely resemble the globins. CcP's are characterized by an invariant distal histidine (His-52) and an invariant arginine (Arg-48) which extend over the heme surface. Unlike the situation observed with metmyoglobin and methemoglobin, the conserved histidine residue is not well situated for hydrogen bonding to the iron-bound water [61], but it is in the proper orientation to hydrogen bond to the distal oxygen of bound peroxide(s).

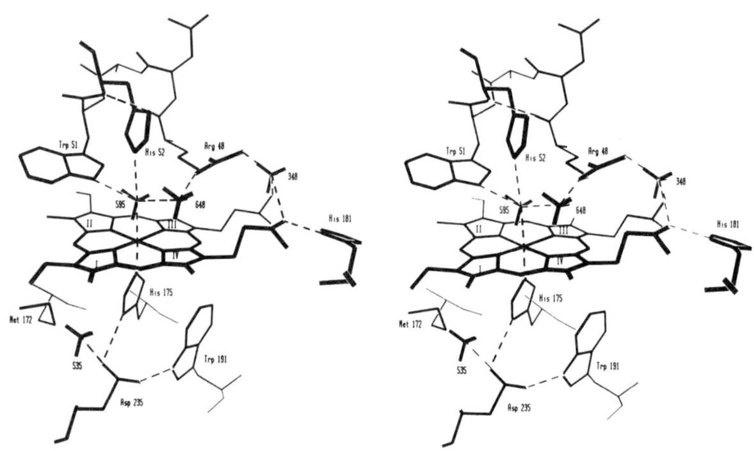

FIG. 4. Stereoscopic view of the active site of CcP [168]. Two water molecules (595 and 648) occupy the active site, providing hydrogen bonds to Trp-51, Arg-48, and His-52. The water-His-52 hydrogen bond is significantly longer (3.2 Å) than optimal hydrogen-bonding distance. (Reproduced from Ref. 168 by permission of Marcel Dekker, Inc.)

Another residue which has received considerable attention is Trp-51, which also lies on the distal side of the heme within 3 Å of the peroxide oxygen. Based on the location of these residues, a mechanism had been proposed which accounts for much of the biochemical and spectroscopic data available [61]. According to this model, the resting ferric CcP binds one peroxide (ROOH or HOOH) as the peroxyl anion, the distal O_2 proton having been transferred to the His-52. The hydroperoxide O-O bond then undergoes heterolytic cleavage, in a manner analogous to similar reactions of horseradish peroxidase [63] or catalase [64]. The guanidinium group of Arg-48 stabilizes the developing negative charge on the distal oxygen atom as heterolytic scission proceeds. However, the resulting compound I-type intermediate, usually referred to as compound ES with CcP, is clearly structurally different from the compound I of the other peroxidases. This intermediate, like compound I, is two oxidation states above the resting enzyme, and therefore capable of oxidizing two equivalents of cytochrome c. The first redox reaction between these proteins would yield oxidized cytochrome c and a species formally equivalent to compound II of the peroxidases [65]. However, several lines of spectroscopic evidence indicate that compound ES does not contain a porphyrin-centered radical cation [66,67], in contrast to the well-documented ferryl porphyrin radical cation of horseradish peroxidase and catalase. With these latter two enzymes, the intermediate has been studied by Mössbauer spectroscopy, magnetic susceptibility, resonance Raman spectroscopy, electron paramagnetic resonance spectroscopy, and UV-visible spectroscopy [68-70]. Typically, however, these peroxidases do not show a strong, sharp EPR signal, but rather a broader signal due to the ferryl-porphyrin radical cation of compound I ($Fe^{4+} = O^{+\cdot}P$). CcP, on the other hand, demonstrates a very intense, sharp radical signal at g = 2.004 after reaction with peroxides, corresponding to one unpaired electron per molecule of compound ES. The oxidation state of the resting, ferric enzyme is apparently changed to the ferryl (4^+) state in compound ES, as indicated by the disappearance of the paramagnetic iron signal. Thus, compound ES has traditionally been described as ($Fe^{4+} = O^{\cdot}R$), where the second oxidizing equivalent is centered on a protein residue rather than the porphyrin [70].

Some investigators found Trp-51 to be a likely candidate for the localized radical, and noted that HRP contains a phenylalanine in an analogous position of the heme pocket. Phenylalanine would not be expected to form a radical of the same stability [71]. However, the g values observed with electron paramagnetic resonance (EPR) and the hyperfine splittings observed with ENDOR [72] were not consistent with an aromatic free radical or a peroxyl radical as suggested by others [73]. Based on this, Hoffman et al. [72] proposed that the EPR signal represented a sulfur-based radical, possibly from a dimeric thioether radical cation $(R_2SSR_2)^{\cdot+}$. In support of this proposal it was noted that CcP contains a methionine, Met-172, at 3.7 Å below the proximal heme surface. Furthermore, other peroxidases contain an unoxidizable serine at the analogous position [71,73].

Not surprisingly, Trp-51 and Met-172 were early targets for site-directed mutagenesis. When Met-172 is replaced with serine or cysteine, a stable reversible, peroxide-dependent intermediate is obtained with optical properties characteristic of compound ES. This intermediate exhibits the sharp free-radical signal and hyperfine structure observed with the wild-type enzyme. In addition, the mutant CcP is fully active with respect to cytochrome c oxidation. These results prompted the conclusion that "either Met-172 does not play a role in the enzyme intermediate, or if it does, efficient alternative mechanistic pathways are available in the absence of this residue" [59].

The mutation of Trp-51 has also been accomplished with the expectation of detecting a structural or catalytic role for this active site aromatic residue. Trp-51 was changed to a phenylalanine and this mutant was also found to exhibit a sharp EPR resonance at g = 2.004 [60]. Based on other EPR resonances, it was apparent that this protein resulted in a change in the ferric spin-state equilibrium.

The mutant demonstrates a shift to predominantly high-spin iron as compared to the wild-type which exists as a mixture of low- and high-spin forms. Furthermore, the mutant was capable of forming an apparently normal compound ES, but this intermediate was less stable than the corresponding species formed with the wild-type enzyme. These results led to the conclusion that Trp-51 is not the site of the free-radical signal associated with compound ES.

It has now been proposed that there are two (or more) distinct radical sites in the peroxide-generated intermediate [60], characterized by the narrow g = 2.004 EPR signal and an axially symmetrical signal which is broadened to appear as "wings" on the narrow signal when the temperature is raised to 77°K. These wings are greatly reduced in the Trp-51 and the Met-172 mutants. A current explanation of these results may involve a radical which exists on two spatially distinct protein sites, where both of these residues play a role in the delicate balance of the active site electronic structure of compound ES, but neither residue by itself is absolutely essential for the functional structure of this intermediate. The recently published x-ray crystal structure of compound ES suggests that residues Met-230, Met-231, and Trp-191 may provide the "radical cluster" [74]. Site-directed mutagenesis will undoubtedly be used to change these residues in the near future. Further discussion of cytochrome c peroxidase spectroscopic and mutagenesis efforts is the subject of Chap. 13 in this volume.

5. CYTOCHROME b_5

The de novo synthesis of the rat hepatic cytochrome b_5 [9] gene represented the first total synthesis of a metalloprotein gene directed for expression in *E. coli* and provided the *prima facia* evidence for the ability of this bacteria to express gram levels of soluble holoprotein with heme incorporation. The design of the synthetic mammalian b_5 gene is shown in Fig. 1. Codon usage was matched to *E. coli* transfer RNA pools [7,75] and the *Pseudomonas* consensus ribosome-binding sequence [6] was placed seven bases prior to the initiator ATG. In addition to these features, restriction sites were designed at convenient locations to facilitate mutagenesis via the cassette method. This system has proved to be extremely useful in providing the basis for genetic manipulation of cytochrome b_5.

Cytochrome b_5 performs several important electron transfer-related functions in mammalian cells. The hepatic form of cytochrome b_5 plays an essential role in two microsomal electron transport reactions, coupling to the fatty acid desaturation pathway and the cytochrome P-450 oxygenase system [76-84]. This microsomal cytochrome b_5 is composed of a small hydrophobic carboxy terminal domain and a larger heme-containing core that can be released by tryptic cleavage. The soluble tryptic form of cytochrome b_5 has been crystallized and the structure determined by x-ray diffraction to 2 Å resolution. A soluble form of cytochrome b_5, nearly identical in primary sequence to the trypsin-released hepatic enzyme, has been purified from erythrocytes [85,86]. This cytochrome b_5 has been shown to mediate electron transfer between methemoglobin reductase and methemoglobin, and thus performs the important physiological role of maintaining the reduced O_2-binding hemoglobin pool [86].

The genes for both the hepatic membrane form of cytochrome b_5 and the soluble, tryptic form for which the crystal structure is known [87] have been constructed [9]. Both forms have been purified from *E. coli,* with the soluble form expressed at levels up to 10% of total cell protein. Initial work concentrated on the soluble cytochrome b_5 for which the wealth of spectroscopic and kinetic characterization gives some guidance to site-directed mutagenesis studies. The following section is divided according to the functional classes of mutations generated with this form of cytochrome b_5.

5.1. Sulfhydral Mutants

The sulfhydral moiety of cysteine has long been utilized as a natural tag for monitorable probes due to its unique chemical properties and reactivities. The absence of cysteine in the wild-type synthetic cytochrome b_5 makes this residue a logical choice for the placement of unique chemical reactivity at desirable locations on the molecular surface. The use of site-directed mutagenesis to place a functional chemical group at a specific locus on the protein surface has only been recently utilized [88].

The utility of such an approach is well represented in a recently completed characterization of cytochrome b_5 association reactions [89]. Modification and crosslinking studies have strongly implicated a set of conserved carboxylate residues on the cytochrome b_5 molecular surface as being important in establishing complex formation with several redox partners [90-92]. These groups are localized around the exposed cytochrome b_5 heme cleft resulting in an asymmetry of charge distribution directly around the proposed location of electron transfer [85]. Computer-modeling studies have included specific electrostatic contacts with these carboxylates and the solvent-exposed heme propionic side chain in their docking of cytochrome b_5 with cytochrome c [93], hemoglobin [94], and myoglobin [95].

Previous attempts to demonstrate the binding of cytochrome b_5 to sperm whale metmyoglobin and cytochrome P-450cam were unsuccessful [96,97]. Site-directed mutagenesis was used to introduce a location for a spectroscopic probe on the periphery of the aforementioned binding surface, designed to detect association without sterically interfering with the process (Fig. 5). Two cytochrome b_5 mutants were engineered to produce sites for specific labeling with a fluorescent dye that allowed precise definition of complex formation. Thr-65 was chosen as a suitable point of replacement with cysteine since it represented a conservative amino acid change and was located on the border of the region of negative potential. To demonstrate that binding was localized at that surface, a second mutant protein was produced with a cysteine introduced on the opposite face of the molecule at amino acid position 8. That these conservative amino acid replacements were noninvasive was demonstrated by the lack of a para-

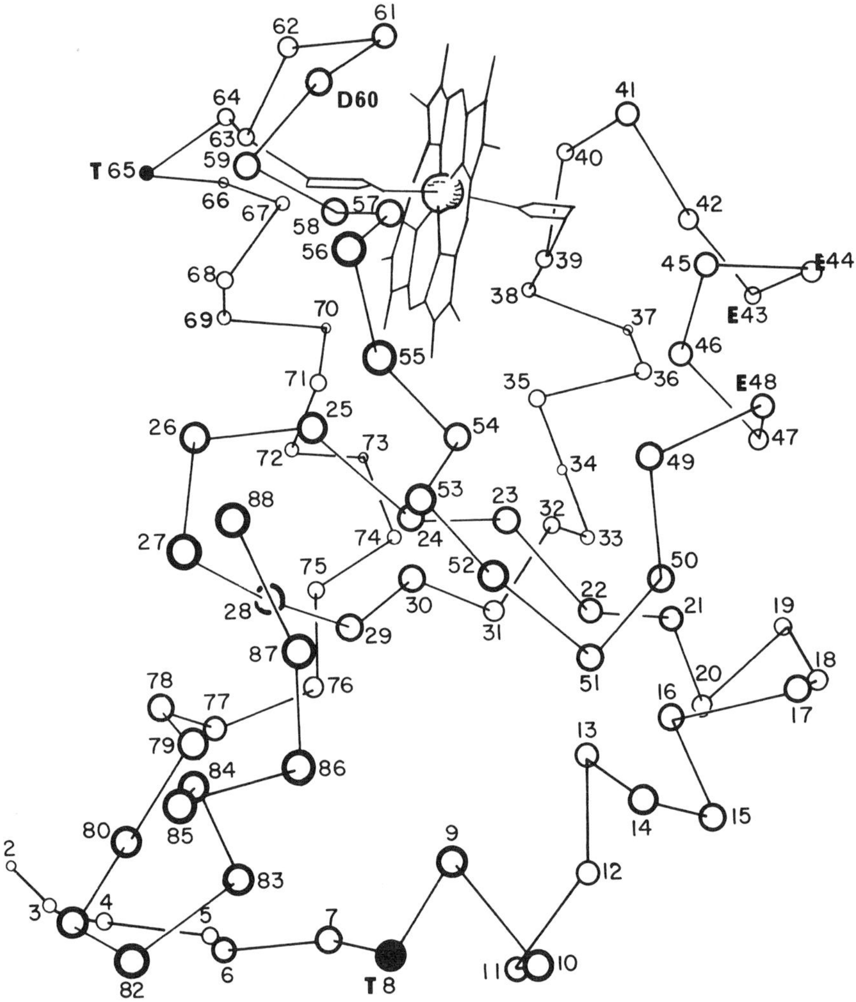

FIG. 5. Cytochrome b_5. The conserved acidic amino acid residues surrounding the heme cleft are indicated. Also shown are the positions of the two introduced cysteine residues.

magnetic ^1H NMR difference spectrum between wild-type and mutant proteins.

The engineering of a reactive thiol onto the surface of cytochrome b_5 made possible the specific labeling of the protein with

TABLE 1

Summary of Dissociation Constants Derived from Fluorescence Binding Titrations

Protein	pH	K_d (μM)
Cytochrome P450cam	7.15	0.65 ± 0.05
Metmyoglobin	6.50	1.1 ± 0.1
	7.15	1.5 ± 0.2
Cytochrome c	7.15	0.5 ± 0.05

Note: The titrations were performed in 1 mM phosphate buffer with T65C b_5. The dissociation constants were extrapolated from either a plot of saturation vs. ln L_{free} for cytochromes P450cam and c, or from a modified Eadie plot for myoglobin. Results from $n \geq 3$ trials are given with standard deviations.

the environmentally sensitive fluorescent probe acrylodan [98]. Acrylodan-derivatized cytochrome b_5 was then used to monitor binding to metmyoglobin, for which conflicting reports of reactivity with cytochrome b_5 existed [95,96], and to cytochrome P-450cam, for which the previous lack of a spectral handle made this the first determination. The results of the binding titrations are given in Table 1.

Cytochrome b_5 has been shown to affect several microsomal cytochrome P-450 activities [79-84] and to function as an effector molecule in the 5-exo hydroxylation of camphor with cytochrome P-450cam [99]. The interaction with cytochrome P-450cam is thus particularly interesting since the crystal structure has been determined to 1.7 Å resolution [100]. Potential cytochrome b_5-binding sites were computer-modeled and the highly conserved P-450 proximal heme-binding loop [101] is found to contain a small cleft at the nearest accessible surface where the basic amino acid side chains of Arg-72, Arg-112, Arg-364, and Lys-344 are arrayed. A striking complementarity of salt bridge contacts with the conserved cytochrome b_5 carboxylates can be modeled which also gives rise to a tight van der Waals interaction between the two proteins. The presence of conserved cytochrome P-450cam Phe-350 between the heme redox centers is particularly intriguing, and could serve as an intermediate in the electron transfer pathway analogous to cytochrome c Phe-87. The availability of genetic handles for both cytochrome b_5 [9] and cytochrome P-450cam [6] make this a potentially rich system for both protein-protein association and electron transfer studies.

In addition to quantifying the binding of cytochrome b_5 to metmyoglobin and P-450cam, the acrylodan emission spectral properties of metmyoglobin were found to differ from those of cytochrome P-450cam and cytochrome c. Acrylodan is an extremely sensitive environmental probe due to its large excited state dipole, and the observed increase in emission enhancement and blue shifts for myoglobin must reflect a more hydrophobic environment relative to that in the cytochrome c and cytochrome P-450 complexes. This can be interpreted as an increase in the averaged binding overlap near position 65, although the possibility of either a myoglobin-induced b_5 conformational change or an increase in secondary myoglobin-b_5 interactions near the acrylodan label cannot be ruled out. Computer modeling of the myoglobin-cytochrome b_5 complex as previously suggested does not result in an increase in protein-protein overlap relative to the cytochrome c model. The increase in averaged binding contact could be due to either a single binding orientation different from that of cytochromes c and P-450cam, or reflect an ensemble of orientations more heavily weighted with species overlapping residue 65. Thus the cytochrome b_5 anionic surface may not be best viewed as a continuous negative patch, but rather as collection of binding sites capable of discriminating between binding partners. The accessed surface would depend on the steric and electrostatic nature of the partners as well as solution conditions. Another approach that can dissect the molecular nature of protein-protein association uses a series of surface charge mutations and is described in the following section [102].

The ability of acrylodan-labeled cytochrome b_5 to detect cytochrome P-450cam association has been utilized in a related study [103]. Cytochrome P-450cam is reduced physiologically by putidaredoxin, an $Fe_2S^*_2Cys_4$ iron sulfur protein. Competitive binding studies indicate that mammalian cytochrome b_5 and putidaredoxin react at the same cytochrome P-450 surface. These studies imply that the modeled P-450cam surface in the cytochrome b_5 complex is also used by putidaredoxin. These structures suggest the first molecular description of a P-450 protein-protein interaction.

In addition to fluorescently labeling the various thiol-derivatized heme proteins generated by site-directed mutagenesis, the unique chemical reactivity of cysteine can be exploited in other manners. The sulfhydrals of proteins have long been used as sites of attachment for the nitroxide series of spin labels [104]. The cysteine mutants of cytochrome b_5 have been successfully labeled with an alkyl iodide nitroxide derivative. A partially immobilized nitroxide EPR spectrum was observed. The spin probe is potentially useful in a number of areas. The paramagnetic nitroxide center perturbs local nuclear magnetic resonances and can therefore be used to assign high-resolution NMR spectra to particular residues. The ability to place the free radical at defined points on the cytochrome b_5 (or myoglobin) surface makes this a potentially valuable tool in the detailed two-dimensional NOE assignment of solution structure, particularly for the diamagnetic protein.

Two other efforts have utilized the b_5 sulfhydral mutants. The first of these takes advantage of the observation that disulfide dimers form spontaneously or may be induced using copper phenanthroline oxidation [88]. There are several permutations of potential dimers formed by the different connectivities of the two cytochrome b_5 mutants and a myoglobin sulfhydral mutant. This is potentially an excellent model system for biological electron transfer because it involves proteins of known crystal structure. The heme-edge to heme-edge distances can be varied over a large range and similarly the potential for examining intervening media effects exists with the ability to engineer amino acid side chains into or out of the reaction path. Myoglobin-cytochrome b_5 heterodimers may be particularly amenable to this study because myoglobin can be reconstituted with a photoreactive porphyrin center to initiate the electron transfer event [36].

Finally, attempts to add a hydrophobic tail to the soluble form of cytochrome b_5 have been initiated in the hope of obtaining an artificial membrane protein. These experiments again rely on the presence of a unique thiol as an attachment point for a hydrophobic molecule. These molecules are interesting in a biochemical sense because they allow delineation of the importance of heme orientation in a reconstituted membrane system. In a slightly different focus there is also interest in the possibility of obtaining a two-dimensional crystal structure at an air-water interface with these proteins. Such an application may have important ramifications in the desired development of bioarray and biosensor technologies.

5.2. Surface Charge Mutants

The elusiveness of x-ray crystal structures for diprotein complexes has made the direct experimental determination of macromolecular binding geometries difficult. Inspection of the independent crystal structures for cytochromes b_5 and c revealed a striking localization of charged amino acid residues around the exposed heme edge in both proteins [87,105]. The potential complementarity of these residues led Salemme to propose a computer model of the cytochrome c-cytochrome b_5 dienzyme complex based largely on the electrostatic stabilization afforded by salt-bridged lysine and carboxylate side chains [93] (Fig. 6). Subsequent experiments aimed at providing experimental evidence for this model focused on the chemical modification of cytochrome c lysine residues and cytochrome b_5 glutamate and aspartate side chains [106-108]. These modified proteins generally exhibited reduced association affinities and electron transfer activities. This motif has been extended to other diprotein complexes for which detailed computer models have been predicted [94,95,109].

Protein-protein recognition and complex formation must play an essential role in the establishment of biological specificity and optimal enzymatic activity. A recent series of surface charge mutations with the synthetic rat liver cytochrome b_5 system has extended the experimental study of protein association within the cytochrome b_5-cytochrome c diprotein complex [102]. In this study, mutations

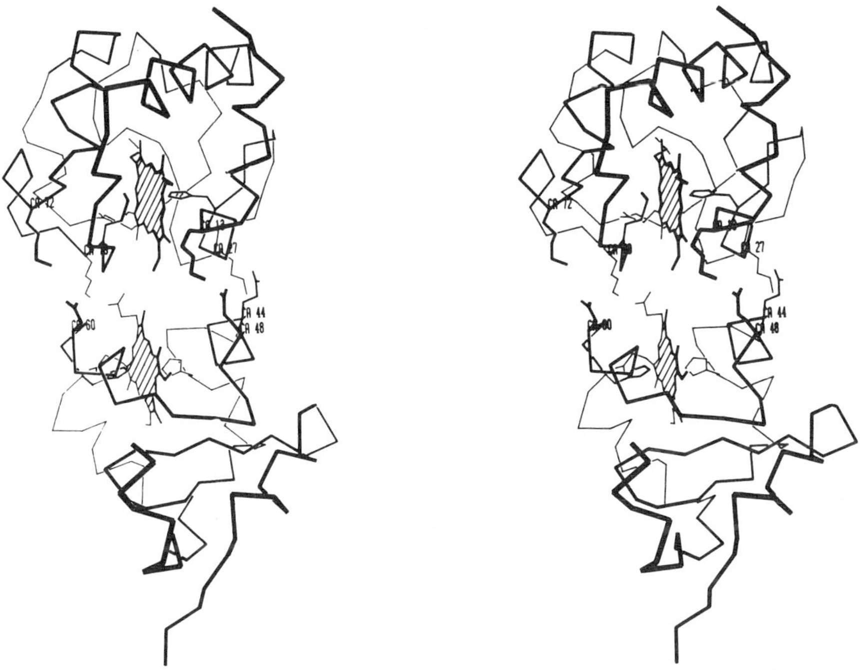

FIG. 6. Stereoscopic view of the proposed cytochrome b_5-cytochrome c complex [168]. The electrostatic contacts are indicated. (Reproduced from Ref. 168 by permission of Marcel Dekker, Inc.)

replaced specific glutamic and aspartic residues with glutamine and asparagine, respectively. The mutant proteins represent noninvasive, best scenario probes of the intramolecular architecture of this complex since they are nearly isosteric with their charged counterparts and retain the ability to form weak hydrogen bonds with their lysine counterparts of cytochrome c, thus potentially maintaining the stereochemistry of the complex.

The specific carboxylate residues implicated in salt bridge linkage with cytochrome c are listed in Table 2 and the mutant constructions are given in Table 3. These mutations, with the substitution of ferriprotoporphyrin IX dimethyl ester, gave a systematic dissection of the individual carboxyl contributions to complex stability. As expected, each of the charge mutations led to a decrease

TABLE 2

Proposed Surface Charge Interactions in Cytochrome b_5 with Several Redox Proteins

Cyt b_5	Cyt c	Hb	Myo	Cyt b_5 (pH 8)	Cyt c (pH 8)
E48	K13	K59	—	E48	K72
E44	K27	K61	K50	E13 Q13[a]	K79
E43	K25	K65	K47	—	—
D60	K72	K95	K98	D60	K87
HEME-COO-	K79	K66	—	E56	K86

[a]In the modeling studies, bovine cytochrome b_5 was used. All of the residues referred to above in the bovine b_5 are conserved in our rat liver sequence with the exception of residue 13, which is a Gln in the rat liver sequence as opposed to a Glu in the bovine sequence.
Abbreviations and symbols: Cyt c, horse heart cytochrome c; Hb, subunit of horse methemoglobin; Myo, bovine heart myoglobin.
Source: Reprinted from [102] by permission.

in complex affinity. An important experimental observation gleaned from the dissociation constants and binding free energies listed in Table 3 is that individual salt bridges do not necessarily contribute the same stabilization to complex stability—all salt bridges are not created equal in an energetic sense.

The significance of this finding was realized through the use of a high-pressure technique [97] that provided information on the volume changes associated with complex dissociation. Le Chatelier's principle governs the association equilibrium such that high hydrostatic pressure should favor the net decrease in system volume produced by solvent electroconstriction around exposed charges in the dissociated proteins [110]. The pressure-induced dissociation of the complex was indeed observed and exhibited a dissociation volume change of -122 ml/mol, which is in good agreement with that predicted by the summation of four to five salt bridges contributing approximately -20 ml/mol each [111].

TABLE 3

Cytochrome b_5-c Association

Species	K_d ($M \times 10^6$)	$-\Delta G$ (kcal/mol)[e]	$-\Delta V$ (ml/mol)[e]
WT[a]	0.40	8.69	122
E48Q[b]	0.80	8.28	87
D60N[b]	1.00	8.15	77
E43,44Q[b]	1.10	8.09	85
WT-DME[c]	1.05	8.12	80
E48Q-DME	1.30	8.00	60
D66S (control)[d]	0.41	8.68	117
WT (pH 8)	1.47	7.92	105
WT-DME (pH 8)	0.22	9.04	96

[a]WT, wild-type cytochrome b_5.
[b]E48Q, D60N, and E43,44Q refer to charge mutants of cytochrome b_5 where the first letter is the symbol of the original amino acid, the numbers refer to the position of the residue, and the last letter is the symbol for the new amino acid in that position. Therefore, in the mutation E43,44Q, the glutamates at positions 43 and 44 have both been changed to glutamines.
[c]DME refers to ferriprotoporphyrin IX dimethyl ester which has been reconstituted into apocytochrome b_5.
[d]As a control, D66 was changed to S. If this residue, which is near the proposed interfacial domain, were sequestered from solvent, one would expect a large decrease in the magnitude of ΔV, which is not observed.
[e]Standard error of the mean are ±0.08 kcal/mol in ΔG and ±5 ml/mol in ΔV for replicate determinations.
Source: Reprinted from Ref. 102 with permission.

Most interesting is the closely observed linear free-energy relationship between the $P\Delta V$ work and the change in total Gibbs energy for complex formation with the different carboxylate mutations. The cytochrome b_5-cytochrome c association has previously been shown to be largely entropically driven with estimated thermodynamic values of $\Delta S = 33 \pm 11$ eu and $\Delta H = 1 \pm 3$ kcal/mol [112]. The linear relationship between ΔV, presumably to a large extent a measurement of the degree of ionic solvation at the protein-protein interface, and ΔG, the overall association free energy, suggests that a common process is dominating both terms. The loss of solvent structure around the isolated ionic charges concomitant with complex formation would be reflected in both functions. In the case of ΔV, the difference in electrostriction between isolated and complexed charges would be related to the degree of isolation from solvation. The linear free-energy relationship suggests that differential solvation, as reflected in the dissociation volumes, accounts for the measured ΔG differences in binding free energy. Similarly, ΔG contains a contribution from the positive ΔS associated with loss of water structure at the solvent-excluded protein-protein interface. Thus these experiments provide important evidence that the specificity and stability of the cytochrome b_5-cytochrome c complex may reflect the degree of solvent exclusion induced at each salt bridge and is consistent with the prior thermodynamic description of the complex. As previously discussed, this description is consistent with the Chothia-Janin model of protein-protein complex formation [113]. The cytochrome b_5-cytochrome c complex doesn't appear to fulfill the criteria for the Ross and Subramanian two-step model [114], although the reaction of cytochrome b_5 with hemoglobin does, possibly pointing to interesting and subtle differences at the complex interface since they are both proposed to involve similar salt bridge contacts. Alternatively, these state functions may be reflecting the energetics of binding-induced conformational changes [96] which differ according to the protein pair utilized.

The two double-charge mutants exhibit special free-energy relationships. The E43,44Q mutation does not follow the linear correlation between Gibbs free energy and volume change. Detailed computer modeling suggests that the Glu-43-Lys-25 salt bridge is partially solvent-exposed [93]. This would lead to a small P ΔV contribution due to electrostriction while some salt bridge enthalpic and entropic contributions to ΔG were maintained. Thus the linkage seen in the other mutants would be broken. The E48Q-DME double mutant falls cleanly on the experimental line suggesting that solvation factors contribute proportionately to ΔG and P ΔV. However, they are non-additive suggesting an interaction free energy that may reflect increased interface solvation, or an electrostatic interaction between Glu-48 and the heme propionate [115].

In summary, the high-pressure approach has proved to be a sensitive probe of the salt bridge contacts in the cytochrome b_5-cytochrome c complex. The observed dissociation volumes agree well with those predicted for the number of salt bridges proposed in the wild-type and mutant complexes. Furthermore the linear relationship between ΔV and ΔG for these complexes has provided the first direct experimental evidence for the suggested entropic drive resulting from loss of solvent structure at the complex interface. A more microscopic characterization of the cytochrome b_5-cytochrome c interface geometry is being currently undertaken with NMR-based studies using specifically labeled ^{13}C charge mutants.

5.3. Axial Ligand Mutants

A central goal of bioinorganic chemistry is to define the role of metal center ligands in defining the chemical reactivity and spectroscopy of the active site. The need for investigations in this area is particularly evident in the heme proteins where one finds a variety of metal ligands including cysteine (cytochrome P-450 and chloroperoxidase), histidine (myoglobin, hemoglobin, horseradish peroxidase, etc.), and tyrosine (catalase). Each of the above examples represents a five-coordinate iron-porphyrin macrocycle, with an additional vacant or easily replaceable sixth position available for oxygen, peroxide, or other exogenous ligand. The cytochromes, however, are usually six-coordinate, low-spin, heme complexes.

Site-directed mutagenesis offers an ideal means to assess the role of heme ligands in metalloprotein function. In the case of cytochrome b_5, for example, the normally bis-imidazole-ligated heme iron was converted to an enzyme active in catalytic oxidative demethylation of N,N-dimethylaniline by replacement of one of the histidine ligands with methionine [116]. It is clear that this methodology offers great promise in understanding the unique features of metal center electronic structure that define particular metalloprotein reactivities.

6. CYTOCHROME P-450

The wealth of data concerning the expression and mutagenesis of heme proteins that has been presented in previous sections will attest to the claim that "the opportunity to tinker with the amino acid sequence of enzymes by using site-directed mutagenesis is proving to be irresistable" [117]. Among the heme proteins which fall into this category are the cytochromes P-450, a family of thiolate-ligated, b-type cytochromes which catalyze a vast array of biological oxidations including alkyl hydroxylations, aryl and olefinic epoxidations, and heteroatom dealkylations [118,119]. Typically, the P-450 reaction cycle begins with substrate binding to the ferric enzyme followed by electron transfer from an NADPH P-450 reductase or a redoxin. Molecular oxygen then binds to the reduced heme prior to a slow, second electron transfer which initiates O-O bond scission and affords a poorly characterized oxidizing intermediate. In light of the overwhelming body of literature concerning these ubiquitous enzymes, it is striking that few data are available from site-directed mutants. This is presumably partly the result of the fact that many of the P-450's documented are integral membrane proteins and their expression in host organisms has lagged behind that of the soluble proteins expressed in bacteria. In this section results obtained with a few such membrane P-450's will be summarized along with results obtained with the cytosolic cytochrome P-450cam.

The mammalian liver microsomal P-450's demonstrate a wide substrate specificity with varying degrees of induction resulting from exposure to diverse chemical compounds. Several lines of evidence indicate that these P-450's are anchored in the membrane bilayer [120 and references therein] and interact with a peripheral membrane protein which contains both FAD and FMN prosthetic groups. This P-450 reductase shuttles reducing equivalents to the cytochromes, sequentially, oxidizing NADPH in the process. Recently, recombinant methods have been used to create a fusion protein from these individual components [121]. A hybrid cDNA was constructed which fused the 5' end of the reductase cDNA to the 3' end of rat P-450c cDNA. The resulting fusion protein was expressed in a P-450-deleted yeast strain and demonstrated activity in the O-deethylation of 7-ethoxycoumarin. The purified fusion protein, which was isolated from the microsomal fraction, was subjected to trypsin treatment, resulting in separation of the P-450 and reductase into the microsomal and soluble fractions, respectively. This suggests that the P-450 domain was embedded in the membrane bilayer, with the reductase moiety anchored to it in the cytosol. These results clearly demonstrate the feasibility of engineering a bifunctional protein from the microsomal monooxygenase components, yet the authors note that based on spectral quantitation of the microsomal heme content, the fusion protein is expressed at only one-sixth the level of the individual P-450c gene in a similar yeast expression system.

A second issue addressed with the rat microsomal P-450's involved an attempt to understand the differences in substrate specificity of two different rat P-450 isozymes, P-450c and P-450d [122]. In this work, the authors conceptualize P-450's as consisting of three domains: (1) an N-terminal section associated with membrane insertion, (2) a central, core region which dictates substrate specificity, and (3) a C-terminal region which promotes protein-protein interaction with the P-450 reductase enzyme. The authors constructed three chimeric P-450's by replacing the central substrate-binding region, the C-terminal region, or both, of the P-450c gene with the corresponding P-450d regions. These chimeric genes were again expressed in yeast, with fascinating results. The yeast cultures containing the first two-thirds of the P-450c gene with the C terminus of the P-450d isozyme exhibited the substrate specificity of P-450c with reduced activity in NADPH consumption, whereas a fusion of the N-terminal domain of the P-450c gene with the 3' two-thirds of the P-450d gene possessed the substrate specificity of P-450d. This suggests that the central core region of these P-450 isozymes dictates the observed substrate specificity. These results provide an exceptional source of optimism for those investigators interested in exchanging large segments of various proteins and engineering new P-450's which retain the functional requirements for monooxygenase activity.

In addition to these structure-function studies on liver microsomal P-450's, work has been recently initiated to investigate the putative signal peptides of these monooxygenases by genetically engineering fusion proteins which possess the N terminus of rabbit cytochrome P-450 PBc2 [123]. The hydrophobic N terminus of this protein, 25 amino acids, was substituted for the signal peptide, 21 amino acids, of bovine preproparathyroid hormone (pre-pro-PTH). The native N terminus of pre-pro-PTH is established as a signal peptide which mediates membrane translocation of the nascent pre-pro-PTH peptide and its subsequent processing in a signal recognition particle (SRP)-dependent manner [124,125]. When fused to the N terminus of P-450 PBc2, however, no proteolytic processing of the in vitro translation product was observed, and the peptide was not translocated across microsomal membranes, when assayed in a wheat germ cell-free translation system with exogenously added microsomal membranes. Instead, the unprocessed peptide was embedded in the membranes, suggesting that the P-450 PBc2 signal peptide functions as an insertion-stop translocation signal. Unlike most signal peptides, the P-450 PBc2 signal does not contain basic amino acids, and in fact is preceded by the acidic residue aspartate. When the signal sequence of P-450 PBc2 was altered via site-directed mutagenesis such that a leucine and this aspartate were replaced with lysine and arginine, the translational fate of this fusion protein was significantly altered. This change in net charge from -1 to +2 with this signal sequence resulted in the complete translocation of the peptide across the microsomal membranes with some processing of the signal. The uncleaved fusion peptides, though translocated across the membranes and protected from trypsin/protease treatment, remained associated with the microsomal fraction upon centrifugation suggesting that, as with the wild-type signal, the mutant signal acts as a membrane anchor. The authors present a model to explain these results in which the mutant, positively charged signal peptide is looped through the membrane, allowing for cleavage by signal peptidase, with the remainder of the translated product still anchored to the membrane. These results demonstrate important progress in understanding the details of expression of these liver microsomal P-450's.

An additional cytochrome P-450 system which has been successfully expressed to high levels in a host organism is the soluble P-450cam, originally isolated from the soil bacterium *Pseudomonas putida*. This monooxygenase catalyzes the regio- and stereospecific hydroxylation of the monoterpene camphor to afford 5-exo-hydroxycamphor. This reaction is the first committed step in the utilization of camphor as a carbon source for this organism. The functional gene for this cytochrome has been expressed in *E. coli*, with apparent transcriptional control being dictated by the lac promoter. This particular P-450 is often considered the prototypical form since much of the spectroscopic and physical characterization of the heme center and its oxygenated states has been conducted with this enzyme [126,127].

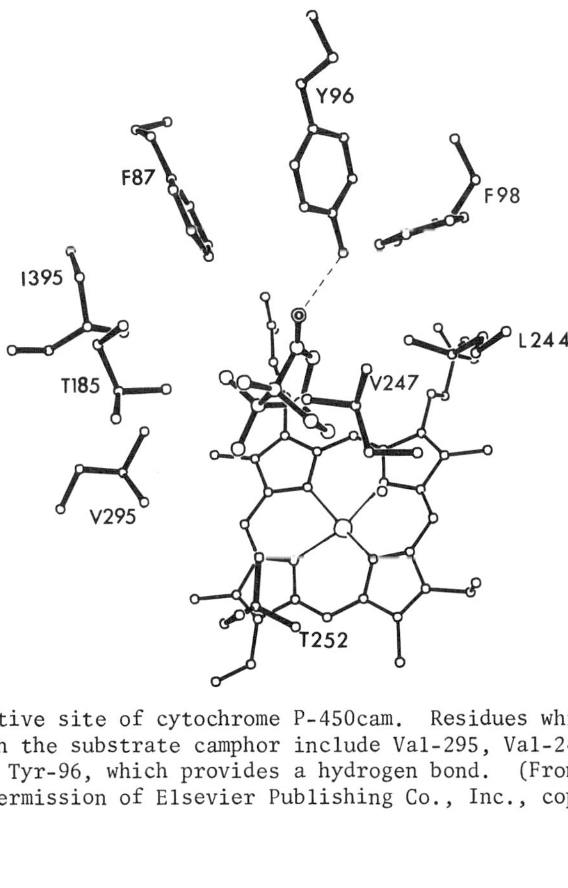

FIG. 7. Active site of cytochrome P-450cam. Residues which are in contact with the substrate camphor include Val-295, Val-247, Leu-244, Phe-87, and Tyr-96, which provides a hydrogen bond. (From Ref. 62, used with permission of Elsevier Publishing Co., Inc., copyright 1984.)

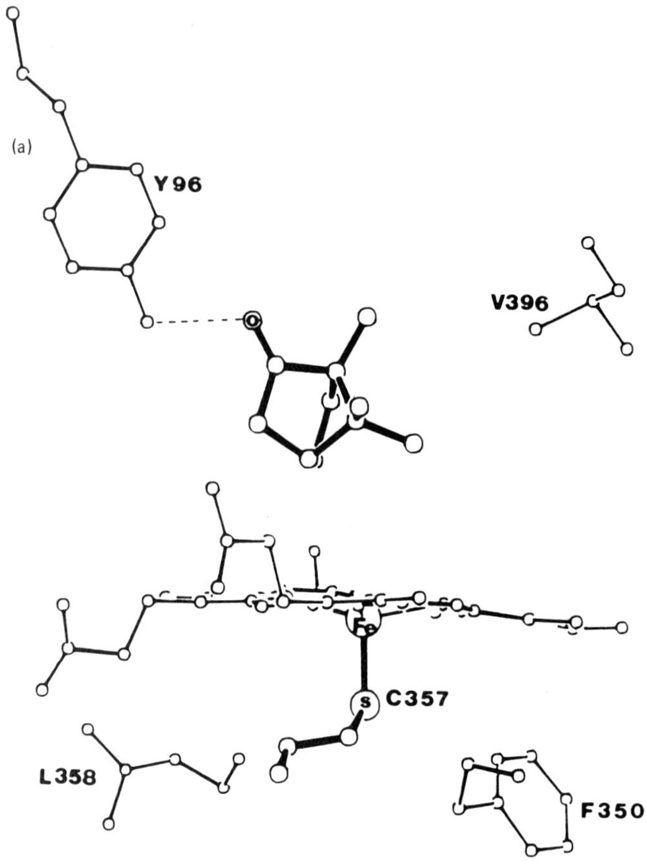

FIG. 8. Comparison of substrate-free and camphor-bound P-450cam. (a) Edge on view of the heme pocket with camphor present. (b) The same perspective in the absence of camphor. A cluster of water molecules is ligated to the heme iron. (From Ref. 62, used with permission of Elsevier Publishing Co., Inc., copyright 1984.)

In addition, the high-resolution x-ray crystal structure of substrate-free, camphor-bound, and inhibitor-bound forms of this enzyme are available, making site-directed mutagenesis particularly attractive as a means of making structural modifications with hypothesized consequences [128-130].

Briefly, the x-ray structures indicate that the camphor molecule is in van der Waals contact with several active site residues

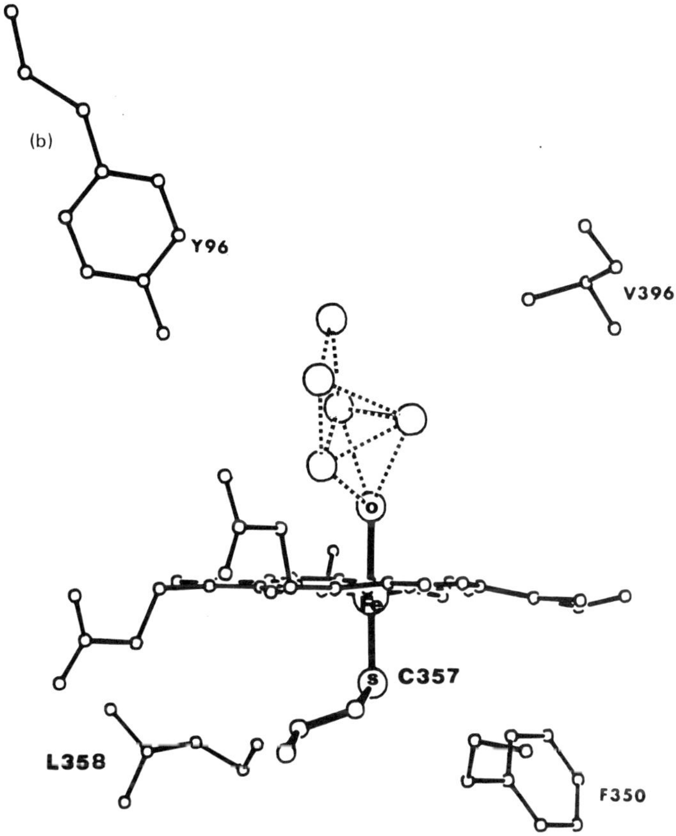

including hydrophobic interactions with Val-295, Val-247, Leu-244, Phe-87, and a hydrogen bond with Tyr-96 (Fig. 7). Comparison of the camphor-bound and substrate-free forms of the enzyme indicates a striking absence of conformational change. Instead, substrate binding results in the expulsion of six water molecules from the active site, which results in a change from low-spin, six-coordinate, aquo-liganded heme to a high-spin, five-coordinate iron system (Fig. 8). This change in spin state is associated with an increase in reduction potential, allowing for electron transfer from the associated electron transfer partners putidaredoxin and putidaredoxin reductase. Recently, with the use of single-stranded mutagenic methods [131], several active site and surface mutants have been constructed. Of these, the most thoroughly characterized to date is the Y96F mutation which destroys the hydrogen bond to the carbonyl moiety of camphor. This change significantly effects the biological specificity of this enzyme in several ways. In addition to an approximate twofold increase in the K_d for substrate binding (1.6 → 3.4 μM), the ferric spin state equilibrium is altered such that saturating levels of camphor do not result in a complete spin state conversion. In contrast to the wild-type P-450cam, which is driven to a 95% high-spin population upon camphor binding, the Y96F mutant only becomes 60% high spin. The rate of NADH consumption of the complete, reconstituted system with the appropriate electron transfer partners is correspondingly decreased from 62 nmol/min/nmol P-450 for the wild type to 29 nmol/min/nmol P-450 for Y96F. Thus, the kinetic control of electron transfer to the cytochrome is compromised when the substrate is not perfectly juxtaposed above the heme iron to exclude water. Furthermore, the decrease in substrate specificity of this enzyme is reflected in the loss of absolute regiospecificity of hydroxylation of the camphor molecule. In addition to the 5-exo-hydroxy alcohol, as much as 8% of the product yield is comprised of other isomeric hydroxycamphors. It is clear that with this particular P-450, nature has managed to couple the control of electron transfer with the control of regio- and stereochemistry. These results also indicate that, with the judicious choice of engineering strategies, it is feasible to construct P-450's which afford new products or metabolize novel organic compounds.

Other P-450cam mutants which have been constructed include V295A, V295I, V247A, F350A, C357H, C357S, and C136S. Preliminary results with these mutants will briefly be mentioned here. The hydrophobic residues lining the heme pocket, Val-295 and Val-247, appear to be intimately involved with protein folding and/or stability. This is based on the observations that these mutants are expressed either not at all or at exceedingly low levels compared to the wild-type protein. This initially surprising result, where relatively conservative changes have been made, directly contrasts the results observed with the chimeric rat P-450's. With P-450cam specific, single-site changes from valine to alanine at position 247 appear to drastically alter the global structure of the enzyme or its folding process to the extent that heme is not incorporated by the protein. With the P-450c and P-450d enzymes entire domains were interchanged with successful expression of active enzyme. These results offer an interesting challenge to the understanding of protein structure.

Other P-450cam mutants which should be discussed include C357H and C357S. As with all P-450's, the fifth axial ligand of the heme iron is a thiolate sulfur of a cysteine residue, which in the case of P-450cam is Cys-357. There is much speculation concerning the role of this thiolate ligand in dictating the enzymatic activity of P-450's vs. other heme proteins which possess nitrogenous or oxygen-bearing fifth ligands [132]. In an attempt to address this question the above-mentioned mutations were made. As with the mutants involving hydrophobic residues of the active site, these axial ligand mutants are not expressed as intact holoenzymes [133]. Although this result may be less surprising than the expression problems found with V295A and V247A, it emphasizes the limitations of site-directed mutagenesis as a technique for conducting structure-function studies in some enzyme systems.

7. HEMOGLOBIN AND MYOGLOBIN

7.1. Background

Hemoglobin (Hb) and the closely related myoglobin (Mb) have been two of the most intensely studied vertebrate proteins. Their physiological importance has been well documented since the early 1900s, and in 1960 the first high-resolution x-ray crystallographic structure of a protein (sperm whale Mb) was reported [134]. The structure of Hb soon followed [135]. Hb, found in erythrocytes, is a tetramer consisting of two identical α chains of 141 amino acids and two identical β chains of 146 amino acids. Each subunit contains one heme with the Fe coordinated to four pyrrole nitrogens and an imidazole nitrogen donated by a globin histidine, termed the proximal histidine, which is the eighth amino acid residue in the F helix (F8). The sixth coordination position of the iron can reversibly bind O_2, CO or NO in the ferrous (Fe^{2+}) state and H_2O, OH^-, F^-, N_3^-, CN^-, and imidazole in the ferric (Fe^{3+}) state. The physiological role of Hb is to transport O_2 from the lungs to the tissues and to return CO_2 (as bicarbonate) to the lungs. Hb is ideally suited for this task because of its ability to bind O_2 cooperatively, i.e., when one subunit binds one molecule of O_2, the O_2 affinity of the other subunits increases accordingly giving rise to a sigmoidal curve. Binding of the first O_2 to the Hb tetramer occurs with a low association constant, but due to the high partial pressure of O_2 in the lungs (pO_2 is approximately 100 mmHg), all four subunits are easily saturated with O_2. This cooperative binding permits the efficient dissociation of O_2 in the tissues (pO_2 is approximately 35 mmHg) since when one molecule of O_2 dissociates, the other subunits lose their O_2 with increasing ease. In addition, protons and O_2 are antagonists. Consequently, at decreasing pH the O_2 saturation curve shifts to the right. This pH dependence stabilizes the deoxy state and is known as the Bohr effect [136]. This is physiologically important since O_2 has a greater tendency to dissociate as muscle acidity increases. Other substances known to stabilize the Hb deoxy complex are CO_2 (found in tissues as bicarbonate), chloride ions, and diphosphoglyceric acid (DPG). The cooperativity observed for Hb arises from the equilibrium between two different structures, the R (relaxed) state, or oxy complex, and the T (tense) state, or deoxy structure. In the absence of O_2 the equilibrium constant for Hb has been described as $K_{eq} = [T]/[R]$. In the presence of O_2 the association constants are commonly expressed as K_T and K_R, where K_T depends on the concentrations of the heterotropic ligands H^+, Cl^-, CO_2, and DPG, whereas K_R varies with H^+ concentrations but remains relatively resistant to Cl^-, CO_2, and DPG [137,138]. The R and T structures differ in their quaternary structure. During the transition between the R and T states, movement at the α chain-β chain interface occurs giving rise to an altered tertiary structure of each subunit such that the O_2 affinity for each subunit is higher than that of individual α or β chains. In the T state the tertiary structure of each subunit is such that the O_2 affinity is decreased compared to individual α or β chains. The effect of H^+, Cl^-, CO_2, and DPG is to physically bind to the tetramer at a site other than the heme iron and stabilize the T structure. More detailed accounts describing R-T transitions, in the presence and absence of ions or small organic molecules, and corresponding ligand affinities are reviewed elsewhere [139-142].

Myoglobin, on the other hand, is a much simpler protein consisting of one globin chain of 153 amino acids with one heme per molecule, somewhat analogous to one subunit of the Hb tetramer. Mb is found in muscle tissue where it functions to store O_2 liberated from Hb and subsequently donates its bound O_2 for mitochondrial oxidative phosphorylation. Like individual Hb chains, Mb has a predominantly α-helical structure forming a "box" designed to protect the heme iron ligand-binding region. The Fe also is coordinated to four pyrrole nitrogens and an imidazole nitrogen donated by the globin proximal histidine (F8). The heme iron of Mb reversibly binds O_2 in a simple bimolecular process that can be described by the law of mass action, giving rise to a hyperbolic O_2-binding curve. Since Mb is monomeric, no cooperative ligand binding is observed and the protein is insensitive to the heterotropic ligands (except H^+) known to effect Hb cooperativity. Owing to its physiological importance, its abundant availability from muscle, its stability, and its high-resolution x-ray structure, sperm whale Mb has been, and still serves as, a popular choice among scientists interested in protein structure-function relationships. (A detailed account of Mb structure and function may be found in Ref. 143.)

The wealth of knowledge accumulated on Hb and Mb over the years inspired the desire to manipulate these related proteins at the genetic level. For many years Hb α and β chains have been compared to naturally occurring mutations with the hope of gaining insight into the role of specific amino acid residues in protein function [142-144]. In addition, comparisons between Hb's and Mb's isolated from other species, with known amino acid changes at evolutionarily conserved positions, have enhanced the understanding of the structure-function relationships of these two proteins. In the last few years more precise control of altered Hb and Mb has been achieved through the efforts of gene cloning and subsequent site-directed mutagenesis. A discussion of gene cloning and protein expression methods for human Hb α and β chains, human Mb, and sperm whale Mb can be found in Sec. 2 of this chapter.

It is known that the rate of Fe oxidation for free heme is very rapid such that a stable O_2 complex cannot be formed, and also that the binding affinity of CO vs. O_2 favors CO by a factor of several thousand [142,145]. In comparison, the binding ratio of CO vs. O_2 for Hb and Mb is roughly 100 times less than free heme complexes, indicating clearly the significance of the globin. X-ray crystal structures have implicated two highly conserved ligand-binding pocket residues, the distal histidine (HisE7) and distal valine (ValE11), as the amino acids most likely responsible for this ligand-binding discrimination (Fig. 9). The HisE7 is thought to provide at least two crucial roles in ligand binding. The first is to stabilize the O_2 complex through hydrogen bonding with the N_ε of the HisE7 imidazole. This hydrogen bonding has been observed by neutron diffraction studies of O_2-bound Mb [146] but not for CO-bound Mb [147]. The second key function of the distal histidine has been thought to provide steric bulk, thus forcing CO to bind in a bent or tilted manner and thereby reduce its binding affinity. Most mutagenesis efforts on Hb and Mb have thus far been designed to answer the question of ligand binding.

7.2. Hemoglobin Mutagenesis

The ability to alter active site amino acid residues that are thought to directly participate in catalysis or ligand binding may be one of the most exciting possibilities of site-directed mutagenesis. Nagai and coworkers altered the active site of human Hb α and β chains and

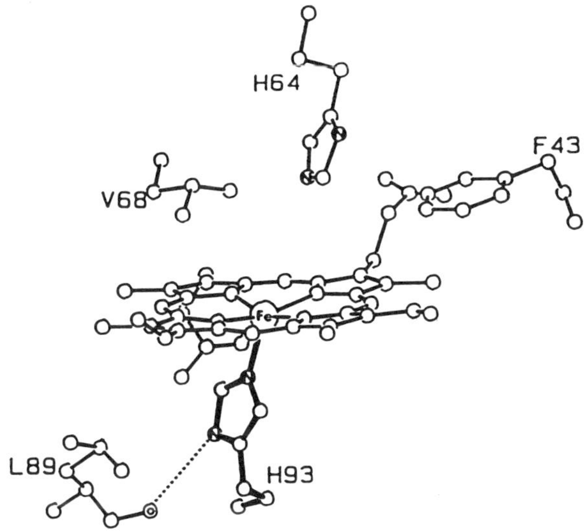

FIG. 9. Active site structure of myoglobin. The highly conserved residues near the ligand-binding site are labeled. These residues are also found in the active site of hemoglobin. (From Ref. 62, used with permission of Elsevier Publishing Co., Inc., copyright 1984.)

characterized the effects of these substitutions on ligand binding. They chose two highly conserved active site residues, the distal histidine His-63 (E7), and Val-67 (E11), which are thought to participate in ligand binding. They substituted Gln, Val, and Gly for β-His-63 (E7) and characterized these mutant proteins by differences in O_2 binding, resonance Raman Fe-CO stretch frequencies, and x-ray crystallography [148]. The β-Val-67 residue only affects O_2 binding when in the T state, whereas in the R state it lies further away from the O_2-binding site [149] and provides no steric interference. Ile-substituted β-Val-67 (V67I) reduces the O_2 affinity of the T state $\alpha_2\beta_2$ tetramer by about 50%, which approximates the theoretical maximum. In the same light, β-V67A shows an increase in O_2 affinity suggesting decreased steric hindrance. β-V67M and β-V67L have little effect on O_2 binding, and although the Leu side chain has the same volume as Ile, its side chain is flexible and disordered. Leu does not extend to the O_2-binding site; therefore, the O_2 affinity is not altered [148]. Steric blocking of the ligand-binding site is in agreement with the observation that CO recombination is also slower with this mutant [148].

The β-H63G mutation leaves a large space in the ligand-binding pocket and, surprisingly, there is no significant movement of the E helix [148]. This mutant protein binds CO 10 times faster than HbA but is unstable, as is β-H63V, thus preventing the determination of O_2 affinities for these two proteins. H63Q shows increased O_2 affinity and although the x-ray crystal structure has been determined to 2.12 Å resolution, this is not sufficient to determine the difference between N_ε or O_ε of the Gln side chain [148]. The Gln side chain is more flexible than His, and Nagai et al [148] suggest that it can donate a hydrogen bond to bound O_2. Resonance Raman data indicate that the Fe-CO vibrational modes are unchanged when Val-67 is substituted by Ile, Ala, Leu, or Met, suggesting that the R-state protein structure is flexible enough to accommodate these residues without distorting the Fe-CO bond [148]. When the His-63 residue is changed to Val, a broad shoulder on the low-frequency side of the Fe-CO stretching line is observed and the C-O stretching line splits to 1951 cm^{-1} and 1971 cm^{-1}. Replacement of His-63 with Gly also creates a split C-O stretching line into 1965 cm^{-1} and 1971 cm^{-1}. They also report that the resonance Raman spectrum of β-H63Q is essentially identical to HbA [148].

These results are not surprising in light of the highly conserved nature of the Val-67 and His-63 residues. Val-67 lies further from the ligand-binding site in α subunits thus permitting Ile in that position without causing serious deleterious effects on protein function [150,151]. However, these data do not explain why Leu is not present in the 67 position more frequently in nature. Since Gly and Val substituted for His-63 are unstable with bound O_2, it is obvious why these residues are not normally permitted. β-H63Q is a change seen in the α subunits of opossum Hb [152], and based on the data of Nagai et al. [148], Gln is thought to be capable of donating a hydrogen bond to bound O_2 in a manner similar to His-63 and therefore can be considered a conservative change.

Nagai and coworkers also reported mutagenesis of human Hb β chains in an attempt to understand the Root effect [153,154]. The Root effect describes an interesting phenomenon observed for some fish Hbs where a drastic decrease in O_2 affinity is observed at low pH [155]. The Root effect is not observed in mammalian Hbs, although a similar yet not so drastic pH effect is observed. The mammalian Hbs show lowered O_2 affinity at low pH but, in contrast to some fish Hbs, cooperative O_2 binding still exists. This phenomenon, known as the Bohr effect [136], is not completely understood at present and is better reviewed elsewhere [137,150,156]. In some fish Hbs, the T state is favored at low pH even after O_2 saturation, thus eliminating any possible cooperative ligand binding [157,158]. Perutz and Brunori have suggested, based on model building, that the salt bridge between the C-terminal His-146 and an Asp or Glu at position 94 in the β chains would be strengthened in fish Hbs, thus stabilizing the T state [159]. In an attempt to reproduce the Root effect using human Hb b-chains, Nagai et al. produced two mutant Hb proteins through the use of site-directed mutagenesis [153]. The first, Hb Nympheas (b-C93S), showed slightly increased O_2 affinity, diminished cooperativity, normal DPG, and slightly reduced alkaline Bohr effects. The second mutant protein, Hb Daphne (b-C93S-b-H143R), showed lower cooperativity, higher O_2 affinity, a slightly reduced alkaline Bohr effect, and an enhanced DPG effect. Neither protein displayed the Root effect. X-ray crystallographic analyses of these mutant proteins revealed that the major structural changes were at the site of substitution [154]. In the Hb Nympheas mutant protein, Asp-94 had moved closer to and formed a hydrogen bond with the Ser-93 hydroxyl resulting in an equilibrium with the normal salt bridge between Asp-94 and His-146 of the β chain. Luisi and Nagai [154] concluded that disruption of this Asp-94 to His-146 salt bridge destabilizes the T state, thereby reducing the Bohr effect. In the Hb Daphne mutant protein, extensive movement of side chains around the H143R substitution was observed [154]. Luisi and Nagai therefore concluded that electrostatic repulsion by the positively charged arginine causes b-Lys-82 to move slightly, resulting in a slight displacement of the corner connecting the E and F helices. The enhanced DPG effect observed for Hb Daphne may be due to the positively charged, long flexible side chain of arginine which could bind DPG more strongly [153]. Luisi and Nagai [154] also suggested that substitution of b-Asp-94 with Glu in addition to the above mutations would be required to produce the Root effect. (Glu is normally at β-chain position 94 in fish Hb's exhibiting the Root effect.) Since the side chain of Glu is longer than Asp, they speculated that the normal salt bridge between the carboxylate residue at position 94 and His-146 would be stabilized.

7.3. Myoglobin Mutagenesis

Sligar and coworkers applied site-directed mutagenesis to a totally synthetic sperm whale Mb gene [8] and substituted the distal histidine, His-64 (E7), to 10 different amino acids (Gly, Val, Phe, Cys, Met, Thr, Asp, Lys, Arg, Tyr). All of these mutant proteins express heme containing Mb in *E. coli* to approximately 10% of the total soluble cell protein. These distal histidine mutant proteins have been characterized in terms of ligand-binding kinetics, low-temperature Fourier transform-infrared spectroscopy and linear dichroism, electron paramagnetic resonance spectroscopy, resonance Raman spectroscopy, NMR, and UV-visible spectroscopy. The proximal histidine (His-93) was substituted with Cys and Tyr in an attempt to change the axial ligation and form cytochrome P-450 (cysteine) and catalase (tyrosine) analogs. Both proximal mutant proteins express in *E. coli* to a high degree but less than 10% are heme containing. Upon purification and reconstitution with hemin, these proteins are more unstable than the distal histidine mutant proteins. Preliminary EPR evidence shows a mixture of states but clearly resolved high-spin rhombic components, with g values near 8 and 6 for the cysteine derivative, and g = 6.4, 5.4 for the tyrosine mutant, suggesting that axial ligand replacement has occurred.

The autoxidation of Mb is known to occur under physiological temperatures and pH through the nucleophilic displacement of bound O_2 by H_2O or OH^- [160-162]. Sligar and coworkers [163,164] measured the autooxidation rates of nine of the His-64 mutant proteins (H64Y does not form a stable O_2 complex) and have found that rates range from 40 to 300 times faster than native sperm whale Mb depending on the substitution for His-64. One interpretation of these results is that the ligand-binding pocket is more accessible to solvent and that lack of a suitable hydrogen bond donor prevents a stable O_2 complex. Consistent with this explanation is the observation that the O_2 affinities measured for H64G, H64V, H64F, H64M, and H64R range from 10 to 150 times lower than native Mb [163]. The potential hydrogen bond donors are probably either too distant (H64T) or not in the correct orientation (H64K, H64R) to donate a hydrogen bond and stabilize bound O_2. Sligar and coworkers [164] also measured the CO-rebinding kinetics of their distal histidine mutants at 24°C following a 15-nsec laser flash. Interestingly, all of the mutant proteins, except H64Y, rebind CO with rates of approximately 10 times faster than native Mb. In addition, they measured the CO affinities for H64G, H64V, H64F, H64M, and H64R and found that the CO affinity has increased approximately 5.3, 5.1, 2.9, 7.0, and 14.0 times, respectively, for these mutant proteins. H64Y rebinds CO at 0.81 the rate of native Mb, but this result is confused by the fact that in the Fe^{3+} state, this protein has the E7Tyr as its axial ligand as determined by EPR (g = 6.64, 5.53) and is five-coordinate [164]. It is not yet clear whether upon reduction and addition of CO the ligation switches back to the proximal histidine, or whether the Tyr remains as the ligand and CO is binding on the proximal side. The steric bulk provided by the distal histidine is clearly important in reducing the affinity for CO, but data on the His-64 mutants suggest that the bulky side chain must be in the proper orientation to reduce CO binding. This stable orientation in native Mb may arise from hydrogen bond and electrostatic interactions between the His-64 imidazole, Arg-45, and a heme propionate. Mutagenesis of the His-64 residue may interrupt this lattice structure and allow easy access of ligands and solvent into the binding pocket. This may help explain the increased bimolecular association rate constants for O_2 and CO. The phenylalanine side chain for the H64F mutant may simply not be in the proper orientation or provide the same lattice structure necessary to decrease CO binding to the same extent as native Mb.

Sligar, Frauenfelder, and coworkers have used low-temperature infrared spectroscopy and found that the three distinct bound C-O stretch frequency bands, ν(C-O), normally seen with native sperm whale MbCO have collapsed into one broad band at 1973 cm^{-1} for the H64G mutant protein. A similar result has been observed for the H64V, H64F, and H64D mutant proteins as well. It is apparent that the presence of histidine at the E7 position gives rise to the multiple CO conformations observed by Frauenfelder and coworkers. For a more detailed description of the analysis of multiple MbCO conformations, see Austin et al. [165] and Ansari et al. [166]. In addition, Frauenfelder, Sligar, and coworkers have employed infrared spectroscopy measurements of the linear dichroism following photoselective flash photolysis at low temperature to measure the angle between bound CO and the heme plane. They observed an angle of approximately 70° for the H64G mutant sperm whale Mb [167], which is in contrast to the current dogma which might predict an angle of nearly 0° when steric bulk from the distal histidine has been removed. Model building suggests that the Val-68 may be providing steric bulk to prevent a 0° CO angle in H64G. Also, the CO-rebinding rates following photodissociation at temperatures ranging from 60 to 160°K for H64G have been measured. This CO-rebinding data reveal nonexponential kinetics and were fit assuming a distribution of conformational substates with different barriers for rebinding [165,166]. Below 180°K in a 75% glycerol-H_2O solution, the ligand is confined to the distal pocket and the protein is frozen in a conformational substate with a unique activation enthalpy for rebinding. A 4 KJ/mol decrease in peak enthalpy is observed for H64G, most probably due to the decreased steric hindrance in the pocket. A reduction in the preexponential is also observed and can be attributed to an increase in the entropy of the system. Pocket volumes were calculated from exponentials and indicated a difference of approximately 84 $Å^3$ for native Mb minus H64G. Interestingly, this number is in good agreement with the difference in side chain volumes between histidine and glycine (approximately 93 $Å^3$).

Resonance Raman spectroscopy has been performed by Champion, Sligar, and coworkers on the met-, deoxy-, and CO-bound forms of the H64G and H64M mutant proteins [167]. No gross structural changes in the heme environment or in the tertiary structure upon substitution at the E7 position is observed. At 295°K, pH 7.0, both H64G and H64M have split νFe-CO at 495 and 508 cm^{-1}. This frequency is quite close to that found for low-pH structures of native Mb, possibly reflecting a more open distal pocket. This finding may help explain the observed increase in autoxidation associated with replacement at the E7 position.

In related work, Boxer and coworkers cloned the human myoglobin gene and employed the Nagai expression system to produce mutant proteins in a bacterial host [2]. Baldwin and coworkers are currently using this system to introduce mutations into the myoglobin gene which stabilize intermediate folding states in a general effort to understand the mechanism of protein renaturation and helix-helix recognition.

ACKNOWLEDGMENTS

This review and the research in our laboratory has been supported by generous grants from the National Institutes of Health. We wish to thank Dr. Thomas Poulos for continuing collaborative endeavors and providing the graphic data on the P-450cam active site. Dr. F. Sherman, Dr. J. T. Hazzard, Dr. B. Hoffman, and Dr. G. Mauk generously provided unpublished work.

REFERENCES

1. K. Nagai and H. C. Thøgersen, *Nature, 309,* 810 (1984).
2. R. Varadarajan, A. Szabo, and S. G. Boxer, *Proc. Natl. Acad. Sci. USA, 82,* 5681 (1985).
3. D. B. Goodin, G. A. Mauk, and M. Smith, *Proc. Natl. Acad. Sci. USA, 83,* 1296 (1986).
4. L. A. Fishel, J. E. Villafranca, and J. Kraut, *Biochemistry, 25,* 351 (1987).
5. G. J. Pielak, A. G. Mauk, and M. Smith, *Nature, 313,* 152 (1987).
6. B. P. Unger, I. C. Gunsalus, and S. G. Sligar, *J. Biol. Chem., 261,* 1158 (1986).
7. T. Ikemura, *J. Mol. Biol., 157,* 389 (1981).
8. B. A. Springer and S. G. Sligar, *Proc. Natl. Acad. Sci. USA, 84,* 8961 (1987).
9. S. Beck von Bodman, M. A. Schuler, D. R. Jollie, and S. G. Sligar, *Proc. Natl. Acad. Sci. USA, 83,* 9443 (1986).
10. H. Koga, B. Rauchfuss, and I. C. Gunsalus, *Biochem. Biophys. Res. Commun., 130,* 412 (1985).
11. E. Margoliash, *Nature, 192,* 1121 (1961).
12. E. Margoliash, *J. Biol. Chem., 237,* 2161 (1962).
13. S. K. Chan and E. Margoliash, *J. Biol. Chem., 241,* 335 (1966).
14. S. K. Chan and E. Margoliash, *J. Biol. Chem., 241,* 507 (1966).
15. K. Narita, K. Titani, Y. Yaoi, H. Murakami, M. Kimura, and J. Vanacek, *Biochim. Biophys. Acta, 73,* 670 (1963).
16. F. R. Sallemme, S. T. Freer, N. H. Xuong, R. A. Alden, and J. Kraut, *J. Biol. Chem., 248,* 3910 (1973).
17. T. Takano and R. E. Dickerson, *J. Mol. Biol., 153,* 79 (1981).
18. N. Tanaka, T. Yamane, T. Tsukihara, and M. Kakudo, *J. Biochem., 77,* 147 (1975).
19. R. Timkovich and R. E. Dickerson, *J. Biol. Chem., 51,* 4033 (1976).
20. R. E. Dickerson, T. Takano, D. Eisenberg, O. B. Kallai, I. Samson, A. Cooper, and E. Margoliash, *J. Biol. Chem., 246,* 1511 (1971).
21. D. Keilin, in "The History of Cell Respiration and Cytochromes", Cambridge University Press, London, 1966.
22. E. Margoliash and A. Schejter, in "Advances in Protein Chemistry" (C. B. Anfinsen, M. L. Anson, J. T. Edsall, and F. M. Richards, eds.), Academic Press, New York, 1966, p. 114.

23. R. E. Dickerson and R. Timkovich, in "The Enzymes", 3rd ed., Vol. II (P. D. Boyer, ed.), Academic Press, New York, 1975, p. 397.

24. M. A. Cusanovich, T. E. Meyer, and G. Tollin, in "Advances in Inorganic Biochemistry", Vol. 7 (G. L. Eichhorn and L. G. Marzilli, eds.), Elsevier, New York, 1988, p. 37.

25. J. R. Winkler, D. G. Nocera, K. M. Yocom, E. Bordignon, and H. B. Gray, *J. Am. Chem. Soc., 104,* 5798 (1982).

26. K. M. Yocom, J. B. Shelton, J. R. Shelton, W. A. Shroeder, G. Worosila, S. S. Isied, E. Bordignon, and H. B. Gray, *Proc. Natl. Acad. Sci. USA, 79,* 7052 (1982).

27. D. G. Nocera, J. R. Winkler, K. M. Yocom, E. Bordignon, and H. B. Gray, *J. Am. Chem. Soc., 106,* 5145 (1984).

28. G. McLendon and J. R. Miller, *J. Am. Chem. Soc., 107,* 7811 (1985).

29. G. McLendon, T. Guarr, M. McGuire, K. Simolo, S. Strauch, and K. Taylor, *Coord. Chem. Rev., 64,* 113 (1985).

30. N. Liang, L. H. Kang, E. Margoliash, P. S. Ho, and B. M. Hoffman, *J. Am. Chem. Soc., 108,* 4665 (1986).

31. K. T. Conklin and G. McLendon, *Inorg. Chem., 25,* 4804 (1986).

32. P. S. Ho, C. Sutoris, N. Liang, E. Margoliash, and B. M. Hoffman, *J. Am. Chem. Soc., 107,* 1070 (1985).

33. J. T. Hazzard, T. C. Poulos, and G. Tollin, *Biochemistry, 26,* 2836 (1987).

34. E. Cheung, K. Taylor, J. A. Kornblatt, A. M. English, G. McLendon, and J. R. Miller, *Proc. Natl. Acad. Sci. USA, 83,* 1330 (1986).

35. G. J. Pielak, K. Oikawa, A. G. Mauk, M. Smith, and C. M. Kay, *J. Am. Chem. Soc., 108,* 2724 (1986).

36. J. L. McGourty, N. V. Blough, and B. M. Hoffman, *J. Am. Chem. Soc., 105,* 4470 (1983).

37. N. Liang, G. J. Pielak, A. G. Mauk, M. Smith, and B. Hoffman, *Proc. Natl. Acad. Sci. USA, 84,* 1249 (1987).

38. J. R. Miller and J. V. Beitz, *J. Chem. Phys., 74,* 6746 (1981).

39. R. Bechtold, C. Kuehn, C. Lepre, and S. S. Isied, *Nature, 322,* 286 (1986).

40. B. M. Hoffman and M. A. Ratner, *J. Am. Chem. Soc., 109,* 6237 (1987).

41. G. McLendon, K. Pardue, and P. Bak, *J. Am. Chem. Soc., 109,* 7540 (1987).

42. N. Liang, A. G. Mauk, G. J. Pielak, J. A. Johnson, M. Smith, and B. M. Hoffman, *Science, 240,* 311 (1988).

43. D. O. Lambeth, K. L. Campbell, R. Zand, and G. Palmer, *J. Biol. Chem., 248,* 8130 (1973).

44. J. T. Hazzard, G. McLendon, M. A. Cusanovich, G. Das, F. Sherman, and G. Tollin, *Biochemistry, 27,* 4445 (1988).

45. I. Ahmad, M. A. Cusanovich, and G. Tollin, *Biochemistry, 21,* 2133 (1982).

46. I. Ahmad, M. A. Cusanovich, and G. Tollin, *Proc. Natl. Acad. Sci. USA, 78,* 6724 (1981).

47. D. M. Hampsey, G. Das, and F. Sherman, *FEBS Lett.,* in press.

48. D. M. Hampsey, G. Das, and F. Sherman, *J. Biol. Chem., 261,* 3259 (1986).

49. G. V. Louie, W. L. B. Hutchcon, and G. D. Brayer, *J. Mol. Biol., 199,* 295 (1988).

50. J. F. Ernst, D. M. Hampsey, J. W. Stewart, S. Rackovsky, D. Goldstein, and F. Sherman, *J. Biol. Chem., 260,* 13225 (1985).

51. D. Holzscho, L. Principio, K. T. Conklin, D. R. Hickey, J. Short, R. Rao, G. McLendon, and F. Sherman, *J. Biol. Chem., 262,* 7125 (1987).

52. S. Tsunasawa, J. W. Stewart, and F. Sherman, *J. Biol. Chem., 260,* 5382 (1985).

53. A. Bachmair, D. Finley, and A. Varshausky, *Science, 234,* 179 (1986).

54. A. Ben-Bassat and K. Bauer, *Nature, 326,* 315 (1987).

55. T. Yonetani, in "Advances in Enzymology", Vol. 33 (F. F. Nord, eds.), Interscience, New York, 1970, p. 309.

56. T. Yonetani, in "The Enzymes", Vol. 13 (P. D. Boyer, ed.), Academic Press, New York, 1976, p. 345.

57. M. Erecinska, P. Oshino, P. Loh, and E. Brockelhurst, *Biochem. Biophys. Acta, 292,* 1 (1973).

58. J. Kaput, S. Goltz, and G. Blobel, *J. Biol. Chem., 257,* 15054 (1982).

59. D. B. Goodin, A. G. Mauk, and M. Smith, *Proc. Natl. Acad. Sci. USA, 83,* 1295 (1986).

60. L. A. Fishel, J. E. Villafranca, M. Maura, and J. Kraut, *Biochemistry, 26,* 351 (1987).

61. T. L. Poulos and J. Kraut, *J. Biol. Chem., 255,* 8199 (1980).

62. T. L. Poulos, in "Advances in Inorganic Biochemistry", Vol. 7 (G. L. Eichhorn and L. G. Marzilli, eds.), Elsevier, New York, 1988, p. 5.

63. A. F. Coulson and T. Yonetani, *Biochemistry, 14,* 2389 (1975).

64. G. R. Schonbaum and S. Lo, *J. Biol. Chem., 247,* 3353 (1972).

65. P. George, *Biochem. J., 54,* 267 (1953).

66. T. Yonetani, *J. Biol. Chem., 240,* 4509 (1965).

67. A. F. Coulson, J. E. Erman, and T. Yonetani, *J. Biol. Chem., 246,* 917 (1971).

68. D. Dolphin, A. Forman, D. C. Borg, J. Fajer, and R. H. Felton, *Proc. Natl. Acad. Sci. USA, 68,* 614 (1971).

69. L. E. Shulze, P. W. Devancy, H. Winkley, P. G. Debrunner, N. Doan, R. Chiang, R. Rutter, and L. Hager, *FEBS Lett., 103,* 100 (1979).

70. T. Yonetani, H. Schleyer, and A. Ehrenberg, *J. Biol. Chem., 241,* 3240 (1966).

71. K. Takio, K. Titani, L. H. Ericsson, and T. Yonetani, *Arch. Biochem. Biophys., 203,* 615 (1980).

72. B. M. Hoffman, J. E. Roberts, C. H. Kang, and E. Margoliash, *J. Biol. Chem., 256,* 6556 (1981).

73. K. G. Welinder, *FEBS Lett., 72,* 19 (1976).

74. S. L. Edwards, N. H. Xuong, R. C. Hamlin, and J. Kraut, *Biochemistry, 26,* 1503 (1987).

75. H. A. deBoer and R. A. Kastelein, in "Maximizing Gene Expression" (W. Resnikoff and L. Gold, eds.), Butterworth, Stoneham, MA, 1986, p. 225.

76. N. Oshino and R. Sato, *J. Biochem., 69,* 169 (1971).

77. P. Strittmatter, L. Spatz, D. Corcoran, J. J. Rogers, and R. Redline, *Proc. Natl. Acad. Sci. USA, 71,* 4565 (1974).

78. T. Shimakata, K. Mihara, and R. Sato, *J. Biochem. (Tokyo), 73,* 1163 (1972).

79. B. Cohen and R. W. Estabrook, *Arch. Biochem. Biophys., 143,* 54 (1971).

80. A. G. Hildebrant and R. W. Estabrook, *Arch. Biochem. Biophys., 143,* 66 (1971).

81. E. T. Morgan and M. J. Coon, *Drug Metab. Dispos., 12,* 358 (1984).

82. P. P. Tamburini, R. E. White, and J. B. Schenkman, *J. Biol. Chem., 260,* 4007 (1985).

83. D. Pompon and M. J. Coon, *J. Biol. Chem., 259,* 15377 (1984).

84. C. Bonfils, C. Balny, and P. Maurel, *J. Biol. Chem., 256,* 9457 (1981).

85. D. Hultquist, L. Sannes, and D. Schafer, *Prog. Clin. Biol. Res., 55,* 291 (1981).

86. D. Hultquist and P. Passon, *Nature, New Biol., 229,* 252 (1971).

87. F. S. Mathews, M. Levine, and P. Argos, *J. Mol. Biol., 64,* 449 (1972).

88. J. J. Falke and D. E. Koshland, *Science, 237,* 1596 (1987).

89. P. S. Stayton, M. T. Fisher, and S. G. Sligar, *J. Biol. Chem., 263,* 13544 (1988).

90. H. A. Dailey and P. Strittmatter, *J. Biol. Chem., 254,* 5388 (1979).

91. P. P. Tamburini, R. E. White, and J. B. Schenkman, *J. Biol. Chem., 260,* 4007 (1985).

92. C. S. Hackett and P. Strittmatter, *J. Biol. Chem., 259,* 3275 (1984).

93. F. R. Salemme, *J. Biol. Chem., 102,* 563 (1976).

94. T. L. Poulos and A. G. Mauk, *J. Biol. Chem., 258,* 7369 (1983).

95. D. J. Livingston, S. J. McLachlan, G. N. La Mar, and W. D. Brown, *J. Biol. Chem., 260,* 15699 (1985).

96. M. R. Mauk and A. G. Mauk, *Biochemistry, 21,* 4730 (1982).

97. M. T. Fisher, R. E. White, and S. G. Sligar, *J. Am. Chem. Soc., 108,* 6835 (1986).

98. F. G. Prendergast, M. Meyer, G. L. Carlson, S. Iida, and J. D. Potter, *J. Biol. Chem., 258,* 7541 (1983).

99. J. D. Lipscomb, S. G. Sligar, M. J. Namtvedt, and I. C. Gunsalus, *J. Biol. Chem., 251,* 1116 (1976).

100. T. L. Poulos, B. C. Finzel, and A. J. Howard, *J. Mol. Biol., 195,* 687 (1987).

101. T. L. Poulos, in "Adv. Inorg. Biochem., Vol 7" (G. L. Eichhorn and C. G. Marzilli, eds.), Elsevier, New York, 1988, p. 1.

102. K. K. Rodgers, T. C. Pochapsky, and S. G. Sligar, *Science, 240,* 1657 (1988).

103. P. S. Stayton and S. G. Sligar, submitted.

104. J. D. Morrisett, in "Spin Labeling: Theories and Applications" (L. J. Berliner, ed.), Academic Press, New York, 1976, p. 274.

105. R. E. Dickerson, T. Takano, D. Eisenberg, O. B. Kallai, L. Samson, A. Cooper, and E. Margoliash, *J. Biol. Chem., 246,* 1511 (1971).

106. L. M. Geren and F. Millett, *J. Biol. Chem., 256,* 4851 (1981).

107. S. Ferguson-Miller, D. L. Brautigan, and E. Margoliash, *J. Biol. Chem., 253,* 149 (1978).

108. C. H. Kang, D. L. Brautigan, and E. Margoliash, *J. Biol. Chem., 253,* 6502 (1978).

109. T. L. Poulos and J. Kraut, *J. Biol. Chem., 255,* 10322 (1980).

110. G. Weber and H. G. Drickamer, *Quart. Rev. Biophys., 16,* 89 (1983).

111. K. Heremans, F. Ceuterick, J. Snauwaert, and J. Wauters, in "Techniques and Applications of Fast Reactions in Solutions" (W. J. Gettins and E. Wyn-Jones, eds.), Reidel, Dordrecht, 1979, p. 429.

112. M. R. Mauk, L. S. Reid, and A. G. Mauk, *Biochemistry, 21,* 1843 (1982).

113. C. Chothia and J. Janin, *Nature, 256,* 705 (1975).

114. P. D. Ross and S. Subramanian, *Biochemistry, 20,* 3096 (1981).

115. L. S. Reid, M. R. Mauk, and A. G. Mauk, *J. Am. Chem. Soc., 106,* 2182 (1984).

116. S. G. Sligar, K. D. Egeberg, J. Sage, D. Morikis, and P. M. Champion, *J. Am. Chem. Soc., 109,* 7896 (1987).

117. R. T. Raines, E. L. Sutton, D. R. Strauss, W. Gilbert, and J. R. Knowles, *Biochemistry, 25,* 7142 (1986).

118. P. Ortiz de Montellano, in "Cytochrome P-450: Structure, Mechanism, and Biochemistry" (P. Ortiz de Montellano, ed.), Plenum Press, New York, 1986, p. 235.

119. S. D. Blake and M. J. Coon, in "Advances in Enzymology, Vol. 60" (A. Meister, ed.), Wiley-Interscience, New York, 1987, p. 35.

120. M. Ingelman-Sundberg, in "Cytochrome P-450: Structure, Mechanism, and Biochemistry" (P. Ortiz de Montellano, ed.), Plenum Press, New York, 1986, p. 124.

121. H. Murakami, Y. Yabasuki, T. Sakaki, M. Shibata, and H. Ohkawa, *DNA, 6,* 189 (1987).

122. H. Ohkawa, Y. Yabusaki, T. Sakaki, H. Murakami, and M. Shibata, in "Cytochrome P-450: New Trends" (R. Sato, T. Omura, Y. Imai, and Y. Fuji-Kunyama, eds.), Yamada Science Foundation, Japan, 1987, p. 103.

123. E. Szczesna-Skopura, N. Browne, D. Mead, and B. Kemper, *Proc. Natl. Acad. Sci. USA, 85,* 738 (1988).

124. D. J. Anderson, K. E. Mostov, and G. Blobel, *Proc. Natl. Acad. Sci. USA, 80,* 7249 (1983).

125. W. T. Wickner and H. F. Lodish, *Science, 230,* 400 (1986).

126. J. A. Peterson, Y. Ishimura, and B. W. Griffin, *Arch. Biochem. Biophys., 149,* 197 (1972).

127. R. I. Murray, M. T. Fisher, P. G. Debrunner, and S. G. Sligar, in "Metalloproteins: I. Metal Proteins with Redox Roles" (P. M. Harrison, ed.), Macmillan, London, 1985, p. 157.

128. T. L. Poulos, B. C. Finzel, I. C. Gunsalus, G. C. Wagner, and J. Kraut, *J. Biol. Chem., 260,* 16122 (1985).

129. T. L. Poulos, B. C. Finzel, and A. J. Howard, *Biochemistry, 25,* 5314 (1986).

130. T. L. Poulos and A. J. Howard, *Biochemistry, 26,* 8165 (1987).

131. J. W. Taylor, J. Ott, and F. Eckstein, *Nucleic Acids Res., 13,* 8764 (1985).

132. J. H. Dawson and M. Sono, *Chem. Rev., 37,* 1257 (1987).

133. B. P. Unger, Doctoral thesis, University of Illinois, 1988.

134. J. C. Kendrew, R. E. Dickerson, B. E. Strandberg, R. G. Hart, D. R. Davis, D. C. Phillips, and V. C. Shore, *Nature, 185,* 422 (1960).

135. M. F. Perutz, M. G. Rossman, A. F. Cullis, H. Muirhead, G. Will, and A. C. T. North, *Nature, 185,* 416 (1960).

136. C. Bohr, K. Hasselbalch, and A. Krogh, *Arch. Physiol., 16,* 402 (1904).

137. K. Imai, in "Allosteric Effects in Haemoglobin", Cambridge University Press, Cambridge, 1982.

138. J. M. Baldwin, *Prog. Biophys. Mol. Biol., 29,* 225 (1975).

139. M. F. Perutz, *Ann. Rev. Biochem., 48,* 327 (1979).

140. M. F. Perutz, G. Fermi, B. Luisi, B. Shaanan, and R. C. Liddington, *Acc. Chem. Res., 20,* 308 (1987).

141. J. M. Rifkind, in "Advances in Inorganic Biochemistry, Vol. 7" (G. L. Eichhorn and L. G. Marzilli, eds.), Elsevier, New York, 1988, p. 155.

142. R. E. Dickerson and I. Geis, in "Hemoglobin", Benjamin Cummings, Menlo Park, CA, 1983.

143. E. Antonini and M. Brunori, in "Hemoglobin and Myoglobin in Their Reactions with Ligands", Elsevier, Amsterdam, 1971.

144. G. Fermi and M. F. Perutz, in "Hemoglobin and Myoglobin: Atlas of Biological Structures" (D. C. Phillips and F. M. Richards, eds.), Clarendon, Oxford, 1981.

145. J. P. Collman, J. I. Brauman, T. R. Halpert, and K. S. Suslick, *Proc. Natl. Acad. Sci. USA, 73,* 3333 (1976).

146. S. E. V. Phillips and B. P. Schoenborn, *Nature, 292,* 81 (1981).

147. J. C. Hanson and B. P. Schoenborn, *J. Mol. Biol., 153,* 117 (1981).

148. K. Nagai, B. Luisi, D. Shih, G. Miyazaki, K. Imai, C. Poyart, A. De Young, L. Kwiatkowsky, R. W. Noble, S. A. Lin, and N. T. Yu, *Nature, 329,* 858 (1987).

149. B. F. Luisi, Doctoral thesis, Cambridge University, 1986.

150. M. F. Perutz, *Nature, 228,* 726 (1970).

151. G. Fermi, M. F. Perutz, B. Shaanan, and R. Fourme, *J. Mol. Biol., 175,* 159 (1984).

152. P. Stenzel, B. Brimhall, R. T. Jones, J. A. Black, A. McLachlan, and D. Gibson, *J. Biol. Chem.*, *254*, 2071 (1979).

153. K. Nagai, M. F. Perutz, and C. Poyart, *Proc. Natl. Acad. Sci. USA*, *82*, 7252 (1985).

154. B. F. Luisi and K. Nagai, *Nature*, *320*, 555 (1986).

155. R. W. Root, *Biol. Bull.*, *61*, 427 (1931).

156. C. Ho and I. M. Russu, *Biochemistry*, *26*, 6299 (1987).

157. A. L. Tan, R. W. Nobel, and Q. H. Gibson, *J. Biol. Chem.*, *248*, 2880 (1973).

158. M. Brunori, *Curr. Top. Cell. Reg.*, *9*, 1 (1975).

159. M. F. Perutz and M. Brunori, *Nature*, *299*, 421 (1982).

160. Y. Sugawara and K. Shikama, *Eur. J. Biochem.*, *110*, 241 (1980).

161. Y. Satoh and K. Shikama, *J. Biol. Chem.*, *256*, 10272 (1981).

162. G. Tajima and K. Shikama, *J. Biol. Chem.*, *262*, 12603 (1987).

163. B. A. Springer, K. D. Egeberg, S. G. Sligar, R. J. Rohlfs, A. J. Mathews, and J. S. Olson, submitted (1988).

164. K. D. Egeberg, B. A. Springer, S. G. Sligar, D. Morikis, and P. M. Champion, submitted (1988).

165. R. H. Austin, K. W. Beeson, L. Eisenstein, H. Frauenfelder, I. C. Gunsalus, *Biochemistry*, *14*, 5355 (1975).

166. A. Ansari, J. Berendzen, D. Braunstein, B. R. Cowen, H. Frauenfelder, M. K. Hong, E. T. Iben, J. B. Johnson, P. Ormos, T. B. Sauke, R. Scholl, A. Schulte, P. J. Steinbach, J. Vittilow, and R. D. Young, *Biophys. Chem.*, *26*, 337 (1987).

167. D. Morikis, P. M. Champion, B. A. Springer, and S. G. Sligar, *Bulletin of the American Physical Society*, *33*, 691 (1988).

168. T. L. Poulos and B. C. Finzel, in "Peptide and Protein Reviews, Vol. 4" (M. T. W. Hearn, ed.), Marcel Dekker, Inc., New York, 1984, p. 115.

13

Exploring Structure-Function Relationships in Yeast Cytochrome c Peroxidase Using Mutagenesis and Crystallography

J. Matthew Mauro, Mark A. Miller, Steven L. Edwards, Jimin Wang, Laurence A. Fishel, and Joseph Kraut
Department of Chemistry B-017
University of California, San Diego
La Jolla, CA 92093

1. INTRODUCTION

1.1. Scope

Yeast mitochondrial cytochrome c peroxidase (CCP) is a 294-residue, heme-containing enzyme, whose unique biophysical properties have made it the subject of a substantial number of investigations. These studies have yielded insight into the mechanisms of enzymic O-O bond cleavage [1,2], macromolecular recognition [3,4], interprotein electron transfer [5-7], and heme-protein interactions [8,9]. Rather than exhaustively examine the published literature, this review will briefly highlight the results of recent work which has utilized the tools of x-ray crystallography and site-directed mutagenesis to explore structure-function questions relevant to CCP.

1.2. Background

1.2.1. Compound I of CCP: Formation, Characteristics, and Breakdown

In a simple formulation of the CCP catalytic mechanism, the enzyme in its resting ferric form binds and reduces H_2O_2; the resulting two-equivalent oxidized enzyme product, compound I, is then reduced to the resting ferric state in two successive one-electron steps by ferrocytochrome c:

$$CCP + H_2O_2 \rightarrow \text{compound I} + H_2O$$

$$\text{Compound I} + \text{cyt-c}(Fe^{2+}) + H^+ \rightarrow \text{compound II} + \text{cyt-c}(Fe^{3+})$$

$$\text{Compound II} + \text{cyt-c}(Fe^{2+}) + H^+ \rightarrow CCP + \text{cyt-c}(Fe^{3+}) + H_2O$$

Although the above scheme is conceptually pleasing, many important mechanistic details remain unknown. For instance, we are unaware of any definitive evidence proving that compound I is actually important in the catalytic pathway, and, in fact, there exists kinetic evidence [10] which suggests that rapid reaction between ferrocytochrome c and iron-bound peroxide in CCP may occur, thus preempting the formation of compound I.

Regardless of its relevance to the catalytic mechanism, the unusual chemical properties of compound I of CCP make it worthwhile to study in its own right. Based on the results of a number of careful spectroscopic studies, it has been proposed that the electronic structure of this species is best represented as $(Fe{=}O)^{4+}\text{-}R^*$, where R^* represents one or more radical centers that are believed to exist within the protein matrix [11-15] and which are responsible for the characteristic EPR behavior of the oxidized enzyme. Intriguing features of preparations of compound I include the stabilities of the oxyferryl and protein radical species ($t_{1/2}$ for spontaneous decay is 2-6 hr at room temperature [16]) and the broad "wings" which flank the narrow isotropic g = 2.00 radical signal in the 90°K EPR spectrum. These "wings" have been attributed to the presence of an axially symmetrical radical center whose EPR signal accounts for approximately 90% of the integrated spin intensity seen in the spectrum [17]. Below 25°K, where the axial signal narrows and dominates the first derivative spectrum, the narrow isotropic signal is still present but is obscured in the EPR powder spectrum by the resolved axial g = 2 feature. Hori and Yonetani [17] thus proposed that two radical species exist in preparations of compound I of CCP.

Compound I is rapidly reduced by stoichiometric quantities of ferrocytochrome c (10^7-10^8 M^{-1} sec^{-1}) [7,18] and much more slowly by other electron donors [19,20]. The rapid rate of reaction of compound

I with ferrocytochrome c has been attributed to a specific and tightly bound CCP:cytochrome c complex or set of complexes which enables electron transfer to occur at an enhanced rate compared with metal complexes having redox potentials and self-exchange rates similar to those of cytochrome c [21]. A hypothetical model for such a complex was previously proposed, based on complementary electrostatic interactions between the two proteins, as predicted from computer graphics modeling studies using x-ray structural data [3]. Although some experimental results have been obtained which support the proposed structure for the electron transfer complex [4,22], no conclusive evidence has yet been reported, and the nature of the intermolecular interactions involved in CCP:cytochrome c electron transfer continues to be the subject of considerable theoretical [23-25] and experimental interest [5,6,7,26,27].

1.2.2. Structure of the CCP Heme Crevice from X-ray Crystallography

Knowledge of the three-dimensional structure derived from high-resolution x-ray diffraction data [28,29] allowed the first rational attempts at explanation of the catalytic and spectroscopic properties of the enzyme, and opened the door to structure-function studies. Ten helices represent over 50% of the protein structure and the heme is held in a pocket made up of residues from two distinct domains. Details of the overall structure have been described previously [29,30]. Figure 1 shows the immediate heme environment, emphasizing some of the residues that we have considered in designing structure-function experiments using crystallography and site-directed mutagenesis. On the distal side of the heme plane, where ligand binding takes place, helix B [28] donates Trp-51, His-52, and Arg-48. Formation of compound I by heterolytic cleavage of bound hydroperoxide has been proposed to depend on His-52 acting as a sequential general acid-base catalyst, with Arg-48 acting to electrostatically stabilize the charge-separated transition state [1]. On the proximal side, Asp-235 links the imidazole of iron-coordinated His-175 with the indole ring

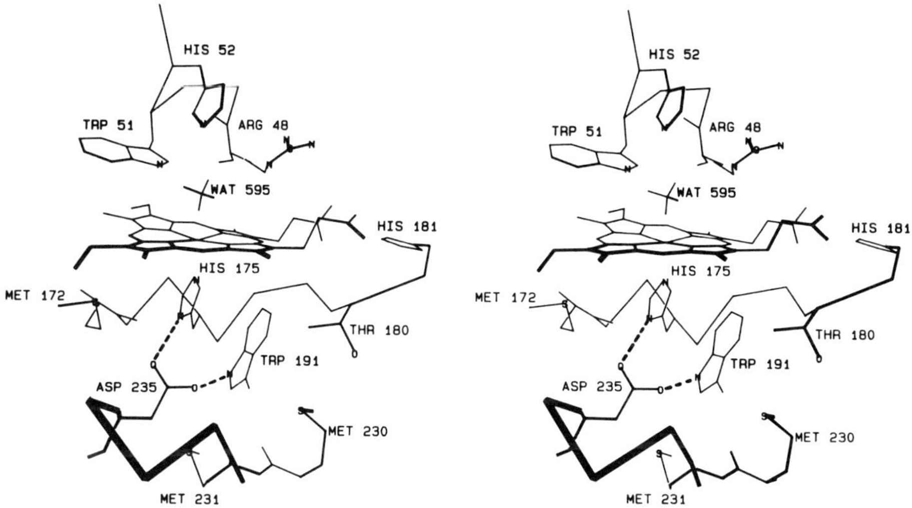

FIG. 1. Stereoscopic view of the heme environment in CCP(MI), the wild-type yeast peroxidase expressed in *E. coli* [34,55]. Hydrogen bonds between the Asp-235 carboxyl, His-175, and Trp-191 are shown by dotted lines.

of Trp-191 by means of hydrogen bonds, with the parallel ring systems positioned in van der Waals contact. Because it seems possible that control of the chemistry of the heme iron center occurs by means of the H-bonding and π-π orbital interactions that exist in this proximal side assembly (together with other nearby structural features described below), considerable effort is being devoted to investigation of structure-function relationships in this region of the CCP molecule.

1.3. Some Structure-Function Questions

A fundamental area of interest is the nature of compound I. Questions that have been used as a basis for designing experiments include the following: Is the structure of the enzyme significantly changed in this highly oxidized state? Can any such structural changes be used to suggest the location(s) of the radical species in compound I? Is the ability to form the radical species obligatory for enzyme activity?

What structural factors contribute to the stability of the radical species and the oxyferryl iron in compound I? Is the rate of electron transfer from cytochrome c enhanced by any special mechanism other than specific and tight binding between the macromolecular redox partners?

Another important area of interest is the modulation by the surrounding protein environment in CCP of the interaction of the heme with ligands. Does the detailed structure on the heme proximal side affect ligand-binding phenomena on the heme distal side? What structural factors are responsible for the pH dependence of ligand-binding behavior exhibited by the enzyme?

2. STRUCTURE-FUNCTION EXPERIMENTS EXAMINING THE NATURE OF COMPOUND I

2.1. Crystal Structure of Compound I of Baker's Yeast CCP

In a direct approach to examination of the structural differences between resting state ferric CCP and compound I, the low temperature (-15°C) x-ray structure of the oxidized intermediate has been determined [31]. For this purpose, crystals of ferric CCP were converted to compound I by soaking in dilute H_2O_2. A complete data set was obtained from six crystals that were irradiated for 4 hr or less per crystal in order to minimize structural contributions from spontaneous decomposition products [16,32].

The refined structure revealed that a number of small changes occur in compound I as compared to the resting state structure determined under identical conditions. The iron in the oxyferryl heme moves about 0.2 Å from its position in the ferric enzyme, from 0.1 Å to the proximal side of the heme plane to a new position 0.1 Å to the distal side of the heme plane. This iron movement is accompanied by a compensatory movement of proximal His-175 so that a 2.0-Å Fe-N coordination interaction is maintained. Additionally, the flexible distal Arg-48 side chain [33] swings to within hydrogen-bonding distance of the ferryl oxygen atom, and the entire heme appears to slide further into its binding pocket by about 0.15 Å.

More distant from the heme, the refined structure showed that the oxygen of Thr-180 had moved by 0.15 Å toward the sulfur atom of Met-230 (see Fig. 1). Close examination of the refined compound I structure as well as the difference Fourier map, which is a very sensitive indicator of small structural changes, revealed that three backbone carbonyl oxygen atoms, in addition to the Thr-180 oxygen atom, had tilted toward the sulfur of Met-230, and we have suggested that this local perturbation may result from the presence of a nearby positively charged radical center [31]. Based on EPR and ENDOR spectroscopic studies, Hoffman et al. [13] had previously proposed that the radical site in compound I may be composed of a pair of interacting methionine sulfurs, and it may be no coincidence that Met-230 lies adjacent to another methionine, Met-231, in the CCP primary sequence. Since these sulfur atoms are at least 6 Å apart in the folded enzyme, stabilization of a radical center cannot result from direct interaction of the two thioether centers, as originally suggested by Hoffman and coworkers [13]. However, an additional feature in this section of the molecule, namely, the position of the indole nitrogen of Trp-191 within 4.3 Å of each of these two sulfur atoms, may be of importance in stabilization of a radical center within the protein (see Fig. 1). Electronic interactions in this S-N-S assembly may be capable of stabilizing a radical, although the chemical feasibility of this idea remains to be demonstrated.

Finally, the effect of the manifold small structural adjustments that accompany compound I formation is the perturbation of a loop of 15 residues connecting proximal His-175 and adjacent Trp-191. This intervening loop extends to the surface of the enzyme where cytochrome c has been proposed to bind [3], and it is possible that this small conformational change influences the detailed characteristics of the CCP-cytochrome c interaction. The reverse process may also be possible, i.e., formation of an intermolecular complex may alter properties of the heme iron center, with the described loop acting as a transducer.

2.2. Using Engineered Mutants to Investigate Compound I

2.2.1. *Expression of Yeast CCP in* E. coli

In order to conveniently obtain large amounts of CCP and specific mutants for physical studies, we have developed a system for production of the enzyme in *E. coli* [34]. The starting point was the cloned pre-CCP gene derived from a standard strain of *S. cerevisiae* [35]. Construction of the plasmid-encoded gene for the mature protein involved replacement of the codons specifying amino acid residues -2 and -1 of the CCP leader sequence [36] with codons specifying Met-Ile, in order to create a BclI restriction site and an initiation codon. This CCP(MI) gene (where MI stands for Met-Ile) was then substituted for the *fol* gene in a pUC8 vector previously used for overexpression of dihydrofolate reductase [37]. The desired codon changes were accomplished using standard oligonucleotide-directed mutagenesis procedures in an M13 bacteriophage/*E. coli* system [38]. A similar system for expression of CCP in *S. cerevisiae* has been described by Goodin et al. [39].

Initial characterization of the *E. coli*-derived CCP(MI) enzyme revealed no significant differences in properties from either the plasmid-encoded enzyme expressed in *S. cerevisiae* or the enzyme isolated from commercial baker's yeast, as judged by catalytic activity or spectroscopic methods [34]. We note, however, that there are two differences between the published chemically determined primary sequence of baker's yeast CCP [36,40] and the plasmid-encoded CCP enzymes studied to date; thus Gly-152 and Ile-53 are found in the plasmid-encoded enzymes vs. Asp-152 and Thr-53 in the enzyme isolated from commercial baker's yeast.

2.2.2. *Directed Mutants Designed to Investigate Compound I: Proximal Met-172 → Ser and Distal Trp-51 → Phe*

A number of single-site CCP mutants have been specifically engineered to investigate structural features which contribute to the observed chemical and spectroscopic properties of compound I. Initial work focused on two residues in very close proximity to the heme which appeared to be reasonable sites for protein-based radical species, namely, proximal Met-172 and distal Trp-51 (Fig. 1). Using a yeast expression system, Goodin and coworkers [39] demonstrated that a Met-172 → Ser mutant had undiminished enzyme activity, a result which indicated that the integrity of this residue is not important in the catalytic mechanism. Met-172 had been previously proposed as a likely locus for a radical based on its location, 3.7 Å below the proximal heme face [29], and on the large g anisotropy of the EPR signal of compound I [13,14]. Compound I of the Ser-172 mutant gave a narrowed g = 2.00 EPR signal at 90°K, indicating that an effect of the mutation had been either (1) to eliminate spin density formerly associated with the methionine or (2) to more subtly change the properties of the radical center. As discussed below, the latter would appear to be the case, and a general theme that emerges from the mutagenesis work conducted to date on CCP is that changes in the detailed characteristics of the compound I EPR spectrum can occur as a result of even small structural disturbances in the heme-binding cleft.

Using the *E. coli* mutagenesis and expression system previously described, a mutant of CCP(MI) at position 51 has also been prepared, in order to test the hypothesis that a site of the unpaired electron observed in compound I is the indole ring of the tryptophan at this position [34]. The indole ring of Trp-51 is about 3.3 Å above and almost parallel to the heme plane on the distal side [28], and we reasoned that substitution of a phenylalanine residue at this position would result in minimal structural perturbations while replacing the indole ring with a much less readily oxidizable benzene ring. It was predicted that this mutation would result in a distal structural motif in the new enzyme analogous to that of the plant peroxidases [41], which have a phenylalanine residue in or close to this position, based on homologous primary sequence information. In the two-equivalent oxidized intermediates of the plant peroxidases, it has been proposed that the presence of the less readily oxidizable distal benzyl side chain may result in preferential oxidation of the heme, resulting in formation of a porphyrin π-cation radical [42-44], in contrast to the protein-based radical species which is believed to exist in compound I of CCP.

Most significantly, the CCP(MI,F51) mutant enzyme is two to four times as active as the parent CCP(MI) enzyme in steady-state assays [34]. Essentially identical kinetic results were obtained by Goodin and coworkers in assays of the same mutant expressed in a yeast system [45]. In addition, these investigators found an increase in the pH maximum for steady-state catalysis in the F51 mutant. The 90°K EPR spectrum of H_2O_2-oxidized CCP(MI,F51) displays a strong, sharp resonance at g = 2.004; however, the wings flanking the signal were narrowed in a manner analogous to the 90°K spectrum of the H_2O_2-oxidized Ser-172 mutant previously described. Based on these results, it was initially concluded that the indole ring of Trp-51 has a role in determining the pH dependence of enzyme catalysis as well as the detailed magnetic properties of the protein radical species, but that it is not required for enzyme activity. Furthermore, UV-visible and EPR studies of CCP(MI,F51) revealed that the lifetime at 23°C of compound I of the mutant is only about 1.4% of the lifetime observed for compound I of CCP(MI) or baker's yeast CCP. Thus, Trp-51 is not necessary for the formation of a radical species in H_2O_2-oxidized CCP(MI) but does exert a strong stabilizing influence. A final significant insight gained from initial studies on the F51 mutant is that replacement of distal Trp-51 with phenylalanine does not suffice to produce a heme protein which spectroscopically mimics the plant peroxidases; therefore, additional structural factors must account for the observed differences between CCP and the plant peroxidases.

Additional low-temperature EPR studies performed by Goodin et al. [45] further illuminated the effects of these single-site mutations on the magnetic properties of the radical species present in the compound I forms of the Ser-172 and Phe-51 mutant enzymes. Thus, although 90°K EPR spectra of both oxidized mutants were essentially missing the broad-wing features which are indicative of the presence of an axially symmetrical radical center, when recorded at 10°K, EPR spectra showed the presence of a signal characteristic of an axially symmetrical species. From these experiments it was concluded that the effect of either mutation had been to alter the temperature dependence of the magnetic properties of the enzyme rather than to abolish either of the radical centers. It was judged that the most likely reason for the similar effects of mutations on opposite sides of the heme was alteration of the zero-field splitting of the ferryl center in either mutant; interaction between the radical center(s) and the perturbed metal coordination center might be expected to result in changes of the temperature dependence of the observed EPR properties of the mutant enzymes.

2.2.3. An Additional Mutant Designed to Investigate Compound I: Proximal Trp-191 → Phe

The position of the indole ring of Trp-191 (Fig. 1), which lies parallel to and in contact with the imidazole ring of proximal His-175, together with its orientation with respect to the thioether sulfur atoms of Met-230 and Met-231, suggested that this residue may be important in formation and stabilization of the radical center(s) of compound I [29,31]. The indole nitrogen of Trp-191 also participates in a network of proximal side hydrogen bonds which includes the side chains of His-175 and Asp-235, and it seemed possible that these interactions could contribute to the stabilization of the oxyferryl heme in compound I [29,46-48]. It was reasoned that replacement of Trp-191 by phenylalanine would disrupt precise proximal side π-π and hydrogen-bonding interactions while causing minimal structural perturbations.

Initial studies of the Trp-191 → Phe mutant revealed that the magnetic characteristics of the H_2O_2-oxidized enzyme had been substantially modified and that the catalytic efficiency of the enzyme had also been profoundly decreased. Thus, 90°K EPR investigations conducted on peroxide-oxidized CCP(MI,F191) show that the mutation affects the shape of the EPR signal centered at g = 2.00 as well as the room temperature stability of the chemical center responsible for the radical signal. As in both the Ser-172 and Phe-51 mutants previously examined, the g = 2.00 signal in the 90°K first derivative EPR spectrum of the peroxide-oxidized F191 mutant is narrowed considerably [49]. However, in marked contrast to the previously examined mutant enzymes, the broad axial EPR features do not emerge at 10°K, and it therefore seems likely that the indole ring at position 191 is either the source of the dominant axially symmetrical signal exhibited by the parent enzyme or is important in formation or stabilization of an as yet unidentified radical center located elsewhere in the CCP molecule.

It is of interest that Trp → Phe replacements on either side of the heme plane (proximal position 191 or distal position 51) result in radical species in the H_2O_2-oxidized mutant enzymes having lifetimes which are only a few percent of the lifetimes of radical species in compound I of the parent enzyme. This suggests that the stability of protein-based radical species in compound I may be due to electron delocalization which occurs in the relatively π-orbital-rich environment that exists in the CCP heme cavity [47].

The Trp-to-Phe replacement at position 191 results in an enzyme with a dramatically diminished ability to catalyze the oxidation of ferrocytochrome c by H_2O_2 as judged by steady-state assays, with CCP(MI,F191) having a k_{cat} which is about 1/3,000th of that of the parent CCP(MI) at pH 7. This striking result attests to the precise structural requirements in this section of the CCP molecule that are necessary for efficient enzyme function. Initial efforts have been directed toward examination of some of the more obvious possible reasons for the low activity of the mutant enzyme. Measurement of the ferric/ferrous redox potential for the F191 mutant revealed an $E_{m,7}$ of -202 mV compared to -194 mV measured for CCP(MI) [50], a result which would appear to rule out a large change in the redox potential of the enzyme as the primary reason for the observed loss of enzyme activity. However, direct measurement of the enzymically relevant Fe(IV)/ferric redox potential would provide more convincing evidence of this.

Preliminary stopped-flow studies of the reaction of CCP(MI,F191) with H_2O_2 at pH 6 reveal biphasic kinetics, indicating that two major species present in solution react with peroxide at different rates [51]. Furthermore, the reaction with both species shows saturation at high H_2O_2 concentrations, in contrast to baker's yeast CCP which shows strictly bimolecular behavior under the same conditions. One interpretation of these results is that equilibration of H_2O_2-reactive and unreactive forms of the mutant enzyme may limit the rate of this reaction step. However, based on this model, the calculated equilibration rate is a relatively rapid 120 sec^{-1} and 37 sec^{-1} for either phase of the reaction, which indicates that reaction with H_2O_2 is probably not the rate-limiting factor in the impaired steady-state reaction sequence employed by the mutant enzyme.

Rapid kinetic measurements of the reduction of the semistable oxyferryl heme of peroxide-oxidized CCP(MI,F191) ($t_{1/2}$ = ~5 hr at pH 6 and 25°C) by horse heart ferrocytochrome c have been performed using two different methods, both of which indicate that transfer of reducing equivalents from ferrous cytochrome c to the oxyferryl center of the mutant enzyme occurs very slowly. Laser flash photolysis studies were conducted using a recently developed system [7] in which flavin semiquinones rapidly and selectively reduce the relatively exposed ferric heme of cytochrome c in preformed CCP compound 1-cytochrome c complexes. Using this method, which allows determination of the rate of electron transfer from Fe^{2+} cytochrome c to the oxyferryl center of CCP, very little reduction of peroxide-oxidized CCP(MI,F191) was observed on a time scale as long as 1 sec. Experiments using stopped-flow and conventional spectrophotometric techniques to more fully characterize this slow reaction are still in progress. Thus far, these studies appear to be in accord with the results of the laser flash experiments, and values for electron transfer rate constants are consistent with the conclusion that the step in the overall catalytic pathway which is most strongly affected by the Trp → Phe mutation at position 191 is electron transfer between the peroxide-oxidized enzyme and bound ferrocytochrome c [51].

The most likely reasons for the diminished catalytic capability of the F191 mutant are (1) failure to form a properly oriented electron transfer complex or set of complexes between the mutant CCP and cytochrome c, (2) the existence of new conformational constraints in the mutant which impede the reduction of the oxyferryl iron center to the ferric state, or (3) the fact that the indole ring of Trp-191 is an important element in some specific electron transfer pathway that exists between the heme centers of these two heme proteins. Evidence that the surface-to-surface interactions required to form CCP(MI,F191)-cytochrome c complexes are probably not strongly perturbed comes from both kinetic and x-ray crystallographic studies. The apparent K_m for cytochrome c for the residual CCP activity exhibited by the mutant enzyme is essentially unchanged from that of parent CCP(MI) [50], indicating that binding of the redox partners is not strongly disturbed, and it would seem unlikely that the minor structural perturbations that occur at the protein surface as a result of the mutation (see x-ray results below) would be sufficient to cause serious protein-protein recognition defects.

When ferrocyanide ion, $Fe(CN)_6^{4-}$, was used as the reductant in steady-state assays of peroxidase activity at pH 7, k_{cat} for CCP(MI, F191) was approximately one-fifth the k_{cat} of the parent peroxidase assayed under identical conditions. This finding indicates that the much more dramatic (~3,000-fold at pH 7) effect of the mutation on the oxidation of ferrocytochrome c may be fairly specific to the macromolecular redox partner, yet there is no evidence that the effect is due to molecular recognition (i.e., binding specificity) problems. To reconcile these observations, one may postulate that binding of cytochrome c results in structural adjustments in the wild-type CCP molecule which enhance the rate at which the oxyferryl center accepts an electron. In this scenario, the F191 mutant enzyme would be unable to undergo the necessary structural changes to allow rapid electron transfer to occur. Effects on ferrocyanide oxidation would be expected to be saml1 since this relatively nonspecific substrate would not be able to utilize the same rate-enhancing mechanism as the natural cytochrome c reducing substrate.

Another possibility for the diminished enzyme activity of CCP(MI,F191) is that the indole ring of Trp-191 is an important link in a specific intermolecular electron transfer pathway between the heme iron centers of CCP and cytochrome c, and that the benzene ring in the F191 mutant is unable to perform in a similar manner. Such a role for tryptophan has recently been proposed in experimental [52, 53] and theoretical [54] studies using ruthenium-modified myoglobins, and it may be that natural protein redox systems have evolved which utilize specifically positioned and oriented aromatic ring systems as "line-of-flight" electron transfer enhancer elements. The studies conducted to date on CCP(MI) and CCP(MI,F191) are not sufficient to confirm or refute this idea; further work will be necessary before any firm conclusions may be drawn.

2.3. Crystal Structures of CCP(MI,F51) and CCP(MI,F191)

The molecular structure of *E. coli*-produced CCP(MI) is nearly identical to that of CCP isolated from baker's yeast [55]. However, different unit cell parameters that result from one or a combination of the primary sequence variations in CCP(MI) as compared to the baker's yeast enzyme (see 2.2.1 above) necessitated that the structure be resolved by molecular replacement techniques [56]. A full discussion of the crystallography would be beyond the scope of this chapter; therefore, we include here only a brief description of the effects of the two Trp → Phe substitutions discussed above, taking the newly solved 2.2-Å CCP(MI) structure as a reference point.

The structure of the F51 mutant reveals that this Trp → Phe replacement is quite well tolerated by the surrounding protein environment [55]. The refined structure shows that the major structural adjustment resulting from the mutation is movement of helix B down toward the heme by about 0.3 Å (Fig. 2). While the side chain of His-52, which projects from helix B, correspondingly moves about 0.3 Å toward the heme iron, fixed water molecule 595 remains 2.7 Å from the iron, outside the transition metal coordination sphere, even though the anchoring hydrogen bond from the indole N of Trp-51 is missing in the mutant. We note that recently performed resonance Raman studies [57] indicate that a high proportion of six-coordinate ferric heme is present in this mutant, a result which is in conflict with the x-ray analysis. It is therefore possible that the iron

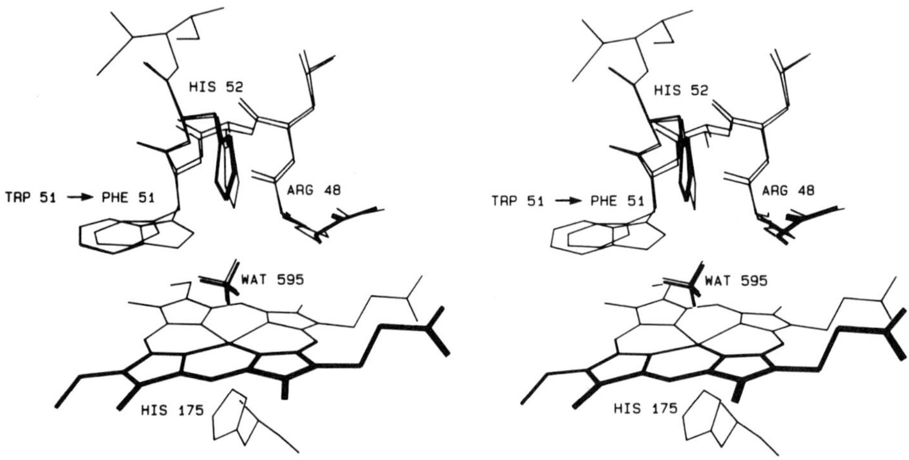

FIG. 2. Comparison of the distal side structures of CCP(MI) and CCP(MI,F51) [55]. The parent structure is displayed with depth cues and the mutant without depth cues to aid visual inspection.

coordination state is different in the crystalline and solution forms of the mutant enzyme. The reasons for the approximately fourfold increase in steady-state peroxidase activity for the F51 mutant are not immediately obvious from the x-ray structure.

The Trp → Phe mutation at position 191 on the proximal side of the heme results in a moderate degree of spatial rearrangement in the surrounding protein [55] (Fig. 3). The phenyl ring in the mutant occupies the same position as the six-membered part of the indole ring system in the parent; the accompanying movement of the extended backbone strand which holds the new phenylalanine is quite large (~1 Å), demonstrating the flexibility of this section of the molecule. Structural adjustment of the adjacent protein in response to this rearrangement in the vicinity of the mutation is best described as a generalized collapse inward toward the heme, as evidenced by the 0.5-Å displacement of an underlying section of three-stranded anti-parallel β sheet. Small adjustments are propagated to the protein surface of the proximal domain, and certain surface residues such as

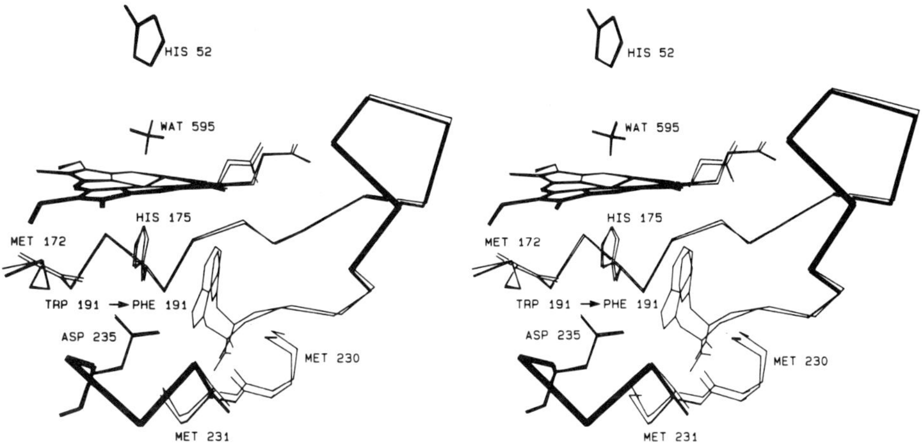

FIG. 3. Comparison of the proximal side structures of CCP(MI) and CCP(MI,F191) [55].

Asp-217, whose side chain has previously been implicated in the hypothetical electron transfer complex of Poulos and Kraut [3], have moved by as much as 0.4 Å. In the original formulation of the model complex, however, adjustments of the positions of surface residues on the order of the actual displacements seen in the x-ray structure of CCP(MI,F191) were allowed in order to optimize electrostatic and hydrogen-bonding interactions between CCP and cytochrome c, and it is expected that similar optimization of surface elements would be possible in the mutant enzyme. Based on the x-ray structure for the mutant, formation of electron transfer complexes characterized by a number of complementary surface-surface interactions between the macromolecular redox partners is unlikely to be seriously perturbed by the mutation at position 191; this conclusion is supported by the kinetic evidence thus far gathered. Consequently, the diminished catalytic competence of the F191 mutant is probably due to other factors as outlined above.

3. INTERMOLECULAR ELECTRON TRANSFER

A remarkable feature of compound I of CCP is its ability to accept electrons efficiently from ferrocytochrome c, despite the apparent inability of the CCP heme edge to interact directly with the heme of the cytochrome c molecule. The previously described hypothetical CCP-cytochrome c complex [3], as well as experimental observations [27,58], suggest that the heme edges in the complexed proteins remain separated by at least 17 Å; this separation is enforced by the presence of intervening protein structure contributed primarily by CCP. The ability to transfer electrons efficiently over relatively large distances is a feature of a number of heme proteins [41,53,59,60], and it is of considerable interest to establish the mechanisms by which this transfer can occur. One possibility, as discussed above, is that aromatic side chains can act as rate-enhancing elements in line-of-flight electron transfer between heme centers [54]. Another mechanism which has previously been proposed to account for efficient electron transfer between ferrocytochrome c and compound I of CCP postulates that electrons are channeled between the hemes by means of a specific through-bond π orbital and hydrogen-bonding pathway [3]. By examining the hypothetical CCP-cytochrome c complex, Poulos and Kraut [3] identified His-181 of CCP (Fig. 1) and Phe-87 of cytochrome c as likely members of such a conduit, since the side chains of both residues lie nearly in the planes of their respective hemes and fit into an extended pathway which has incremental separations of no greater than 7 Å between components.

Studies of the participation of Phe-87 in electron transfer have been conducted using mutants of yeast cytochrome c in which position 87 is occupied by tyrosine, serine, or glycine [6,61]. Steady-state oxidation of these mutants catalyzed by CCP is decreased by only 20-80% [61], indicating that Phe-87 is not an obligatory participant in physiological electron transfer. In contrast to the steady-state results, the rate of reduction of a Zn-CCP π-cation radical by ferrocytochrome c in the preformed bimolecular complex is 10,000-fold slower when Phe-87 is replaced with serine [6]. These observations show that the presence or absence of an aromatic residue at this position in cytochrome c can dramatically influence rates of electron transfer, although the exact nature of the redox process taking place is important in determining the outcome of experiments.

Since the proposed conduit for electron transfer also involves His-181 of CCP, we have initiated experiments to evaluate the role of this residue in electron transfer from cytochrome c to CCP. For this purpose, we have created a mutant of CCP(MI) in which His-181 is replaced with glycine. Preliminary characterization of the mutant enzyme has shown that under standard assay conditions at pH 6.0, k_{cat} for the oxidation of ferrocytochrome c is decreased only about four-fold relative to CCP(MI) [62]. The presence of this substantial residual activity in the Gly-181 mutant indicates that, like Phe-87 of cytochrome c, His-181 is not absolutely required for electron transfer between the cytochrome and CCP(MI). Further experiments to more fully characterize the effect of this mutation on the rate of electron transfer within the ferrocytochrome c-CCP-compound I complex are in progress.

4. STUDIES OF LIGAND BINDING USING ENGINEERED CCP MUTANTS

4.1. CCP-CO Binding Kinetics

The heme-containing proteins comprise a set of biomacromolecules whose functions are quite diverse, and it is of great interest to try to correlate the functional differences among these proteins with their heme environments. Important tools that have been used to establish correlations include x-ray crystallography, proton NMR [63-68], resonance Raman [48,69], Mössbauer [15], and EXAFS [47,70] spectroscopy, and, in fact, each of these methods has been used to compare the heme-binding site of CCP to those of various globins. Additional studies of the binding kinetics of ligands to the ferrous proteins, especially with CO, have been useful in revealing functional variations due to differences in the protein surrounding the heme [71].

The CO-binding kinetics of ferrous CCP [9,71] and plant peroxidases [72,73] differ significantly from vertebrate and invertebrate [9] globins, with both the CO association and dissociation rates of the globins ($k_a = 10^6$-10^7 M^{-1} sec^{-1}; $k_d = 10^{-2}$ sec^{-1}) being at least 100-fold faster than the corresponding rates for the peroxidases at pH 6 and 25°C ($k_a = 10^3$-10^4 M^{-1} sec^{-1}; $k_d = 10^{-4}$ sec^{-1}). Explaining these differences constitutes an interesting structure-function exercise, since the iron protoporphyrin IX prosthetic group is common to all these proteins, and therefore differences in ligand-binding phenomena must be due to the protein environment of the heme.

Previous characterization of CO binding to ferrous baker's yeast CCP using flash photolysis techniques has shown that the association rate is strongly dependent on pH between pH 7 and 9 [71]. Furthermore, two solution forms of CO-CCP are reversibly interconvertible in this pH range by a cooperative two-proton process with an apparent pK_a of 7.5. In ferrous CCP(MI), the association of CO with the more acidic enzyme form (Soret λ_{max} = 423.4 nm) occurs at a rate similar to that seen in the plant peroxidases ($k_a = 2 \cdot 10^3$ M^{-1} sec^{-1}) [62]; the alkaline form (Soret λ_{max} = 420.4 nm) binds CO at a rate which is at least 50 times faster ($k_a = 1.5 \cdot 10^5$ M^{-1} sec^{-1}). Additionally, we have found that the rate of CO dissociation from CCP(MI) is also pH-dependent, increasing approximately 10-fold (from $k_d = 0.7 \cdot 10^{-4}$ to $k_d = 8.0 \cdot 10^{-4}$ sec^{-1}) as the pH is raised from 7 to 8.4 [62].

4.2. Postulated Structural Factors Governing CO-Binding Rates

Studies of heme model compounds have demonstrated that dramatic alterations in the rates of CO binding and dissociation can be achieved by changing either the distal or the proximal heme environment [72-77]. Based on these studies and the heme-cleft structural differences known to exist between the CCP and the globins, we have identified some key residues both proximal and distal to the CCP heme plane which could be expected to affect ligand-binding characteristics.

The most obvious difference between the peroxidases and the globins is the presence of an arginine residue on the heme distal face in the peroxidases (Arg-48 in CCP; [29,41]), in a position where the globins have relatively nonpolar residues such as leucine or phenylalanine. This positioning of the large and positively charged arginine is expected to have both electrostatic and steric effects on ligands entering or leaving the heme pocket. On the opposite side of the heme plane, an important way which CCP differs from the globins is in the interaction of a buried carboxylate group with Nδ1 of the iron-coordinated proximal histidine; in myoglobin, the corresponding interaction consists of a much weaker H bond between the proximal histidine and a backbone carbonyl oxygen atom. The stronger proximal His-175/Asp-235 hydrogen bond in CCP is expected to increase the anionic character of the imidazole of His-175 relative to that of the globin proximal imidazole; model studies suggest that the effect of increased anionic character of the proximal imidazole is manifested functionally by decreased CO association rates [77]. Thus, it seemed reasonable to propose that this effect may be responsible for the difference in CO recombination rates observed between the heme peroxidases and the globins.

4.3. Studies of CO Binding Using Ferrous CCP(MI) Mutants

In order to directly investigate these structure-function questions, the visible spectroscopic and CO-binding behavior of mutants of CCP(MI) which have been specifically altered at distal position 48 and proximal position 235 have been characterized.

Spectroscopic characterization of the Fe^{2+}-CO complex of a mutant of CCP(MI) with leucine at distal position 48 revealed that two pH-dependent forms of the mutant enzyme continue to exist, in analogy to the CCP(MI) parent enzyme [62]. However, the pK for the transition of the Soret maximum of the CO complex is decreased in the mutant to 6.2 from 7.5 in CCP(MI), indicating that Arg-48 in CCP(MI) is important in controlling the acid-alkaline transition. Consistent with the spectroscopic results, the rates of CO association are rapid in the more alkaline form of the Leu-48 mutant, with a slow-binding component appearing in assays conducted at pH 6. Importantly, the CO association rate constant for the alkaline form of the mutant enzyme is $5 \times 10^5\ M^{-1}\ sec^{-1}$, which is still quite similar to that of the alkaline form of CCP(MI). Thus, the presence of arginine at position 48 in the parent peroxidase apparently elevates the pK_a of some group or groups which control the transition from an acidic to alkaline enzyme form, but the arginine does not ultimately control the CO association rate in CCP. Finally, the rate at which CO dissociates from the Leu-48 mutant has increased compared to the parent peroxidase ($k_d = 2 \cdot 10^{-3}\ sec^{-1}$) and is independent of pH, indicating that no simple correlation may be drawn relating the effects of this mutation on the association and dissociation processes.

In order to evaluate the importance of the anionic character of the proximal imidazole, we have investigated the kinetics of interaction of CO with a mutant of CCP(MI) having an Asp → Asn substitution at position 235. Interestingly, this replacement, which we believe significantly changes the coordination chemistry of the heme iron (based on resonance Raman [57] and x-ray results [55]), does not have a large effect on either the pH dependence of the shift of the Soret maximum or on the CO-binding and dissociation rates compared to the parent enzyme. We believe that proximal His-175 in the Asn-235 mutant very likely has decreased anionic character, due to the absence of a strong proximal His-Asp hydrogen bond, yet the effects on the CO-binding phenomena we have examined are minimal, contrary to the results predicted from model studies where proximal-side effects were significant [77]. It is concluded that distal-side structure is the dominant factor in control of CO binding in CCP, and studies are in progress to further define the residues responsible both for the pH-dependent transition in ferrous CCP-CO complexes and for the slower rate of CO binding to peroxidases compared to globins.

5. CONCLUDING REMARKS

Some important questions concerning the properties of yeast CCP have already been answered using site-specific mutagenesis combined with spectroscopic, enzymological, and other biophysical methods of analysis. With respect to compound I, studies of mutant proteins have demonstrated that Trp-51 and Met-172 are almost certainly not radical sites, while the presence of Trp-191 appears to be critical for cytochrome c peroxidase activity and in determining the magnetic characteristics of the oxidized enzyme. Additionally, the results of a straightforward experiment in which glycine replaced histidine at position 181 with only small effects on CCP activity has cast serious doubt on the validity of a proposed model for electron transfer [3]. Finally, strong evidence that CO binding in ferrous CCP is determined largely by distal-side effects has been demonstrated using specifically designed mutants. Although explanation of the new physical and enzymological features of CCP mutant enzymes using x-ray crystallography and other biophysical probes remains a major challenge, continued application of these types of experiments promises to rapidly and substantially expand our knowledge of heme protein chemistry.

ABBREVIATIONS

CCP	yeast cytochrome c peroxidase
CCP(MI)	yeast chtochrome c peroxidase producod in *E. coli*
EPR	electron paramagnetic resonance
ENDOR	electron nuclear double resonance
$E_{m,7}$	midpoint potential at pH 7
NMR	nuclear magnetic resonance
EXAFS	extended x-ray absorption fine structure

ACKNOWLEDGMENTS

This work was supported by NRSA Postdoctoral Fellowship PHS GM10292-02 to J. M. M.; Hemoglobin and Blood Protein Training Fellowship 5 T32 AM07233-12 to M. A. M.; Hemoglobin and Blood Protein Training Fellowship 5 T32 AM07233-11 to L. A. F.; and NSF DMB 85-11656-01 to J. K.

REFERENCES

1. T. L. Poulos and J. Kraut, *J. Biol. Chem., 255,* 8199 (1980).
2. J. E. Frew and P. Jones, *Biochim. Biophys. Acta, 742,* 1 (1983).
3. T. L. Poulos and J. Kraut, *J. Biol. Chem., 255,* 10322 (1980).
4. B. Waldemeyer, R. Bechtold, H. R. Bosshard, and T. L. Poulos, *J. Biol. Chem., 257,* 6073 (1982).
5. E. Cheung, K. Taylor, J. A. Kornblatt, A. M. English, G. McLendon, and J. R. Miller, *Proc. Natl. Acad. Sci. USA, 83,* 1330 (1986).
6. N. Liang, G. J. Pielak, A. G. Mauk, M. Smith, and B. M. Hoffman, *Proc. Natl. Acad. Sci. USA, 84,* 1249 (1987).
7. J. T. Hazzard, T. L. Poulos, and G. Tollin, *Biochemistry, 26,* 2836 (1987).
8. T. Yonetani and T. Asakura, *J. Biol. Chem., 243,* 4715 (1968).
9. M. P. Mims, A. G. Porras, J. S. Olson, R. W. Noble, and J. A. Peterson, *J. Biol. Chem., 258,* 14219 (1983).
10. J. K. Yandell and T. Yonetani, *Biochim. Biophys. Acta., 748,* 263 (1983).
11. T. Yonetani, H. Schleyer, and A. Ehrenberg, *J. Biol. Chem., 241,* 3240 (1966).
12. B. A. Wittenberg, L. Kampa, J. B. Wittenberg, W. E. Blumberg, and J. Peisach, *J. Biol. Chem., 243,* 1863 (1968).
13. B. M. Hoffman, J. E. Roberts, T. G. Brown, C. H. Kang, and E. Margoliash, *Proc. Natl. Acad. Sci. USA, 76,* 6132 (1979).
14. B. M. Hoffman, J. E. Roberts, C. H. Kang, and E. Margoliash, *J. Biol. Chem., 256,* 6556 (1981).
15. G. Lang, K. Spartalian, and T. Yonetani, *Biochim. Biophys. Acta, 451,* 250 (1976).
16. J. E. Erman and T. Yonetani, *Biochim. Biophys. Acta., 393,* 350 (1975).
17. H. Hori and T. Yonetani, *J. Biol. Chem., 260,* 349 (1985).

18. T. Yonetani and G. S. Ray, *J. Biol. Chem., 241,* 700 (1966).

19. A. F. W. Coulson, J. E. Erman, and T. Yonetani, *J. Biol. Chem., 246,* 917 (1971).

20. W. L. Purcell and J. E. Erman, *J. Am. Chem. Soc., 98,* 7033 (1976).

21. R. A. Marcus and N. Sutin, *Biochim. Biophys. Acta., 811,* 265 (1985).

22. H. R. Bosshard, J. Bänziger, T. Hasler, and T. L. Poulos, *J. Biol. Chem., 259,* 5683 (1984).

23. S. H. Northrup, J. C. L. Reynolds, C. M. Miller, K. J. Forrest, and J. O. Boles, *J. Am. Chem. Soc., 108,* 8162 (1986).

24. S. H. Northrup, J. O. Boles, and J. C. L. Reynolds, *J. Phys. Chem., 91,* 5991 (1987).

25. V. R. Lum, G. D. Brayer, G. V. Louie, M. Smith, and A. G. Mauk, in "Protein Structure, Folding and Design 2" (D. L. Oxender, ed.), Alan R. Liss, New York, 1987, p. 143.

26. J. T. Hazzard, S. J. Moench, J. E. Erman, J. D. Satterlee, and G. Tollin, *Biochemistry, 27,* 2002 (1988).

27. H. Koloczek, T. Horie, T. Yonetani, A. G. Maniara, and J. M. Vanderkooi, *Biochemistry, 26,* 3142 (1987).

28. T. L. Poulos, S. T. Freer, R. A. Alden, S. L. Edwards, U. Skogland, K. Takio, B. Eriksson, Ng. H. Xuong, T. Yonetani, and J. Kraut, *J. Biol. Chem., 255,* 575 (1980).

29. B. C. Finzel, T. L. Poulos, and J. Kraut, *J. Biol. Chem., 259,* 13027 (1984).

30. T. L. Poulos, in "Heme Proteins", Vol. 7 (G. L. Eichhorn and L. G. Marzilli, eds.), Elsevier, New York, 1988, p. 1.

31. S. L. Edwards, Ng. H. Xuong, R. C. Hamlin, and J. Kraut, *Biochemistry, 26,* 1503 (1987).

32. A. F. W. Coulson and T. Yonetani, *Biochem. Biophys. Res. Commun., 49,* 391 (1972).

33. S. L. Edwards, T. L. Poulos, and J. Kraut, *J. Biol. Chem., 259,* 12984 (1984).

34. L. A. Fishel, J. E. Villafranca, J. M. Mauro, and J. Kraut, *Biochemistry, 26,* 351 (1987).

35. S. Goltz, J. Kaput, and G. Blobel, *J. Biol. Chem., 257,* 11186 (1982).

36. J. Kaput, S. Goltz, and G. Blobel, *J. Biol. Chem., 257,* 15054 (1982).

37. J. E. Villafranca, E. E. Howell, D. H. Voet, M. S. Strobel, R. C. Ogden, J. N. Abelson, and J. Kraut, *Science, 222,* 782 (1983).

38. M. Smith, *Ann. Rev. Genet., 19,* 423 (1985).

39. D. B. Goodin, A. G. Mauk, and M. Smith, *Proc. Natl. Acad. Sci. USA, 83,* 1295 (1986).

40. K. Takio, K. Titani, L. H. Ericsson, and T. Yonetani, *Arch. Biochem. Biophys. 203,* 615 (1980).

41. T. L. Poulos and B. C. Finzel, *Pept. Protein Rev., 4,* 115 (1984).

42. D. Dolphin, A. Forman, D. C. Borg, J. Fajer, and R. H. Felton, *Proc. Natl. Acad. Sci. USA, 68,* 614 (1971).

43. J. E. Roberts, B. M. Hoffman, R. Rutter, and L. P. Hager, *J. Biol. Chem., 256,* 2118 (1981).

44. C. E. Schultz, R. Rutter, J. T. Sage, P. G. Debrunner, and L. P. Hager, *Biochemistry, 23,* 4743 (1984).

45. D. B. Goodin, A. G. Mauk, and M. Smith, *J. Biol. Chem., 262,* 7719 (1987).

46. I. Fujita, L. K. Hanson, A. F. Walker, and J. Fajer, *J. Am. Chem. Soc., 105,* 3296 (1983).

47. M. Chance, L. Powers, T. Poulos, and B. Chance, *Biochemistry, 25,* 1266 (1986).

48. S. Hashimoto, J. Teraoka, T. Inubushi, T. Yonetani, and T. Kitagawa, *J. Biol. Chem., 261,* 11110 (1986).

49. C. P. Scholes, Y. Liu, L. A. Fishel, M. F. Farnum, J. M. Mauro, and J. Kraut, *Isr. J. Chem.,* in press (1988).

50. J. M. Mauro, L. A. Fishel, J. T. Hazzard, T. E. Meyer, G. Tollin, M. A. Cusanovich, and J. Kraut, *Biochemistry, 27,* 6243 (1988).

51. J. E. Erman, personal communication.

52. R. J. Crutchley, W. R. Ellis, Jr., and H. B. Gray, *J. Am. Chem. Soc., 107,* 5002 (1985).

53. S. L. Mayo, W. R. Ellis, Jr., R. J. Crutchley, and H. B. Gray, *Science, 233,* 948 (1986).

54. A. Kuki and P. G. Wolynes, *Science, 236,* 1647 (1987).

55. J. Wang, Ph.D. dissertation, University of California, San Diego (1988).

56. R. A. Crowther, in "The Molecular Replacement Method" (M. G. Rossman, ed.), Gordon and Breach, New York, 1972, p. 173.

57. G. Smulevich, J. M. Mauro, L. A. Fishel, A. M. English, J. Kraut, and T. G. Spiro, *Biochemistry, 27,* 5477 (1988).

58. J. J. Leonard and T. Yonetani, *Biochemistry, 13,* 1465 (1974).

59. M. R. N. Murthy, T. J. Reid, A. Sicignano, H. Tanaka, and M. G. Rossman, *J. Mol. Biol., 152,* 465 (1981).

60. T. L. Poulos, B. C. Finzel, and A. J. Howard, *J. Mol. Biol., 195,* 687 (1987).

61. G. J. Pielak, A. G. Mauk, and M. Smith, *Nature, 313,* 152 (1985).

62. M. A. Miller, J. T. Hazzard, J. M. Mauro, S. L. Edwards, P. C. Simons, G. Tollin, and J. Kraut, *Biochemistry,* in press (1989).

63. J. D. Satterlee and J. E. Erman, *J. Biol. Chem., 258,* 1050 (1983).

64. G. N. LaMar, J. S. DeRopp, V. P. Chacko, J. D. Satterlee, and J. E. Erman, *Biochim. Biophys. Acta, 708,* 317 (1982).

65. J. D. Satterlee and J. E. Erman, *Biochim. Biophys. Acta, 743,* 149 (1983).

66. J. D. Satterlee, J. E. Erman, G. N. LaMar, K. M. Smith, and K. C. Langry, *Biochim. Biophys. Acta, 743,* 246 (1983).

67. J. D. Satterlee, J. E. Erman, G. N. LaMar, K. C. Smith, and K. C. Langry, *J. Am. Chem. Soc., 105,* 2099 (1983).

68. S. J. Moench, T. M. Shih, J. E. Erman, and J. D. Satterlee, *Biophys. J., 53,* p. 27A (1988).

69. J. A. Shelnutt, J. D. Satterlee, and J. E. Erman, *J. Biol. Chem., 258,* 2168 (1983).

70. M. Chance, L. Powers, C. Kumar, and B. Chance, *Biochemistry, 25,* 1259 (1986).

71. T. Iizuka, R. Makino, Y. Ishimura, and T. Yonetani, *J. Biol. Chem., 260,* 1407 (1985).

72. D. Kertesz, E. Antonini, M. Brunori, J. Wyman, and R. Zito, *Biochemistry, 4,* 2672 (1965).

73. B. A. Wittenberg, E. Antonini, M. Brunori, R. W. Noble, J. Wittenberg, and J. Wyman, *Biochemistry, 6,* 1970 (1967).

74. T. G. Traylor, *Accts. Chem. Res., 14,* 102 (1981).

75. T. G. Traylor, N. Koga, and L. A. Deardurff, *J. Am. Chem. Soc., 107,* 6504 (1985).

76. T. G. Traylor, S. Tsuchiya, D. Campbell, M. Mitchell, D. Stynes, and N. Koga, *J. Am. Chem. Soc., 107,* 604 (1985).

77. M. A. Stanford, J. C. Schwartz, T. E. Phillips, and B. M. Hoffman, *J. Am. Chem. Soc., 102,* 4492 (1980).

Author Index

Numbers in parentheses are reference numbers and indicate that an author's work is referred to although his name may not be cited in the text. Underlined numbers give the page on which the complete reference is listed.

C

D

E

F

H

I

L

N

O

P

Q

R

S

W

X

Y

Z

Subject Index

B

E

F

N

O

P

S